An Introduction to Scientific Computation and Programming

Daniel T. Kaplan

THOMSON

BROOKS/COLE

Australia • Canada • Mexico • Singapore • Spain • United Kingdom • United States

Publisher: *Bill Stenquist*
Editorial Assistant: *Julie Ruggiero*
Technology Project Manager: *Burke Taft*
Executive Marketing Manager:
 Tom Ziolkowski
Marketing Assistant: *Jennifer Gee*
Advertising Product Manager: *Vicky Wan*
Editorial Production Project Manager:
 Kelsey McGee
Composition Buyer: *George Brown*

Print/Media Buyer: *Doreen Suruki*
Permissions Editor: *Joohee Lee*
Production Service:
 RPK Editorial Services
Copy Editor: *Rose Kernan*
Cover Designer: *Denise Davidson*
Cover Image: © *Digital Vision*
Cover Printing, Printing and Binding:
 Transcontinental–Louiseville
Compositor: *Atlis Publishing Services*

For more information about our products, contact us at:
Thomson Learning Academic Resource Center
1-800-423-0563

For permission to use material from this text, contact us by:
Phone: 1-800-730-2214
Fax: 1-800-730-2215
Web: http://www.thomsonrights.com

Library of Congress Control Number: 2003113084

ISBN 0-534-38913-9

Brooks/Cole—Thomson Learning
10 Davis Drive
Belmont, CA 94002
USA

Asia
Thomson Learning
5 Shenton Way #01-01
UIC Building
Singapore 068808

Australia/New Zealand
Thomson Learning
102 Dodds Street
Southbank, Victoria 3006
Australia

Canada
Nelson
1120 Birchmount Road
Toronto, Ontario M1K 5G4
Canada

Europe/Middle East/Africa
Thomson Learning
High Holborn House
50/51 Bedford Row
London, WC1R 4LR
United Kingdom

Latin America
Thomson Learning
Seneca, 53
Colonia Polanco
11560 Mexico D.F.
Mexico

Spain/Portugal
Paraninfo
Calle Magallanes, 25
28015 Madrid
Spain

*In memory of my father,
Alan H. Kaplan, a gentle
advocate of logical reasoning.*

Preface

Scientists and engineers have always been among the leading innovators in computation: the designers of the first electronic computers, the developers of the first high-level computer language (FORTRAN), and the creators of the Internet and the World Wide Web. Scientists have also been innovators in computer education, perhaps motivated by the importance of computers in modern science and the ongoing need to train young scientists and engineers to become proficient in developing the new computational methods and ideas of their disciplines.

This book continues the tradition of innovation in the computer education of scientists and engineers by providing an integrated approach. It addresses not only the topic of programming but also the important methods and techniques of scientific computation (graphics, the organization of data, data acquisition, numerical methods, etc.) as well as the design and organization of software.

In almost all scientific programming texts, the emphasis is on numeric data. A major theme of this book is that numbers, despite their central role in scientific computation, are not the only important objects of computation. Scientists have to represent on the computer the real-world or theoretical entities that are the subject of their scientific work. This goes well beyond the built-in data types of computer languages (floating point numbers, character strings, etc.) to include complex entities like census databases, bridge designs, voice recordings, and x-ray photographs. Today's students have grown up using computers to perform sophisticated searches of distributed databases, compose and play musical sounds, and edit photographs and videos. A modern text on scientific computation must be based on examples that have a similar level of realism and richness.

Another way in which this book attempts to be innovative involves finding the appropriate middle path between two extreme approaches used in teaching about computers: *teach-a-package* and *teach-a-language*. The teach-a-package approach emphasizes the computations that can be performed with a particular piece of software such as EXCEL, MINITAB, or SPPS. This approach is appropriate in teaching subjects such as introductory statistics where there is a limited set of computations that need to be performed (e.g., the *t*-test, analysis of variance [ANOVA], and regression). Unfortunately, it fails to give students the skills needed to express their own ideas or to work outside of a limited domain.

The *teach-a-language* perspective is that students should work with a general-purpose system (e.g., C++ or Java) that is sufficiently flexible to build anything that needs to be built. This provides training in programming but is so time consuming that there is little opportunity left to learn about computation. For example, one of the most important operations for scientists is the production of scientific graphics; however, using C++ or Java few people are able to generate scientific graphics with less than a couple of months' experience.

The approach taken here is to teach general-purpose language skills and concepts and to take advantage of existing software so that realistic computations can be performed after just a few days or weeks of class time. Examples in the book include text processing, database searching, the synthesis of false-color images and the building of user interfaces. Given the right software and the right language skills, impressive feats of computation can be accomplished by beginners.

The book grows out of my own experiences over two decades working with scientists and students of science—chemists, economists, electrical and biomedical engineers, mathematicians, molecular biologists, neuroscientists, physicists, physicians, statisticians, and others—both in research and in formal classroom settings. My goal is to provide readers with the skills and concepts needed to be able to use the computer expressively in scientific work.

Most students using this book will take one and only one course on computation and programming. It's necessary, but not sufficient, to cover the basics of programming: how to express algorithms, and how to use programming-language constructs such as conditionals and loops, and ways to organize, save, and retrieve information. But today's students will be working in a web of software of startling complexity. An important part of their training must be in the principles of software development that help them to create programs that exploit and work with a large body of existing software. If modern scientists are to see further, they must, like Newton, stand on the shoulders of giants. In the case of software, this often means using existing software rather than building from the ground up.

Studying computation is similar in many ways to learning a foreign language. It's not enough to read about the principles of communication and grammar; one has to practice and use them actively. This is why beginning language courses are always about a particular language (e.g., English, Chinese or French). There is little point in studying languages abstractly except at an advanced level.

This book uses the computer language MATLAB. The use of a particular language allows us to be specific and to illustrate abstract concepts with working examples. MATLAB has many important advantages as a first language for scientists: It has a relatively simple grammar and is interpreted rather than compiled; it provides an integrated development environment that includes computation, graphics, user interface design and debugging; and it is powerful enough to handle realistic computations. Many operators of importance to scientists are built-in. Beyond its uses in education, MATLAB is a powerful professional-level programming environment that is widely used in academic, commercial, and government centers.

MATLAB is flexible sufficiently so that it can be used for general-purpose projects, and MATLAB's package-like aspects make it highly efficient for scientific computation. Almost all of the concepts that one learns using MATLAB are transferable to languages such as C++ and Java. Inevitably, though, there are important operations, such as the detailed manipulation of graphics and graphical user interfaces, where learning how to perform the operation in MATLAB provides little in the way of transferable skills. I have tried to minimize the use of such operations and, when they are needed, to emphasize general principles over language-specific issues. For example, in discussing user interfaces, the general concept of a "callback" function is introduced; the means of creating them in MATLAB is touched on only lightly.

I use this book in a one-semester course at Macalester College called "Introduction to Scientific Programming." The course attracts students from all scientific disciplines and levels, including many first-year students, most of whom have no previous programming experience and a mathematical background consisting of only a single semester of calculus. The examples used in the book are designed to be accessible to such students, while remaining compelling to more advanced students.

The basic material on programming is contained in Chapters 1 through 8, which can be covered in six to nine weeks depending on the intensity of the course and the enthusiasm of the students. Depending on the interests and orientations of the students and the instructor, the remaining chapters can be covered in a variety of ways. For example, a course with an emphasis on programming might supplement the first eight chapters with Chapters 9 through 12: scope, events, data organization, and recursion.

I find that students are motivated strongly by graphical user interfaces. Despite my warnings that they are among the most difficult types of programs to write, students want to learn this skill as soon as possible.

Chapters 9 and 10 provide an introduction to simple but useful graphical user interfaces. The goal is not to reproduce the highly refined user interfaces of mass-market programs but to enable scientists to collect those sorts of input most effectively provided with a mouse or with keystrokes.

General science students are well served by the material in Chapters 13 and 14. Such students are keen to perform computations on real-world objects; sounds and images provide an excellent motivation and illustrate general principles. Chapters 15 and 16 introduce basic concepts and methods of what is traditionally called "scientific computing." The material in that chapter—solving equations, optimization, interpolation—is accessible to most students because it is limited to one dimension. Concepts can be illustrated simply and concretely with a graph.

A course oriented to mathematicians might cover the first eight chapters, then Chapters 12 (Recursion), 15, 16, and 17. Chapter 17 extends the concepts and methods of Chapter 16 to multiple dimensions. This is suitable for students with a comparatively advanced background including linear algebra and serves as a transition to a follow-up numerical methods course.

Each chapter includes a number of short exercises that provide a means to learn and reinforce the material in that chapter. In later chapters, there is a shift of focus to short projects of a more open-ended nature. These projects integrate the various skills needed for scientific computation. For example, scientific graphics is only partly about drawing lines and points and the operators needed to do this; it's important to be able to store and read in data and therefore to manipulate files and their contents, translating them into physical units as appropriate. Project 9.10 involves all of these facets in the task of plotting an electrocardiogram according to the clinically accepted format. Other projects introduce diverse areas of science and technology, but all are self-contained, requiring no previous exposure to the relevant area of science.

I would like to acknowledge my colleagues G. Michael Schneider, Tom Halverson, Libby Shoop, Stan Wagon, Lenny Smith, Leon Glass, Sarah Little, and Steve Panizza who contributed to this book in various ways, and to thank Priya Arora, Nelson Coates, Issidor Iliev, Nazim Osmancik, and the other students who worked with successive drafts of this book over several years in CS 20 (now CS 121) at Macalester College. Special thanks are due my wife, Maya Hanna, and daughters Tamar, Liat, and Netta who provided support and encouragement for this work.

Contents

About the Author

Danny Kaplan teaches applied math, statistics, and scientific computing at Macalester College in St. Paul, Minnesota. His research concerns the uses of nonlinear dynamics ("chaos theory") in the analysis of time series, with a particular emphasis on variability in physiological systems. He has a B.A. in physics from Swarthmore College, an M.S. in engineering economic systems from Stanford University, and a Ph.D. in biomedical physics from Harvard University. He has taught at M.I.T. and McGill University and worked as a signal processing engineer for a medical instrumentation company. He has also previously worked as an economist at the Department of Energy modeling energy consumption and conservation.

He is the co-author of *Understanding Nonlinear Dynamics*. He and his wife, Maya, have three daughters.

Visit **www.macalester.edu/~Kaplan** for
Color versions of the figures in this text

CHAPTER 1

What Is
a Computation?

This book is about how to use computers to build your own specialized tools for carrying out work in science and technology. The computer, virtually unknown in its present form 50 years ago, by now has emerged fully as a universal tool of science and technology, as well as business, communication, and even entertainment. The computer has achieved this universal status in large part because it is a general-purpose tool that can be configured to perform specialized tasks. We accomplish this configuration most often by writing software: instructions that tell the computer how to be the tool we want it to be.

1.1   Computation as Transformation

Much of the reason why the computer is so useful and adaptable is that it is an *information tool*. We are all familiar with physical tools like screwdrivers and hammers that amplify and concentrate forces and thereby allow us to change the configuration of physical things. Similarly, an information tool is one that allows us to change the configuration of information. This may be hard to understand because "information" itself is such an elusive concept.

Let us use this idea of transforming information to offer a definition of a computation:

> A ***computation*** is a transformation from one or more inputs to an output.

This definition doesn't specify what is an input or an output, but perhaps you can anticipate that both of them are information in some form.

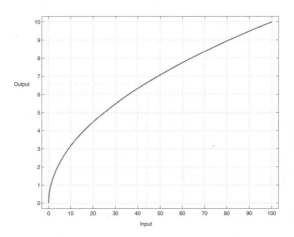

Figure 1.1. A graphical means of computing square roots.

We are concerned in this book with a particular modern technology for carrying out computations: software for the electronic computer. You will see many examples of such computations in the coming chapters. But for now, let us emphasize that computations are carried out all the time and everywhere by myriad other means than electronics. Some of these computations have evolved naturally and others involve human technologies of various sorts. The following are some noncomputerized examples of computations.

The graph in Figure 1.1 shows an ordinary plot of $y = \sqrt{x}$. The graph is part of a simple technology for carrying out a computation, finding the square root of a number. The input is x; the output is y. We all learn how to use such graphs to carry out a transformation from x to y: (1) Given the numerical value of x, find the corresponding point on the x-axis. (2) Trace a vertical line from this point up to intersect the curve. (3) From the intersection point, trace a horizontal line to meet the y-axis, and (4) read off the position of the meeting point on the y-axis. For example, an input of $x = 44$ leads to an output of $y \approx 6.6$.

Each year, U.S. citizens are required to compute their taxes. The inputs are numbers representing income, the number and ages of children, and so on. The output is the amount of tax owed. The government provides a detailed description of the algorithm to be followed in the shape of the form shown in Figure 1.2. Actually, the government does not require that its citizens do the computation; they are obliged only to provide the input information. The government will carry out the computation based on the inputs and send the output in the form of a bill or refund check.

Look at Figure 1.3. It's likely that a specific name occurs to you. If not, probably you recognize the picture as that of a woman or a face. There is a computation involved here too. The input is the collection of small squares

Form 1040 (2000) — Page 2

Tax and Credits				34	
	34	Amount from line 33 (adjusted gross income)		34	
	35a	Check if: ☐ **You** were 65 or older, ☐ Blind; ☐ **Spouse** was 65 or older, ☐ Blind.			
		Add the number of boxes checked above and enter the total here . . . . ▶ 35a ☐			
Standard Deduction for Most People	b	If you are married filing separately and your spouse itemizes deductions, or you were a dual-status alien, see page 31 and check here . . . . ▶ 35b ☐			
	36	Enter your **itemized deductions** from Schedule A, line 28, **or standard deduction** shown on the left. **But** see page 31 to find your standard deduction if you checked any box on line 35a or 35b or if someone can claim you as a dependent		36	
Single: $4,400	37	Subtract line 36 from line 34		37	
Head of household: $6,450	38	If line 34 is $96,700 or less, multiply $2,800 by the total number of exemptions claimed on line 6d. If line 34 is over $96,700, see the worksheet on page 32 for the amount to enter .		38	
Married filing jointly or Qualifying widow(er): $7,350	39	**Taxable income.** Subtract line 38 from line 37. If line 38 is more than line 37, enter -0- .		39	
	40	**Tax** (see page 32). Check if any tax is from **a** ☐ Form(s) 8814 **b** ☐ Form 4972 . . .		40	
Married filing separately: $3,675	41	Alternative minimum tax. Attach Form 6251		41	
	42	Add lines 40 and 41 . . . . ▶		42	
	43	Foreign tax credit. Attach Form 1116 if required	43		
	44	Credit for child and dependent care expenses. Attach Form 2441	44		
	45	Credit for the elderly or the disabled. Attach Schedule R . .	45		
	46	Education credits. Attach Form 8863	46		
	47	Child tax credit (see page 36)	47		
	48	Adoption credit. Attach Form 8839	48		
	49	Other. Check if from **a** ☐ Form 3800 **b** ☐ Form 8396 **c** ☐ Form 8801 **d** ☐ Form (specify)	49		
	50	Add lines 43 through 49. These are your **total credits**		50	
	51	Subtract line 50 from line 42. If line 50 is more than line 42, enter -0- . . . . . ▶		51	

Figure 1.2.   The algorithm for computing income taxes for individuals in the United States.

(pixels) of various shades of gray. The output is some other representation of the input's information: perhaps "Mona Lisa" or "woman" or "face."

The algorithm by which this transformation is performed is by no means completely known, although computers have been taught to compare a picture of a face to a database of such pictures.

Figure 1.4 shows graphs of short segments of speech recorded by a microphone. If you were to play these sounds one at a time through a speaker, you would hear EE for the top sound and OH for the bottom sound. The way that humans accomplish this transformation from sound input to an output as a perceived vowel is not unique. Unlike image recognition, computerized speech recognition is within the range of today's technology; successful computer algorithms have been developed for carrying out a sound-to-perception computation that can translate sound to text.

Figure 1.3.   A low-resolution image consisting of 1008 pixels.

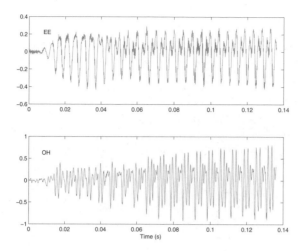

Figure 1.4. Speech sounds as recorded by a microphone. Top: EE. Bottom: OH. The signals are in the files ee.wav and oh.wav.

1.2 Computation as Reaction to Events

The definition of a computation as a transformation from inputs to outputs ignores the issue of where the inputs come from. These inputs often come from measurements.

A measurement (for instance the length of an object, the pH of a chemical solution, the electrical voltage difference between two points in a circuit, the time between heart beats, or the IQ of a test subject) assigns a value, often numerical, to some object or event. As such, the measurement is a translation from an input—the object or event—to an output; a measurement is itself a computation.

Key Term   Measurements are performed using sensors or *transducers*.[1] A mercury thermometer, for example, is a transducer from temperature to position of the mercury meniscus. A speedometer transduces vehicle speed into the position of a pointer. In both cases, the numerical value of the measurement can be read from a graduated scale. Even a computer keyboard is a kind of transducer, measuring which key is pressed at what time.

Some of the earliest computational devices were designed to transduce from the physical alignment of the earth and the sun to a reading of time. The antique sundial shown in Figure 1.5 is considerably more complicated than the decorative variety found in gardens today. The complication stems from the need for accurate measurement of time in the days before high-

---

[1]The word *transduce* is from the Latin for "across" (trans) and "lead" (ducere). It is closely related to the word *translate,* which comes from "across" and "carried" (latus).

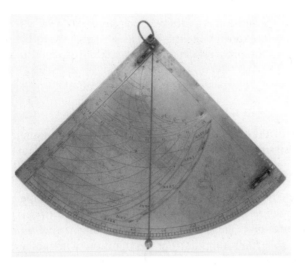

Figure 1.5.   A sixteenth-century sundial. By aligning the sight holes with the sun, the string transduces the altitude of the sun onto a set of several scales that indicate the time of day. The appropriate scale is picked based on the day of the year. (*Source:* Science Museum/Science & Society Picture Library. London, SW72DD. Reprinted with permission.)

A transducer from temperature to position.

precision clocks. For an accurate reading, it is necessary to do more than transduce the angle of the sun in the sky; the day of the year must also be known. Given these two inputs with reasonable precision, the time of day can be computed to within a few minutes. This computation is performed using the scales engraved on the device.

Measurements are often used to monitor a situation and respond appropriately to events. For example, one of the earliest uses of electronic computers was in aircraft control [1], where scores of measurements are made simultaneously by radar. The system needs to react to changes in its environment and perform the computation to produce the appropriate output, perhaps a directive for an airplane to turn or change altitude.

An important concept is the *state* of a system. For an air traffic control system, for example, the state includes the measured and desired positions and altitudes, headings, destinations, and so on of the aircraft currently in flight. In a computer, the state is the current configuration of the information stored: the contents and arrangement of the various forms of computer memory.

The idea of a computation as a reactive response to measurements motivates another perspective on the definition of a computation:

*A **computation** is a transition from an old state to a new state.*

This definition is particularly appropriate when designing systems that respond to external events. But, as we shall see in the next section, it also describes the progression of an algorithm from one step to the next.

1.3 Algorithms

Some of the computations described at the beginning of this chapter may seem so simple as not to be worth calling computations. But this is only because we are so well educated, spending several years learning about numbers so that we can easily translate a number into a point on a graph's axis and do the reverse translation of a point into a number.

Imagine the task of training a young child to carry out these "trivial" computations. The child comes with a built-in ability or learns in infancy to perform certain computations: recognition of images and language. We need to train the child to perform other computations that are built on these: to recognize letters and digits, to compare numbers. With these new computations, we can teach the child still more complicated computations: arithmetic (e.g., adding two numbers), translating a number into a point on a graph, and so on.

The process is much the same in programming a computer: We build new computations out of ones that already exist. The computer hardware comes with certain built-in capabilities. Using these as building blocks, others are added by software: Compare two numbers, scan over a list, look up in a table, and so on. By the time the computer arrives on your desk, it has been programmed to do a huge variety of computations, from square roots to database queries. Sometimes the precise computation you need is already programmed. If so, you need only set up the inputs and receive the outputs. If not, your job is to construct your computation out of those that already exist.

Key Term

An *algorithm* is a set of directions for carrying out a computation in terms of other, simpler computations. By "simpler computations" we mean merely that we already know how to perform these computations.

One of the things that makes it difficult to learn to write computer algorithms is that most of our skills for giving directions were formed in an environment of interacting with people, not automata. We are familiar with the types of tasks people can perform and the types of ambiguity they can handle. To produce computer algorithms, we need to learn what capabilities the computer already has and how to build our directions out of these and only these.

Our experiences giving directions to humans can be of limited value in constructing computer algorithms; computers are quite different from humans. Often the difference is taken to be a matter of speed: Computers seem amazingly fast, performing millions or billions of arithmetic operations every second. But the computer's speed advantage comes about because we judge speed based on tasks that computers are best at and people are poor at: arithmetic, copying information, comparing for exact matches. For many tasks where people excel (e.g., processing visual and sound information, comparing for inexact or approximate matches and analogies), computers are quite slow.

Key Term

For our purposes, the most important difference between people and computers is that computers are *automata:* machines that do things automatically. Computer automata have the great advantage of reliability and repeatability and precise memory. This means that if we program a computation by following a certain algorithm, we can be confident that the algorithm will be followed exactly. As a consequence, complicated computations can be built out of simple ones.

People, in contrast, are not particularly good at following instructions: We are distractible and forgetful, we make mistakes, we entertain ourselves by performing variations on a theme. This makes it difficult to build complicated computations out of simple ones; if the simple computations are unreliable, then the complicated ones will be highly error prone.

To illustrate the point of how poor memory leads to difficulty and error, attempt the simple task of adding 7398 to 5134 in your head. (Stop and try it.) Most people find this extremely challenging because it is hard to remember the components of the answer as they are derived; we have trouble keeping several numbers in memory at once. To solve our human memory problems, we have developed certain technological aids: for example, the abacus and writing. With paper and pencil, we can perform the addition easily by laying out the problem typographically so that digits to be combined are adjacent and there is room to store intermediate results.

But though our memory may be faulty, as people we are capable of judgment and generalization. Judgment allows us to identify error and deal with ambiguity. Generalization lets us see similarities between things that are not identical. Cats and dogs are both animals. A handwritten 3 and a 8 are both numbers, equivalent to the printed 3 and 8. Computers can be trained, with difficulty and usually with mediocre results, to carry out some computations (such as handwriting recognition) that people perform with unconscious ease.

Expressing Algorithms

Consider a simple task: Compare two Arabic numerals (for example, 23 and 97) to decide which is the bigger number. We can think of this as a computation that takes two numbers as inputs and returns as output a single number.

How do we express an algorithm for this operation? Here's one possible algorithm: Count the digits in each numeral. If one of the numerals has more digits than the other, that numeral is larger. If both numerals have the same number of digits, then compare the leftmost digits in the two numerals. The numeral with the larger leftmost digit is the output. If the two leftmost digits are identical, then continue on with the next digit to the right, and so on. If the digits are the same in all the places, the two numerals are identical and so either numeral can be the output.

This algorithm breaks down the task of comparing two numerals into simpler computations: Count the number of digits, compare two digits, find the leftmost digit, and so on. These simpler computations are linked together; we've used phrases like "if . . . then" and "continue on." We have also done something more subtle, since in order to say whether one numeral has more digits than the other, we need to be able to compare two numbers: the number of digits in each numeral. Such circularity (defining

Key Term

an algorithm in terms of itself) leads to a seemingly strange but extremely useful technique for writing algorithms called *recursion*.

Often there are many different algorithms for performing the same computation. A case has been reported of a man who could not tell in the normal way which of two numbers is larger. Instead, he would count from 1 until reaching one of the numbers [3]. This would identify the smaller number. Although this algorithm works in theory and is effective for very small integers, it is impractically slow for large numbers; it would be a week's work to tell which is the smaller of 436,279 and 328,592.

When expressing an algorithm to a person, the goal is often arranging things so that the algorithm makes sense. We want the person to understand what is happening. Until a level of understanding is reached, humans have great difficulty following directions. We usually need to have a specific example, something to follow along so that we can make sense out of the directions. We then generalize from the example to a procedure we can use on any inputs. To illustrate, consider how you would teach someone the algorithm for adding two multidigit numbers using paper and pencil, as illustrated in Figure 1.6. You would almost certainly use an example, probably a series of examples.

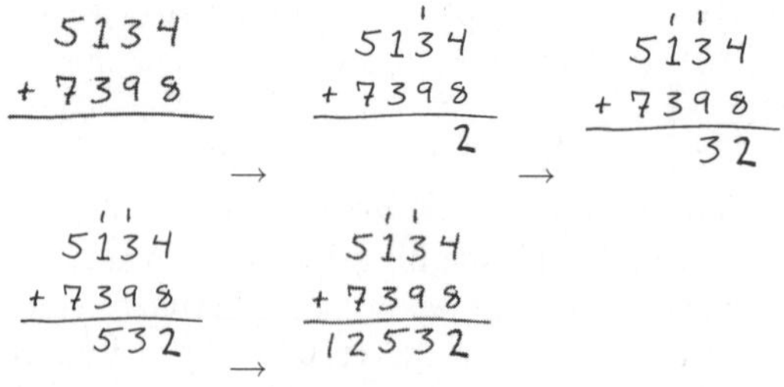

Figure 1.6. Several snapshots of the paper-and-pencil algorithm for adding two multidigit numbers.

While there are techniques for programming computers using examples, the most common and important way of describing algorithms involves naming the objects involved in the computation and giving a series of steps. At each step, directions are given for how to change the named objects. The

Key Term

named objects are part of the *state of the computation;* at each step the state is updated. The state also contains information about which is the following step or whether the computation is complete. When it is complete, one or more of the components of the state becomes the output of the computation.

We'll start with an extremely simple example: how to add three numbers when we already have a capability for adding two numbers. The inputs are three numbers; the output is the sum of the three. The following algorithm breaks this computation down into a sequence of simpler computations:

Add Three Numbers

Step	Computation	What to Do Next
Setup	Call the inputs A, B, and C.	Continue with step (1).
(1)	Compute A + B. Call this X.	Continue with step (2).
(2)	Compute C + X. Call this X.	Return X as the output. STOP.

The add-three-numbers algorithm has been laid out typographically as a set of steps. Within each step, the state of the computation is updated. After this, move to the next step as indicated by the What to Do Next column. Note that both steps (1) and (2) involve another algorithm: that for adding two numbers.

This tabular English-language presentation of the algorithm admittedly is awkward. In the next chapter, we will start to use a computer language for implementing algorithms that allows a more concise and fluent presentation of algorithms.

In the preceding algorithm, the information about what to do next seems hardly worth bothering with. It's common sense that from each step you go on to the next step and then stop after the last step.

Algorithms often have a more interesting structure in which the sequence of steps depends on the results of previous computations. That is, the next step depends on the state of the computation. Take, for instance, an algorithm for performing the computation that takes a list of numbers as an input and returns as output the smallest number in the list.

Find the Smallest Number in a List

Step	Computation	What to Do Next
Setup	Call the input list L. Call the position in the list P. Set P = 1 (the position of the first number in the list).	Continue with step (1).
(1)	Find the number in position P in the list L. Call this number S.	If there are more numbers in L, continue to step (2). Otherwise, the output is S. STOP.
(2)	Increase P by 1. Find the number in position P in the list L. Call this X.	Continue with step (3).
(3)	Let S be the smaller of X and S.	If there are more numbers in L, continue to step (2). Otherwise, the output is S. STOP.

To illustrate how the algorithm works, let's follow it step-by-step using as input the list 7 1 5. The chart shows the state of the computation *after* each step. Think of the chart as analogous to the frames of a movie; each row is a snapshot of the state of the computation at the end of a step.

Find the Smallest Number with Input List 7 1 5

Step in Algorithm	State				Next Step
	L	Position P in List	S	X	
Setup	7 1 5	1			Continue with step (1).
(1)	7 1 5	1	7		Continue with step (2).
(2)	7 1 5	2	7	1	Continue with step (3).
(3)	7 1 5	2	1	1	Back to step (2).
(2)	7 1 5	3	1	5	Continue with step (3).
(3)	7 1 5	3	1	5	Output S. STOP.

Output: 1

Just as the find-the-smallest-number computation was built out of other, simpler computations, so we can use find-the-smallest number as part of a larger computation. For example, we can write a computation that takes as input a list of numbers and returns as output another list with the input values in ascending order. Here's one simple but inefficient algorithm for doing this.

Sort a List of Numbers

Step	Computation	What to Do Next
Setup	Call the input list L and create an empty holder for another list called X.	Continue with step (1).
(1)	Find the smallest number in L. Call this S. Append S to the end of list X.	If there are more numbers in L, continue to step (2). Otherwise, the output is X. STOP.
(2)	Delete S from L.	Continue with step (1).

To illustrate, here's the sequence of values of L, S, and X as we step through the algorithm for an input list 8 5 1 2:

Sort Input List 8 5 1 2

Location in Algorithm	State			Next Step
	L	S	X	
Setup	8 5 1 2			Step (1)
(1)	8 5 1 2	1	1	Step (2)
(2)	8 5 2	1	1	Step (1)
(1)	8 5 2	2	1 2	Step (2)
(2)	8 5	2	1 2	Step (1)
(1)	8 5	5	1 2 5	Step (2)
(2)	8	5	1 2 5	Step (1)
(1)	8	8	1 2 5 8	Step (2)
(2)		8	1 2 5 8	Step (1)
(1)		8	1 2 5 8	L is empty, so output X and STOP.

Output: 1 2 5 8

1.4   From Algorithms to Software

The art of doing computations consists in large part of figuring out how to create an algorithm for a desired computation in terms of other, simpler computations. This new computation then becomes part of the repertoire; we can use it to construct other, more advanced computations.

If we are going to use an automaton to execute the algorithm, we need to set the algorithm in a form that the automaton can follow: When working with computers, we need to express the algorithm in a computer language. Learning such a language is the subject of the rest of the first part of this book.

Learning to create effective algorithms is another, more subtle matter. This relies on your creativity and ingenuity. The second part of this book describes and illustrates computations that are of general use in science and technology. Algorithms are given for a variety of computations such as solving equations, searching through data for matches, finding maxima and minima, simulating random processes, and so on. These computations may be directly useful in your own scientific and technical work or may inspire new ideas for computations. At a minimum, they may provide a useful repertoire of computations for developing new algorithms, or, by increasing your understanding, help you to use existing algorithms more effectively.

Typically, we use a computation to study some object. For example, an engineer might compute the stress in a component of a bridge. To perform such computations, we need somehow to provide the object as an input. Since we can hardly feed a bridge in through the floppy disk slot of a computer, we need somehow to construct on the computer a *representation of the object.* How to build such representations—constructions on the computer that behave analogously to the real-world objects we wish to study, analyze, and synthesize—will also be a major theme of the following chapters.

Key Term

Another important topic is *modularity:* how to package our algorithms so that other people—and ourselves!—can use them without needing to worry much about what goes on inside them. Algorithms, when constructed carelessly or sloppily, can interact needlessly with other algorithms or parts of the computer. The result is behavior that is much more complex than what was intended. Perhaps you have seen similar behavior when installing new software on a computer; sometimes some unrelated piece of software or hardware stops functioning. We all know that computers are complex devices; complex beyond any individual's ability to understand them completely. Surprisingly, even the software you write by your own hand can quickly become so complex that you can't understand it or manage it. If houses were like many computer programs, painting your wall might well cause the roof to leak! Opening a window might back up the plumbing! Unless appropriate care is taken, your own software can become so complex that you won't be able to get it to work. By taking a disciplined approach to organizing and packaging your computations, you will be able to manage the complexity of the computational world you have created.

1.5 Exercises

Discussion 1.1:
Describe in writing an algorithm for comparing two decimal numerals; for instance, 786.89 and 621.3.

Discussion 1.2:
Describe in detail the algorithm you use for adding two numbers in Arabic notation.

Discussion 1.3:
Describe in detail an algorithm for adding two numbers in Roman notation; for example,

$$\begin{array}{r} \text{DCXCI} \\ + \quad \text{CDXXX} \\ \hline \end{array}$$

Discussion 1.4:
Follow the find-the-smallest-in-list algorithm on page 10 for the input 41762.

Exercise 1.1:
The date-display system of a digital clock receives an input each midnight that triggers the date to advance. Describe the state of the date-display system and give an algorithm for updating it in response to the midnight event. Don't forget leap years!

Explain what was deficient about the state of systems that suffered from the Y2K problem.

Exercise 1.2:
Describe the algorithm you use for looking up a person's telephone number in the phone book. The input is the person's name; the output is the corresponding phone number.

Exercise 1.3:
Describe the algorithm you would use if you had a telephone number and needed to look up the corresponding name in a telephone book.

Exercise 1.4:
The square-root computation based on Figure 1.1.1 works only for $1 \leq x \leq 100$, since the graph only covers x in that interval. Describe another algorithm that builds on the graph-based square-root algorithm but that will work for any $0 \leq x$. (*Hint:* $\sqrt{100^k x} = 10^k \sqrt{x}$).

Exercise 1.5:

Describe an algorithm to find the second smallest number in a list of numbers.

Exercise 1.6:

Describe an algorithm for a computation that takes two inputs, a list of numbers and an integer n and returns the nth smallest element in the list.

CHAPTER **2**

Invoking a
Computation

He was so learned that he could name a horse in nine languages;
so ignorant that he bought a cow to ride on.
— Benjamin Franklin

When we first study a natural language such as English or Chinese, we usually learn single words and then ways to combine these words to form simple expressions: *Hello. How are you? Where is the pen? The book is on the table.* Only later do we study how to combine these simple expressions into more complex forms involving more elaborate grammar.

In the next several chapters, we will study a language intended for communicating with computers. This language, MATLAB, is designed particularly for describing algorithms in sufficient detail and precision that the computer can perform the computation we desire. Although computer languages are much less complex than natural languages, we will start our study with the simplest of expressions: ones that convey an instruction to perform a specific computation, the algorithm, which already is known to the computer. This is called "invoking a computation" but also is known less formally as "running a command" or "executing a command."

Before you can invoke a computation in MATLAB, you need to start up a program called the MATLAB "interpreter." You start this program in the familiar way, by clicking on an icon or whatever procedure you are used to on your computer. The MATLAB interpreter then displays a prompt, usually something like ≫. You type text at the prompt and, after pressing the ENTER key, MATLAB interprets your text as the invocation of a computation: the interpreter is the genie that performs your commands.

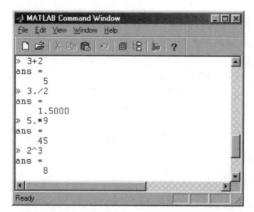

Figure 2.1. The MATLAB command window displays a prompt ($\gg$) at which expressions can be typed to invoke a computation. On different computers, the window's appearance may differ.

2.1 Expressions and Commands

Here is an example:

```
≫  3 + 2
   ans: 5
```

We asked the computer to add 3 and 2. The computer did so and told us the answer. (Since the typographical requirements of a book are different from those of the computer screen, the examples shown in this book are printed in a somewhat different form than they would appear on the computer screen. The book uses an `ordinary typewriter` font to display a command and a `typewriter italic` font to display the response. Figure 2.1 shows how things look on the computer screen.)

Key Term   An *expression* is a statement that provides all of the information needed to invoke a computation. Here are some other examples of expressions and the output of the resulting computation:

```
≫  3 ./ 2
   ans: 1.5000
≫  5 .* 9
   ans: 45
≫  2^3
   ans: 8
≫  cos( 3.14159 )
   ans: -1.0000
```

```
≫ log10( 100 )
  ans: 2
```

To execute a command, you type an expression followed by ENTER. MATLAB responds by *evaluating* the expression and returning the output, which is often called the *value* of the expression.

Whenever we invoke a computation, we need to provide three pieces of information:

1. What is the operator?
2. What are the inputs?
3. What should be done with the output?

You can see from the previous examples that there are two different styles of specifying the operator and inputs. One of the styles is similar to the standard arithmetic notation learned by youngsters. The notation is this: The two numbers to be operated on are written on either side of a special symbol. The symbol itself tells which operation to perform. This is called the *infix* style of notation and is used mainly for the common arithmetic operations. Some of the symbols might be unfamiliar at first: 3.*2 rather than 3×2; 3./2 rather than $3 \div 2$; 3.^2 rather than 3^2.

The second style for specifying the operator is called *functional-style notation*. In this style, the name of the operator (e.g., cos or log10) is followed by one or more arguments enclosed in parentheses. When there is more than one argument, the arguments are separated by commas as in

```
≫ plus(3,2)
  ans: 5
≫ twoCircles(0,1,2,1,0,3)
  ans: -1.9490 0.5510
```

There is no fundamental difference between the functional-style notation and the infix notation: (3+2) is precisely equivalent to plus(3,2).

Students with tradition-minded teachers learn that in an expression such as $3+6$, the 3 is called the addend and the 6 the augend. Perhaps it's silly to identify the two inputs with different names. After all, $3+6$ is the same as $6+3$. But this commutativity isn't true in subtraction. In $7-4$, the 7 is called the minuend and the 4 the subtrahend; the distinction between the two inputs is important.

Terms such as *minuend* and *augend* are esoteric and obsolete. A more general and useful term is *argument*, which refers to each of the inputs of a computation. The computation to be performed is called the *operation*. In some operations, such as subtraction, the two arguments play different roles. We therefore keep track of which argument is which by ordering them: In $7-4$ the 7 is the first argument and the 4 is the second argument.

Key Term **Key Term**

Key Term

Key Term

Key Term

All of the preceding examples are *simple expressions* that involve a single operation. The operations used in the preceding examples are addition, division, multiplication, exponentiation, the cosine, and the base-10 logarithm. MATLAB displayed the output of each computation in the command window as shown in Figure 2.1. Often, though, the output is needed as the input to another computation. This is done using *compound expressions* in which one or more of the arguments to an operation is itself an expression. For instance,

```
≫  (3+2) ./ 5
   ans: 1
```

The two arguments to the division operator are the expression (3+2) and the number 5. In order to perform the division, MATLAB first evaluates the expression (3+2), using that output as the first argument to the division operator. Other examples are as follows:

```
≫  3 + 15./ 5
   ans: 6
≫  sqrt(3.^2 + 4.^2)
   ans: 5
≫  2.^2 + 1
   ans: 5
≫  2.^(2 + 1)
   ans: 8
≫  10.^log10(592.3)
   ans: 592.3
```

These few examples may provide all the information you need to deduce how to invoke any arithmetic computation that might occur to you.

All of the preceding examples are *well-formed expressions*: They completely and properly describe a computation according to the rules of the MATLAB language. *Ill-formed expressions* are those that are meaningless to the interpreter. Using such expressions causes an error message to be generated by the interpreter. For example,

```
≫  (3+2./5
???: (3+2./5
        |
     A closing right parenthesis is missing.
     Check for a missing ")" or a missing operator.
≫  logg(10)
   logg(10)
   Undefined function or variable 'logg'.
```

Even well-formed expressions can generate warnings or errors when the computation cannot be meaningfully carried out:

```
>> 1./0
    Warning: Divide by zero.
    ans = Inf
```

A Matter of Style:

Some expressions may appear ambiguous to a human reader (for instance, does 3+15./5 mean $3 + \frac{15}{5}$ or $\frac{3+15}{5}$?). MATLAB, like other computer languages, applies simple rules to eliminate any potential for ambiguity. For example, there is a rule that multiplication and division have a higher precedence than addition or substraction, so that in 3+2./5, the division occurs first and the addition second, resulting in an output 3.4. In interpreting the expression 1./3./2, MATLAB applies a rule that says that when dealing with operations of equal precedence, the leftmost operation is applied first. Thus, 1./3./2 corresponds with $\frac{1/3}{2}$ and not $\frac{1}{3/2}$.

Due to the rules, all well-formed expressions are completely unambiguous to the MATLAB interpreter. The problem is that the unambiguous intentions of the human reader or writer of a command may not match the unambiguous interpretation of the command by the computer. Even if you master the rules that the computer uses to interpret commands, keep in mind that some future reader of your programs, who perhaps has experience with another language with different rules, may misunderstand even a correctly formed statement.

Here's a simple rule for avoiding problems:

When any potential for ambiguity exists, resolve it with parentheses!

2.2 Changing State: Assignment

In the previous chapter, we saw how algorithms set up a state and modify it at each step. The state was defined as the information in the computer's memory. The interpreter provides facilities for the manipulation of the information in memory. The memory is organized into *variables* each of which has two components: a name and a value. To create a variable, you use a special command that always has the same form

Key Term

name   =   expression

Key Term

This is called an *assignment* statement; it associates a value to the variable. The value is not the expression itself, but the result of evaluating the expression. For example, you can create a variable named a that holds the value of $\sqrt{2}$:

```
>> a = sqrt(2)
 a: 1.4142
```

Once the value has been assigned to the variable, you can access the variable at any time by using the variable name in an expression. For instance, the variable name itself is a complete expression:

```
>> a
 a: 1.4142
```

The variable name can also be used as part of an expression, as in

```
>> a.^2
 ans: 2.0000
```

To change the value of a variable, use an assignment statement with the new value

```
>> a = 1
 a: 1
```

Note that MATLAB displays the value as well as the variable name in response to the assignment command. It's usually the case that you aren't interested in this information. You can *suppress the display* by using a semicolon (the character ;) at the end of the command; for instance,

```
>> a = 1;
```

Suppressing the display just saves space on the screen; the change of state—the assignment of a value to a variable—is still performed. In later chapters, we'll encounter computations where assignments are being done thousands or millions of times. In such situations, suppressing the display becomes essential.

A variable name can be any sequence of characters and numbers, but it must not begin with a number or contain any punctuation characters other than the underscore (_). Some examples of valid names are `fred`, `codeBlue`, `cheddar_cheese`, `type3`. Assignment to an invalid name generates an error message. The allowed number of variables is practically unlimited; you can have as many different variables as can fit into the memory of your computer.

Variables serve several important purposes:

- Variables are one of the main ways that "state" is stored on the computer. Assignment is the means by which you update the state of the computer. For example, suppose you want to make a counter that will keep track of how many times a particular event has happened. You can initialize the counter with the assignment statement:

```
>> n = 0;
```

Whenever the event happens, you would want to execute the command

```
>> n = n + 1;
```

This command might seem bizarre to you at first. There is no possible number n for which n equals $n + 1$, the meaning of the mathematical statement $n = n + 1$. But remember that the assignment command is not a statement of numerical equality. (In Chapter 3 we will encounter the == operator, which does test for numerical equality.) Instead, assignment takes the value on the right-hand side of the = character and makes that the value of the variable named on the left-hand side of =. The command n=n+1 means "increase the value of n by 1." It is a change of state.

- Variables allow you to use the same value, without recomputation, in more than one place. For example, suppose you want to compute the gravitational force of the earth on the moon. If the moon is at a location (x, y, z) relative to the center of the earth, Newton's law of gravitation says that the gravitational force has the three components $F_x = \frac{-GmMx}{(x^2+y^2+z^2)^{\frac{3}{2}}}$, $F_y = \frac{-GmMy}{(x^2+y^2+z^2)^{\frac{3}{2}}}$, and $F_z = \frac{-GmMz}{(x^2+y^2+z^2)^{\frac{3}{2}}}$, where G is the universal constant of gravity, m is the mass of the moon, and M is the mass of the earth. The computation of these force components is greatly streamlined if you create variables to hold the common parts of the three forces.

```
>> G=6.67e-11;
>> m=7.38e22;
>> M=5.98e24;
```

The notation 6.67e-11 is used by the computer to avoid the superscripts involved in standard scientific notation: It means 6.67×10^{-11}. Note that no units have been included with the numbers when entered as variables. We'll return to this topic in Chapter 15; for now, we note that the units of G are Nm^2kg^{-2}, while those of m and M are kg.

The quantities F_x, F_y, and F_z have many terms in common. We can pull them out into a single variable used as an intermediary.

```
>> tmp = G*m*M/((x^2 + y^2 + z^2)^(3/2));
```

Now the forces are simply:

```
>> Fx = -x*tmp;
>> Fy = -y*tmp;
>> Fz = -z*tmp;
```

One great advantage of this style is that it reduces redundancy. Redundancy is a problem because it introduces the possibility of inconsistency and error. In the preceding, by computing the value of

$\frac{GmM}{(x^2+y^2+z^2)^{\frac{3}{2}}}$ in only one place, we arrange things so that if there is a mistake (and there invariably are mistakes made when writing computer programs) we only need to find and fix it in one place.

- Variables allow you to write general expressions. For instance, everyone knows that in a right triangle with perpendicular sides of length a and b, the length of the hypotenuse c can be computed using the Pythagorean formula:

$$c = \sqrt{a^2 + b^2}$$

This relationship holds for every right triangle (in Euclidean geometry). Using variables allows us to write the computation in an abstract way, without referring to any particular value of length:

```
>> c = sqrt(a^2 + b^2);
```

This command allows us to compute the hypotenuse of any triangle simply by first assigning values to a and b.

It's important to keep in mind that the = sign is merely a punctuation mark meaning "assignment," in a command where the variable name is on the left side and the value to be assigned is on the right side. This makes the command look a lot like traditional mathematical notation, but the similarity is misleading. In traditional mathematics, the = symbol stands for "equals" and $a = a + 1$ means exactly the same thing as $a + 1 = a$. But in MATLAB, the command

```
>> a + 1 = a;
```

is ill formed because a+1 is not a valid variable name. Similarly, in mathematical notation, $a = b = c$ means that $a, b,$ and c all are equal. In MATLAB,

```
>> a = b = c;
```

is ill formed: The allowed syntax is _____ = _____ . The word
                                          name   value

Key Term   bug is often used to describe flaws in a computer program. Buggy statements in this book are marked with a small icon.

▶

Example: The Fibonacci Numbers

The Fibonacci numbers are an infinite sequence of numbers that start 0, 1, 1, 2, 3, 5, 8, 13, 21, 34, $\cdots$. This is an important sequence, both mathematically and historically. The sequence was described originally in the year 1202 in

The Book of the Abacus by Leonardo of Pisa, otherwise known as Filius Bonacci.

Finding the next number in a Fibonacci sequence is easy: Add the final number to the previous one. With the new number in place, it becomes the final number of the extended sequence and the sequence can be extended once again.

Key Term

To translate this into terms suitable for computer commands, we need to define the *state of the computation.* One way to do this is to create two variables, final and previous, and initialize them to the first two values in the series.

```
≫ previous = 0;
≫ final = 1;
```

Extending this short sequence is a matter of following the rule

```
≫ new = final + previous
   new: 1
```

The variable new is now a component of the state.

To extend the sequence further, we need to update the state. This is done by assignment. What used to be the final value now becomes the previous value

```
≫ previous = final;
```

and the new value becomes the final one

```
≫ final = new;
```

Now, with the updated state, the rule can be applied again to extend the series

```
≫ new = final + previous
   new: 2
```

We can repeat this process of updating the state and computing the new value to extend the sequence as far as we want. Table 2.1 shows the state after each command.

The details of this assignment are critically important. If we had tried the assignment in the opposite order, that is,

```
🐞 ≫ final = new;
🐞 ≫ previous = final;
```

then both final and previous would end up having the same value, the value of new. In this case, the logical flaw—the bug—comes not from the statements individually but from their order with respect to one another.

Table 2.1 The State of a Computation of the Fibonacci Sequence Changes after Each Assignment Expression.

Expression	State		
	previous	final	new
`previous=0;`	0		
`final=1;`	0	1	
`new=final+previous;`	0	1	1
`previous = final;`	1	1	1
`final = new;`	1	1	1
`new=final+previous;`	1	1	2
`previous = final;`	1	1	2
`final = new;`	1	2	2
`new=final+previous;`	1	2	3
`previous = final;`	2	2	3
`final = new;`	2	3	3
`new=final+previous;`	2	3	5
`previous = final;`	3	3	5
`final = new;`	3	5	5

◀

A Matter of Style:

Shakespeare wrote

> *What's in a name? that which we call a rose*
> *By any other name would smell as sweet.*
> — Romeo & Juliet (II, ii, 1–2)

While I am in no position to argue with Shakespeare about style and language, I will point out that the Immortal Bard had very limited experience in writing computer programs.

Two aspects of names are important. First, the rules for the construction of names must be obeyed.

- Names can be either upper- or lowercase or a mixture of both. Different case names are distinct: `fred`, `FRED`, `Fred`, and `fRED` are all distinct variables with nothing in common so far as the computer is concerned.
- Names can contain digits but must not start with them.
- Names can contain the character `_` but no spaces or special characters such as `*`, `.`, `^`, and so on. `Root2` and `sumOfSquares` are valid names, but `2root` and `sum.of.squares` are invalid.
- Some words are reserved for special purposes in the language. These *keywords* cannot be used as variable names. They are `if`, `else`, `elseif`, `end`, `for`, `while`, `break`, `switch`, `case`, `otherwise`, `try`, `catch`, `return`, `global`, `function`, and `persistent`.

Key Term

The second aspect of selecting names has to do with human readers. The way in which you choose names can be an important aspect of how you communicate with other people; the computer, however, doesn't care so long as the name is syntactically valid. The assignment commands

```
>>  goldenratio = 1.61803398;
```

and

```
>>  v78wq9_8 = 1.61803398;
```

create two variables with identical values, but the second variable's name is much harder for a human to remember or use.

Here are some suggestions on how to pick variable names:

- Use names that are mnemonic, that is, remind you what is the role of the variable.
- Names shouldn't be overly long. MATLAB will only use the first 63 characters of any variable name, but a human reader will find it difficult to distinguish long names, and even harder to type them accurately. `VladimirPutinsFavoriteNumber` (only 28 characters) is descriptive but will be hard to use. In Chapter 4, we'll introduce ways to store data hierarchically so that a complete name doesn't have to be given to each piece of information individually.
- Variable names shouldn't be misleading. The following is allowed but is bad style:

```
>>  SquareRootOf3 = sqrt(2);
```

- Some single-letter names (for instance, `i`, `j`, `k`, `l`, `m`, `n`) are, by tradition and common usage, suggestive that the value is an integer. Others—`x` and `y`—suggest noninteger values such as 3.2.
- Don't be unnecessarily creative. Use mundane names for mundane roles (e.g., `counter`, `tmp`, `result`).

2.3 A Variety of Notations

Computer notation is harder to read than the traditional mathematical notation. For example, compare the traditional

$$\frac{GmM}{(x^2 + y^2 + z^2)^{\frac{3}{2}}}$$

to the equivalent infix

```
G*m*M/(x^2 + y^2 + z^2)^(3/2)
```

and the functional

```
rdivide(times(times(G,m),M), ...
       power(plus( power(x,2), ...
             plus( power(y,2), power(z,2))), ...
       rdivide(3/2)))
```

The problem with the traditional notation is that it relies on complicated typographical layouts that are difficult to enter into a computer. Even scientific notation for a number involves special typography (e.g., 6.67×10^{-11}) whereas the computer notation is a plain sequence of typed numbers: `6.67e-11`.

Some traditional notations require special symbols not found on the standard keyboard (e.g., $\sqrt{2}$). Sometimes the operation is implicit in the arrangement of the arguments, as in $8^{\frac{2}{3}}$, and sometimes the multiple arguments are arranged around a special symbol as in $\sqrt[3]{2}$. At more advanced levels in mathematics, the notation becomes richer:

$$\sum_{j=0}^{4} 2^{-j}, \quad \text{which means} \quad 2^0 + 2^{-1} + 2^{-2} + 2^{-3} + 2^{-4}$$

Since the keyboard is the primary means by which we provide instructions to computers, most computer languages are restricted to one of several types of notation that allow expressions to be sequences of typed characters. Some of the notations in actual use in computer languages are

`(+ 3 2)`	Prefix notation	Lisp and Scheme languages
`(3 + 2)`	Infix notation	MATLAB, FORTRAN, C++, and so on
`(3 2 +)`	Postfix notation	RPN calculators
`plus(3,2)`	Functional notation	MATLAB and other languages

At some fundamental level, all of the notations are the same; intrinsically, none is better than the others. All four notations in the previous list convey the same essential information: the name of the operator and the values and order of the arguments.

MATLAB uses a combination of functional-style and infix notation. The infix notation is convenient for arithmetical statements, which are common in scientific programming. The functional-style notation is used for most other purposes. When, in later chapters, you write your own operators, they will always be invoked using the functional-style notation.

There are a variety of other notations that have been introduced to streamline expressions. For example, MATLAB has a collection operator that can be invoked with the [and] brackets and a component-referencing operation involving the use of the period (.). We will encounter these notations, and others, in later chapters. But whatever the notation, it is important to keep in mind that the only purpose of the notation is to organize the basic facts needed to invoke a computation: what is the operator and what are the arguments.

2.4   Parsing

The cognitive processes involved in a human's reading a sentence are not well understood for the most part. How surprising, then, that the interpreter is able to read and understand our commands.

Part of the reason the computer can do this is that well-formed statements adhere to a strict and rather simple grammar that involves operators and arguments. Although the subject of computer-language grammar is a fascinating one—an area of close ties between linguistics, logic, and mathematics—people who use programming languages generally get by just fine without learning the explicit rules of grammar. Much as native speakers of a natural language learn to speak fluently without learning school-room grammar, so programmers seem to master the grammar without learning its rules.

Key Term

Nevertheless, it is important for programmers to understand, at least in outline, certain aspects of how the computer interprets expressions: the way that the interpreter breaks down an expression into its component parts, the operators and the arguments. This breaking down of the whole expression into parts is called *parsing.* Knowing some principles of parsing can help us to understand how to fix simple mistakes.

Key Term

The purpose of parsing is to divide the sequence of characters that constitutes an expression into discrete *tokens.* For example, consider the statement

```
foo = a + sqrt(32.1.*b);
```

This statement has several tokens, foo , = , a , sqrt , (, 32.1 , .* , b ,) , and ; . (I have adopted as a typographical convention placing each token in a ⎡box⎤. This is intended to indicate precisely which characters are in the token.)

The rules for forming tokens are rather complex, but there is a simplification that handles most cases. While you don't need to memorize these rules—the computer does that for you—being aware of them can help to understand problems that arise from ill-formed expressions.

Key Term

- The space between printed characters (such as inserted with the space bar at the center bottom of the keyboard or the TAB key) is called *white space*. White space is one of the delimiters that can mark the beginning or end of a token, but not the only one. White space itself is not a token, and it doesn't matter how much white space is placed between tokens. For example,

```
sqrt(2)
```

and

```
sqrt      (                    2    )
```

are completely equivalent expressions, each consisting of the four tokens ⎡sqrt⎤, ⎡(⎤, ⎡2⎤, and ⎡)⎤.

Key Term

- The newline character that separates lines is a *token*. In MATLAB, this token generally instructs the interpreter that the expression typed before the newline is complete. Thus, putting a newline within an expression causes an error. For instance, attempting to type a command like

```
sqrt( 2
)
```

generates an error message. This is because there are five tokens here: ⎡sqrt⎤, ⎡(⎤, ⎡2⎤, ⎡NEWLINE⎤, and ⎡)⎤. Ordinarily, MATLAB takes the tokens before ⎡NEWLINE⎤ and interprets them, in this case resulting in the ill-formed expression sqrt(2.

If, for typographical reasons, it is desired to place a newline within a statement, the continue-the-line token ⎡...⎤ must be placed before the newline, as in

```
sqrt( 2 ...
)
```

- A variable name must always be a single token.

Key Term

- There are some character sequences that always constitute a token, regardless of what surrounds them. These are called *atomic tokens*. Some of them are ⎡+⎤, ⎡-⎤, ⎡=⎤, ⎡.*⎤, ⎡./⎤, ⎡,⎤, ⎡:⎤, ⎡.⎤, the opening and closing parentheses ⎡(⎤ and ⎡)⎤, the left and right square brackets

$\boxed{[}$ and $\boxed{]}$, left and right curly braces $\boxed{\{}$ and $\boxed{\}}$, double-equal $\boxed{==}$, caret $\boxed{\wedge}$, dot-caret $\boxed{.\wedge}$, $\boxed{\&}$, $\boxed{|}$, $\boxed{\%}$, $\boxed{@}$, the tilde $\boxed{\sim}$, single quote $\boxed{'}$, slash $\boxed{/}$ backslash $\boxed{\backslash}$, and dot-backslash $\boxed{.\backslash}$.

Atomic tokens have their own meanings; for instance, .* identifies the multiplication operator and = identifies the assignment operator. We'll encounter the others in later chapters. Aside from an atomic token's meaning, it serves the same purpose as white space does, delimiting other tokens. For this reason, the characters in an atomic token can never be included within another token. For instance, dessert fanciers might like to create a variable named after a favorite ice cream

```
ben&jerrys = 'delicious';
```

But this is improper. ben&jerrys is actually three tokens—$\boxed{ben}$, $\boxed{\&}$, and $\boxed{jerrys}$—and the name of a variable must always be a single token.

- Any token that starts with a digit (or a period followed by a digit) is taken to be a number. Scientific notation numbers can also include the letter e followed by an integer. Similarly, an i following a number indicates an imaginary number (e.g., -6.7i or 7e11i). Some examples of valid tokens standing for numbers are 3.2, 0.32e1, and .32e1, and .00032e4, all of which stand for 3.2. Here's a statement that generates an error message:

```
≫  3.2.3
   ??? 3.2.3\\
      |\\
   Missing operator, comma, or semi-colon.
```

The statement 3.2.3 is interpreted as three tokens, a number $\boxed{3.2}$, the atomic token $\boxed{.}$, and a number $\boxed{3}$. Since the atomic token $\boxed{.}$ isn't meaningful with numbers, MATLAB complains that an operator is missing. By replacing the meaningless second period with a proper operator on numbers, the error is avoided. For instance, 3.2+3 and 3.2.*3 are both well-formed expressions.

- Variable names are sequences of characters. Since tokens starting with a digit are taken to be numbers, variable names can't start with a digit. This is mainly for historical reasons.

The only character that can appear in a variable name other than letters and digits is the underscore (_). Many programmers like to create variable names out of words, but since white space is a token delimiter, they are forced to use the underscore; for instance, an_interesting_variable_name. This is sometimes called the "bumpy" style. In this book, I will use the "caps" style: anInterestingVariableName.

Key Term

- The single-quote character (') gives an instruction to the parser to collect any characters that follow, including white space, into a single token called a *character string*. The character string is terminated by a matching single-quote character. Although the two single-quote characters that mark the beginning and end of a character-string token are identical typographically, due to their differing roles they are called the "open quote" and "close quote," respectively.

When an ill-formed statement is executed, an error message will be generated. Keeping the parsing rules in mind can help to make sense out of what may be a cryptic message.

But parsing is not everything: The order of tokens determines whether a statement is grammatical. For instance,

```
≫  foo - a = sqrt(32.1*b);
```

is easily broken down into tokens and has the same mathematical meaning as

```
≫  foo = a + sqrt(32.1*b);
```

But the first form is not a grammatical statement, since one of the rules of grammar says that the only tokens allowed on the left-hand side of the assignment operator = are those specifying a variable.

Help!

Every MATLAB operator has associated with it a bit of English-language explanation of how the operator works. You can have MATLAB print this explanation by using the help command:

```
≫  help cos

COS    Cosine.
COS(X) is the cosine of the elements of X.
```

Some commands have simple explanations, some have long and detailed ones.

2.5 Exercises

Discussion 2.1:
What is the difference between e1 and 1e1?

Discussion 2.2:
Explain how the following statements are broken down into tokens and whether they are valid expressions. If the expression is invalid, explain why.

```
>>   e3
>>   3.2;3
>>   3.2,3
>>   3.2x
>>   interestRate% = 3.2
>>   foo3.2
>>   23skidoo = 9;
```

Discussion 2.3:

A microwave oven takes two inputs (aside from the food to be warmed!): the duration of operation and the level of power. It would seem to be a simple matter to enter two numbers, but since different brands of ovens use different syntax, it can be quite confusing to use a new oven. Describe the button-pushing syntax used to enter the two arguments on an oven you are familiar with.

Discussion 2.4:

Explain why the following statement generates an error message:

```
>>   1.3e(1+2)
```

Discussion 2.5:

Write statements that will swap the value of two variables, a and b. To illustrate,

```
>>   a = 1;
>>   b = 2;
```
_____
                Your statements here

After your statements have been executed, the values of a and b should be swapped.

```
>>   a
   ans: 2
>>   b
   ans: 1
```

Discussion 2.6:

In evaluating a compound arithmetic expression, MATLAB obeys the following rules of precedence to determine the order of evaluations of the subexpressions:

First   Parentheses. Any subexpression in parentheses is evaluated before any other operation can be applied to that subexpression. In function evaluations, the arguments are effectively each in parentheses.

Second Exponentiation.

Third Multiplication and division.

Fourth Addition and subtraction.

For example, consider the compound expression:

$$5*(1+1)^2+1$$

The answer given by MATLAB is 21. This can be seen by following successive steps in the simplification of the compound expression.

$5*\boxed{(1+1)}\,^2+1$	Original compound expression.
$5*\boxed{2^2}+1$	The parenthetical expressions have the highest precedence.
$\boxed{5*4}+1$	Exponentiation is higher in precedence that multiplication or addition.
$\boxed{20}+1$	Multiplication is higher than addition.
21	Addition done last.

Some expressions involve operators of the same level of precedence. In this case, the operations are performed from left to right. For example, 1 - 2 - 3 involves doing the leftmost subexpression 1 - 2 first, then taking this result as input to the subexpression to the right, giving **(-1) - 3** and producing a result **-4**.

Evaluate the following expressions using the rules of precedence. Show the steps in each case.

1. 1 - 2 - 3 - 4
2. 1 - 2 ^3 - 4
3. 1 + 2 * 3 ^4
4. 1 / 2 / 3 / 4
5. sqrt(1 + 2 - 3 + 4)
6. 3 - 5 + 4/6 - 8*4^2

Although MATLAB faithfully and precisely follows the rules of precedence, it's wise generally to use parentheses to make your intentions explicit to yourself and any other human reader.

Exercise 2.1:
Find four different reasonable ways of evaluating the ambiguous statement

$$3 - 5 + 4/6 - 8*4^2$$

For each of the ways, rewrite the expression with parentheses to make it unambiguous and give the resulting value.

Exercise 2.2:
Write $(3-(5+2*8))/4$ in functional style using plus, minus, and rdivide. times.

Exercise 2.3:

Write the following arithmetic-notation expressions as both operator(argument) and infix-notation commands. You will have to assign values to the variables a, b, and c before you can evaluate your expressions successfully.

- $a+b+c$
- $a-(b+c)$
- $\sqrt{a^2+b^2}$

Exercise 2.4:

Write the following expressions in infix notation:

- `plus( plus( a, b ), minus( c, d) )`
- `rdivide( plus( sqrt(a), a), plus( power(a,b),b))`
- `power( rdivide(a, rdivide(b,c)), minus(d,power(f,g)))`

3

Simple Types: Numbers, Text, Booleans

We often think that when we have completed our study on"one,"
we know all about "two," since "two is one and one."
We forget that we still have to make a study of "and."
— Arthur Stanley Eddington (1882–1944)

All of the operators we have seen so far involve numbers; they take numerical values as inputs and return them as outputs. Of course, computers deal with nonnumerical information as well: text, images, sound, and so on. In this chapter, we start our tour of diverse sorts of information by taking a look at ways of computing with text and with logical statements of truth and falsity. We also examine in more detail how computers represent numbers.

All of these sorts of information are stored in the computer's memory. We start with some background on how computer memory is arranged.

3.1 The Organization of Computer Memory

Key Term

Almost without exception, modern computers store information using collections of *bits*. Each bit can have one of two values, generally written as 0 or 1, which might equally well be denoted as + and -, • and ○, ON and OFF, or TRUE and FALSE. In the computer hardware, the symbols are represented physically by the presence or absence of electrical charge or magnetic orientation, or other schemes. The word "bit" is short for Binary digIT,

Key Term

where *binary* refers to the two possible values of a single bit. A modern computer has a memory consisting of typically around 1 billion (10^9) bits with additional disk storage of perhaps 10 to 1000 times that number of bits.

A single bit has only two possible configurations: 0 or 1. A collection of 2 bits has four possibilities: 00, 01, 10, 11. A collection of 3 bits has eight possible configurations: 000, 001, 010, 011, 100, 101, 110, 111. More generally, in a k-bit collection, there are $\underbrace{2 \times 2 \times \cdots \times 2}_{k \text{ times}}$, or 2^k, possible configurations.

In most computer hardware, the large number of bits that make up computer memory is split up into small groups containing 8 bits. Such an 8-bit

Key Term

group is called a *byte*. A byte has 2^8 possible configurations: It can be used to represent the numbers 0 to 255 or any other 256 possibilities. Bytes are not the only types of grouping. For instance, a collection of 64 bits is used in MATLAB to represent real numbers, such as 3.14159 or -2.7×10^{26}. Such

Key Term

a collection of bits is called a *double-precision floating point number*, or "double" for short, and is the subject of Chapter 15.

To illustrate how collections of bits can be used to represent numbers, let's consider a small, 4-bit collection, something not actually used in MATLAB. Such a collection has $2^4 = 16$ configurations, and so, it's fairly easy to list all the possibilities. (Doing the same for a normal-sized 8-bit byte would take up 16 times as much paper.)

A simple way to interpret 4-bit collections is analogous to the way we interpret 4-digit numbers, such as 1374, in Arabic notation. But rather than the familiar ones place, tens place, hundreds place, thousands place, and so on, in binary numbers, we have the ones place, the twos place, the fours place, the eights place, ... The Arabic numeral 1101 is one thousand plus one hundred plus no tens plus one. The binary number 1101 is an eight plus a four plus no twos plus a one: altogether thirteen. The following table gives the complete set of such 4-bit numbers, using the convention that the rightmost bit is the ones place.

A Simple Numerical Interpretation of Bit Patterns

Collection of Bits				Expanded as a Sum of Decimal Numbers							Decimal Equivalent
0	0	0	1	_	+	_	+	_	+	1	1
0	0	1	0	_	+	_	+	2	+	_	2
0	0	1	1	_	+	_	+	2	+	1	3
0	1	0	0	_	+	4	+	_	+	_	4
0	1	0	1	_	+	4	+	_	+	1	5
0	1	1	0	_	+	4	+	2	+	_	6
0	1	1	1	_	+	4	+	2	+	1	7
1	0	0	0	8	+	_	+	_	+	_	8
1	0	0	1	8	+	_	+	_	+	1	9
1	0	1	0	8	+	_	+	2	+	_	10
1	0	1	1	8	+	_	+	2	+	1	11
1	1	0	0	8	+	4	+	_	+	_	12
1	1	0	1	8	+	4	+	_	+	1	13
1	1	1	0	8	+	4	+	2	+	_	14
1	1	1	1	8	+	4	+	2	+	1	15
0	0	0	0	_	+	_	+	_	+	_	0

In Arabic notation, we generally omit any leading zeros—we write 351 rather than 0351—but since bits are physical components in computer memory, they must always be in either the 0 or 1 state; they can't be blank.

Using this scheme, one byte with its 256 possible configurations can store any integer in the range 0 to 255; that is, 00000000 to 11111111. To represent numbers larger than 255, we can use collections larger than 8 bits. For example, a 16-bit number can represent the numbers 0 to 65535 since $2^{16} = 65536$. Because of the way that computer hardware is designed, such collections almost always have a size that's a multiple of 8 bits: one byte, two bytes, four bytes, and so on.

The numbers we have been describing so far are all positive. In Arabic numerals, we signify a negative number with a special symbol prefixing the number (for example, −376). This notation is cleverly reminiscent of how negative numbers can be generated from positive numbers using the subtraction operator: It's short for $0 - 376$. Another notation, used mostly by accountants, is to enclose the number in parentheses, [e.g., (376)] or to write the number in red ink. An early notation, used by Indian mathematicians of the sixth century, was to draw a circle around the number [3]. This works typographically because on a piece of paper there's space to write a new symbol or symbols beside the number and there's an opportunity to change the color of ink. In a computer, we face some challenges here. The indication of negativity has to be shoehorned into the number of bits allocated for storage, and a new symbol is not available; each bit is restricted to holding either 0 or 1. And, obviously, bits do not have an ink color. No problem! It's easy to imagine reserving the leftmost bit for an indication of the sign of the number with the remaining bits holding the magnitude of the number. Perhaps we would use 0 to mean + and 1 to mean −. Under this convention, using byte-sized numbers, we would have only 7 bits to hold the number magnitude, giving us the possible magnitudes 0 through 127. Here are some examples for an 8-bit signed number.

Bit Place and Meaning								**Decimal**
Sign	64	32	16	8	4	2	1	**Equivalent**
0	0	0	0	1	0	0	0	8
1	0	0	0	1	0	0	0	−8
1	1	1	1	1	1	1	1	−127
0	1	1	1	1	1	1	1	127
0	0	0	0	0	0	0	0	0
1	0	0	0	0	0	0	0	−0

The sign-bit convention can represent in one byte the numbers −127, −126, ⋯, 126, and 127: altogether 255 different numbers. The discrepancy between this and the 256 possibilities capacity of an 8-bit number is accounted for by the fact that one number, zero, is represented in two different ways, as illustrated in the preceding table.

Although the sign-bit scheme seems simple and exploits an analogy with Arabic notation, it is not the only way to represent negative numbers. For instance, another technique, *twos complement*, is often used because it simplifies the addition algorithm and because it has only a single representation for zero.

Key Term

Floating Point Numbers

Key Term

Floating point numbers, the sort generally used by MATLAB to store numerical values, are stored in a sort of scientific notation. A number such as 6.023×10^{26} consists of a *mantissa* (6.023) and an *exponent* (26). The *base* of the number is 10. Computer floating point numbers are stored with a base of 2. The mantissa and exponent are separate numbers, stored together in a 64-bit package (for double precision): 11 bits for the exponent and the remainder for the mantissa. (There is also a single-precision type, but this is generally not used in MATLAB.)

Numbers are often entered and displayed using scientific notation. For instance, 6.023×10^{26} might be entered as

```
>> avogadrosNumber = 6.023e26
```

The exponent can, of course, be negative,

```
>> plancksConstant = 6.626196e-34
```

Note that the units of Planck's constant—Joule seconds—have not been specified as part of the number.

When displaying numbers, MATLAB shows as many digits as you request. By default, 4 digits to the right of the decimal are shown; for instance,

```
>> plancksConstant
  plancksConstant: 6.6262e-034
```

This is the short display format. The long display format displays more digits:

```
>> format long
```

```
>> plancksConstant
  plancksConstant: 6.626196000000000e-034
```

Regardless of the precision of the display, double-precision floating point numbers are always stored and processed using the entire 64-bit capacity.

The accepted international standard for floating point computation (the IEEE 754 standard) includes two special forms of numbers. Inf, short for infinity, stands for a quantity whose magnitude is larger than can be precisely stored in the available space; for instance,

```
>> 1e400
  ans: Inf
>> 1./0
  ans: Inf
```

NaN, short for "not a number," is the value returned when the result of a numerical computation has no well-defined value; for instance,

```
>> 0./0
  ans: NaN
>> 0.*Inf
  ans: NaN
>> Inf-Inf
  ans: NaN
```

3.2 Text

Numbers are not the only type of data encountered in science. We often need to be able to store names and descriptions. Sometimes we need to manipulate and compare text data. Examples include the following:

- DNA sequences notated using the letters A, C, G and T
- Times and dates such as "July 16, 2001 at 3:02 PM" or "c. 812 BCE"
- Names of elements such as K or Na as well as chemical formulas such as C_6H_{14}
- Text in general: "Call me Ishmael"
- Latitudes and longitudes: 180 degrees 3 minutes 4 seconds E, 44 degrees 56 minutes 8 seconds N, the location at which this text is currently being written
- Identifying names (e.g., *E. coli*)
- Messages and warnings to the user of software (e.g., `Invalid input: Q is not a chemical element symbol`)
- Units attached to numbers (e.g., 7 kg, 56 m, or $6.626196 \times 10^{-34} J \cdot s$)

Key Term

Text data are composed of *characters*, a term that comprises letters, numbers, symbols, and the various no-ink formatting entities, such as the blank space, tab, and so on. Since ordinary commands are also text, there needs to be a way to distinguish text data from command text. The single-quote token $\boxed{\,'\,}$ is used to communicate to the parser that the text that follows is to be regarded as text data and not as text to be parsed into tokens. For instance,

```
>> dnaLetters = 'ACGT';
```

indicates to the parser that ACGT is text data. Without the quotation marks, MATLAB will parse ACGT in the ordinary way and then attempt to find the variable or operator with the indicated name. If it is unable to find one, it will generate an error:

```
>> dnaLetters = ACGT
   ???: Undefined function or variable 'ACGT'.
```

The opening and closing quotes do not become part of the character string: They just signal to the parser where the text begins and ends. For this reason, the quotes are not shown when a string is displayed as a result. For example,

```
>> greeting = 'hello'
   greeting: hello
```

Note also that the double-quote character (") is distinct typographically from two single quotes (' '). The double quote has no special significance to the parser; only the single quote is used to collect text data.

Key Term

The collection of characters in an entity such as 'ACGT' is called a *character string*. The number of characters in a string, the *string length*, can be found with the `length` operator:

```
>> length('How long is this string?')
   ans: 24
>> dnaLetters = 'ACGT';
```

```
≫  length(dnaLetters)
   ans: 4
≫  length('dnaLetters')
   ans: 10
```

Note that in `length(dnaLetters)` it is not the name `dnaLetters` whose length is being counted but the character string stored under the name `dnaLetters`.

The opening quotation mark instructs the command-line parser to collect into a string the characters that follow until a closing quote is found. The quoted characters are stripped of any special meaning they might otherwise have had for the parser; for example, the expression

```
≫  '3+2'
   ans: 3+2
```

just collects the three characters 3, +, and 2. In contrast, leaving out the quotes invokes a different computation

```
≫  3+2
   ans: 5
```

where the arithmetic operation is performed.

The semicolon (;) has no special meaning within a character string.

```
≫  'This ; is taken literally.'
   ans: This ; is taken literally.
```

but outside the string it is a directive to the interpreter:

```
≫  'But the one here means do not print!';
```

There are a few exceptions where the parser does modify the quoted string:

- The quotation mark (') signifies the end of a character string unless it is followed immediately by a second single quote. A pair of quotes in the interior of a character string indicates that an apostrophe is intended:

  ```
  ≫  'I can''t see John''s dog.'
     ans: I can't see John's dog.
  ```

Key Term

- Even within a string, the *newline* character (the character that is produced by the keyboard when you press the ENTER key) still signals to the parser that the typed command line is being handed over for processing. If the string has not been terminated by the closing quotation mark, an error will result even if the continuation token has been used.

```
🐛 ≫ 'hello ...
    ???: 'hello ...
         |
         Missing ' at end of string or misplaced
         transpose operator.
```

Internal Representation of Characters

There are all sorts of ways to represent characters using bits. If you look carefully at a computer screen, you will see that the characters are composed of little dots: pixels. Each pixel is either on or off (at least on a black-and-white screen) and so this visually oriented representation really consists of a collection of bits. Figure 3.1 shows an example.

The visual representation is effective in communicating with humans, but it is somewhat wasteful in terms of storage. The character in Figure 3.1 consists of a block, or "array," of 12×16 bits, altogether 192 bits. Many different patterns could be stored in such a way, some of which resemble characters and most of which don't. With 192 bits we can represent $2^{192} \approx 10^{58}$ different configurations. This is more than we need, even for all the writing systems of the world. It is actually quite bigger than all the characters that have ever been written by all the people ever on earth!

Even worse than the extra capacity of the bit array is the difficulty of identifying and manipulating such characters. Consider what an algorithm would look like that translates characters from uppercase to lowercase. It would have to recognize each of the many possible configurations corresponding to $\boxed{\text{A}}$ and then pick one of the configurations corresponding to $\boxed{\text{a}}$.

How many bits do we really need to store characters? In English, there are 26 letters (a, b, ..., z) as well as the 10 digits and sundry punctuation marks and symbols (e.g., period, comma, colon and semicolon, ?, !, −, #, and so on), perhaps 50 symbols altogether. With just 6 bits we could hold $2^6 = 64$ distinct patterns, giving us some extra capacity, which we could use for special, rarely used symbols. But what about the uppercase letters A, B, ..., Z? One possibility is to add a "case bit"—another bit that signifies whether a letter is intended to be uppercase or lowercase. This is essentially the role played by the SHIFT key on a typewriter. The SHIFT key isn't an ordinary key; it doesn't stand for any particular letter or symbol. Instead, it is a two-position switch that sets whether another pressed key will be upper- or lowercase. A two-position switch is either on or off: It carries one bit's worth of information. The use of a case bit would work well and certainly would simplify the algorithm for translating from upper- to lowercase: Just turn off the case bit.

The seven resulting bits—6 character bits plus a case bit—can store 128 distinct patterns. The 7 bits produce patterns like 0011011 or 1110101 or 1010101 that do not in any way resemble characters, particularly when we

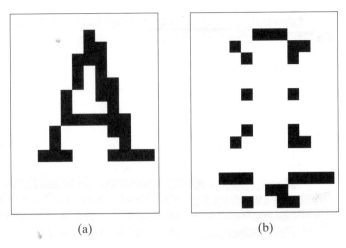

(a)   (b)

Figure 3.1. (a) The character A as a pixel array. (b) A pixel array that doesn't resemble a character.

remember that the 1 and 0 are just the printed notation we use to stand for the presence or absence of electrical charge or magnetic polarization. In principle, we could interpret the 128 seven-bit patterns in any way we like, perhaps having 0000000 stand for *a*, 0000001 stand for *b*, and so on. There are obvious advantages in standardization so that computers can communicate with one another and so that the various subsystems on a computer can work consistently. Imagine the confusion if the e-mail program on your computer interpreted 0010111 as N while the program on the e-mail recipient's computer interpreted 0010111 as Y.

A widely used standard used by MATLAB and many other programs is called ASCII: a 7-bit character code that assigns an interpretation as characters to each of the 128 possible 7-bit patterns. The term ASCII (ASCII) is short for the organization that established the standard: the American Society for Communication and Information Interchange.

Not all of the patterns are printed characters: 0001000 is a sound ("ring the bell"), 0100000 is a space, 0001001 is a tab, 0001101 means "go to the next line," and 0001010 means "bring the carriage back to the left margin of the page." "Carriage?" ASCII was developed when typewriters were dominant. Typewriters have separate mechanisms for advancing to a new line on the page and for moving the print head to the left. The consequences of this still annoy scientific programmers who transfer data between different types of computers. Unix, Windows, and Macintosh computers each use a different way of indicating the end of a line of text and the beginning of a new one on the left side of the page. Some software fails to recognize the differences.

Since computers commonly deal with 8-bit blocks, the 7-bit ASCII characters generally are packaged into 8-bit-long bytes. In an international world, uses have emerged for the information capacity of the additional bit—it gives an additional 128 possible characters beyond those in the ASCII set. Often, these extras are used to represent non-English characters and letters with accents (e.g., á, ö, ñ, and so on). The failure to standardize the meaning of the eighth bit is one reason why internationally transmitted e-mail often contains out-of-place characters. MATLAB packages ASCII characters into 16-bit (2-byte) units: It can therefore potentially handle 65,536 different characters, but only the first 256 are defined in most display fonts.

Key Term

Recognizing the inadequacy of the ASCII code to represent non-English characters, a more extensive standard, *Unicode*, has been developed [8]. Unicode, along with the international standard *ISO 10646* under which Unicode was developed, defines a 31-bit character set. So far, characters have been assigned to only the first 65,534 positions: the first 16 bits. This is adequate to cover the characters used in practically all known languages: from Latin and Greek to Hebrew and Arabic, Chinese, Japanese, Korean Han, Tamil, Thai, Khmer, Bengali, Runic, Cherokee, Yi, and others. The use of Unicode is still not universal, and we will not consider it further in this book.

People often fail to appreciate the severe limitations of using ASCII to represent typography and formatted documents. Although ASCII does include several formatting characters—white-space characters such as tab, newline, and the space between words—it has no built-in mechanism to indicate subscripts or superscripts, font shapes (e.g., italics or bold), underlining, color, and so on. Many word processing programs use a non-ASCII code to allow formatting and font information to be stored; this is why the contents of a word-processed document can be incomprehensible if opened in another brand of word processor.

Some standards for representing formatted documents are based on ASCII and use special words and markings to denote format information. Hypertext Markup Language (HTML) is probably the most widely known of these. This book is being written in another ASCII-based formatting system called LaTeX, which was specifically designed for scientific and technical documents and has excellent facilities for the typographical layout of mathematical notation. Examples of HTML and LaTeX commands are given in Figures 3.2 and 3.3.

It is important to remember that computer commands in almost all computer languages are based on ASCII. This has some consequences:

- You can't use subscripts, superscripts, or non-ASCII characters in commands, variable names, or character strings. For example,

    $\gg$ π = 3.14159;

  is not a possibility and instead has to be written as something like

```
<!doctype html public "-//w3c//dtd html 4.0 transitional//en">
<html>
<head>
<meta http-equiv="Content-Type" content="text/html;
charset=iso-8859-1">
<title>Widgets: A History</title>
</head>
<body>

<h1>A History of Widgets</h1>

<p>Widgets have played an important role in history. 
They should be carefully distinguished from
<i>thingamabobs</i>and
<i>gadgets</i>,
to say nothing of
<i><a href="http://www.silly.org/dict.html">whatyamacallits</a></i>.
</body>
</html>
```

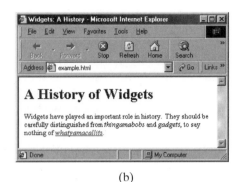

(a)   (b)

Figure 3.2.   (a) A source HTML document. ASCII characters are used both for the contents and to provide information about formatting (for example, section headers and hyperlinks).
(b) The document as rendered by Web browsing software that interprets the formatting commands.

$$\gg \texttt{ pi = 3.14159;}$$

Similarly, we can't write 3^2 but instead write `3.^2` or `power(3,2)`.

- Formatted documents from word processors can't be used directly to hold computer programs. Instead, the documents must be saved in ASCII format. Most word processors have an option for doing this. MATLAB includes an ASCII-based word processor for creating, displaying, and editing programs. Word processor programs intended to create and edit ASCII documents are generally called *text editors*.

```
\centerline{\bf Latex for Mathematical
Typesetting}

\LaTeX\ is a typesetting system for mathematics
and other technical documents. Mathematical
symbols such as $\pi$ and $\xi $
can be specified in {\sc ascii} text,
as can equations, e.g.,
$$\ln x = \int_1^x \frac{1}{x} dx $$

This book is composed using \LaTeX.
```

Latex for Mathematical Typesetting

LaTeX is a typesetting system for mathematics and other technical documents. Mathematical symbols such as π and ξ can be specified in ASCII text, as can equations, e.g.,

$$\ln x = \int_1^x \frac{1}{x}\,dx$$

This book is composed using LaTeX.

(a)   (b)

Figure 3.3.   (a) A source LATEX document. Complex formatting information is provided by ASCII commands.
(b) The LATEX compiler uses the typesetting instructions to translate the document into a non-ASCII format that can be printed. LATEX is widely used in mathematics, physics, and computer science.

Here's an example of how the lack of formatting commands in the ASCII standard can lead to confusion. *The New York Times* reported the following [7]:

> **...a new musical ... called "Proof" ... is about Andrew Wiles, the Princeton professor who explained one of mathematics' most famously lingering enigmas, Fermat's Last Theorem. (It declares that for integer values of *n* greater than 2, no whole number solution exists for *xn* plus *yn* equals *zn*.)**

Reading this, one wonders why Fermat's Last Theorem took more than 200 years to be proven. As stated in the newspaper, the theorem is obviously false. The newspaper equation

$$xn + yn = zn$$

has solutions for any value of *n*, for instance, $x = 1, y = 2, z = 3$.

An accurate statement of the theorem is that

$$x^n + y^n = z^n \text{ has no whole number solutions for integer values of } n.$$

Since ASCII has no built-in superscripts, the mathematical expression needs to be represented in a special way. Using MATLAB notation, for example, we might write the expression as

```
(x.^n + y.^n) == z.^n
```

which has the correct meaning and uses only ASCII characters.

3.3 Collections of Numbers and Plotting

We have seen that character strings can be represented as a sequence of numbers interpreted using the ASCII code. A single variable can hold an entire sequence of characters, making it easy to handle a character string as a unit. Sequences of numbers, even without the ASCII interpretation, are useful in general; indeed, they are possibly the most important type of data used in scientific work.

Key Term A sequence of numbers is often called a *vector*. You can create a vector in a command using a special syntax involving the *collection brackets* [and]:

```
≫ [1, 5, 7, 9]
   ans: 1 5 7 9
```

(In the next chapter, we'll see that the collection brackets are rather more versatile than this.)

An important use for the collection brackets is in assembling data for plotting or analysis. The following table gives measurements of the temperature of water in a coffee mug. The water was heated to boiling in a microwave oven and, after removing the mug from the oven, the temperature was measured at various times.

Time (min)	Temperature (deg C)
0	93.5
1	90.6
2	87.7
3:30	83.9
7	76.6
10	71.3
18	61.5
26	54.8
34	49.2
52	41.0
113	29.1
166	25.5
191	25.0

We can collect these data into two vectors, one called `time` and the other `temp`.

```
≫  time = [0, 1, 2, 3.5, 7];
≫  temp = [93.5, 90.6, 87.7, 83.9, 76.6];
```

(To save space and ink, we entered only the first five values here.) Note that the traditional notation, 3:30, meaning 3 minutes and 30 seconds, has been translated into a decimal form, 3.5 minutes.

Many operators take vectors as arguments. For example, to convert the time measured in minutes into seconds, we can multiply by 60:

```
≫  time.*60
  ans: 0 60 120 210 420
```

Converting the temperature from Celsius to Fahrenheit is done with

```
≫  32 + temp.*9./5
  ans: 200.30 195.08 189.86 183.02 169.88
```

Other operators work in the same way:

```
≫  time +1
   ans: 1 2 3 4.5 8
≫  sqrt(time)
   ans: 0 1.00 1.41 1.87 2.65
≫  time.^2
   ans: 0 1 4 12.25 49
≫  log(time)
   ans: 0 0.69315 1.2528 1.6094 2.1972
```

Those operators that take two arguments (for instance, the arithmetic operators) will work on two vectors so long as they have the same size:

```
≫  time + temp
   ans: 93.5 91.6 89.7 87.4 83.6
```

Although the preceding computation works arithmetically, it makes no sense physically because time and temp are in different physical units: minutes and degrees Celsius.

Key Term

The arithmetic operators +, -, . *, . / and the exponentiation operators . ^ and sqrt are all *element-by-element operators*; that is, they work on the components of a vector one at a time and return another vector with the same number of elements as the input. Other useful operators compute a single number from the vector. As we've already seen, the length operator counts the number of elements in the vector. The sum operator adds up all the elements

```
≫  sum([1 2 3 4])
   ans: 10
```

and the prod operator multiplies them:

```
≫  prod([1 2 3 4])
   ans: 24
```

There are also operators for the cummulative sum and product, cumsum and cumprod:

```
≫  cumsum([1 2 3 4])
   ans: 1 3 6 10
≫  cumprod([1 2 3 4])
   ans: 1 2 6 24
```

There are several operators for computing statistics on vectors: mean computes the arithmetic mean of the elements of the vector, median returns a value that half of the elements are greater than or equal to and the other half less than or equal to, min returns the smallest element in the vector, `max` returns the largest, and `std` calculates the standard deviation of all the elements.

Both beginning and advanced programmers often unintentionally leave out the dot in the element-by-element operators `.*`, `./`, and `.^`. These operators each involve two characters. The corresponding one-character operators, `*`, `/`, and `^`, are for *matrix operations*, a subject we will not need until Chapter 17. An expression that makes perfect sense using an element-by-element operator, for instance,

Key Term

```
≫  [1 2 3].*[4 5 6]
  ans:  4 10 18
```

may generate an error when a matrix operation is used, for example,

```
🐞 ≫  [1 2 3]*[4 5 6]
    ???:  Error using ==> *
          Inner matrix dimensions must agree.
```

Here are some more important numerical operators that work on vectors on an element-by-element basis:

Sign-related operators: `abs`, `sign`

Exponential functions and their inverses: `exp`, `log`, `log10`, `log2`

Trigonometric functions and their inverses: `sin`, `cos`, `tan`, `asin`, `acos`, `atan`, and `atan2` and the hyperbolic functions `sinh`, `cosh`, `tanh`, `asinh`, `acosh`, `atanh`

Fraction-to-integer conversions: `round`, `floor`, `ceil`, `fix`

Remainders and moduli: `rem`, `mod`

Plotting

One of the most important operations using vectors is plotting. MATLAB contains many operators for producing graphics; the most important is the `plot` operator. Plot is simple to use but also flexible and powerful. The command

```
≫  plot(time, temp,'-o')
```

makes a graph (see Figure 3.4) with the time variable on the horizontal axis and the temperature variable on the vertical axis. From the plot we see: The temperature is going down, rapidly at first but later at a slower rate.

The `plot` operator does not return any value, but it has the useful side effect of drawing a graph. `Plot` graphs the corresponding elements of the two vectors one against the other: in this case 90.6 versus 1, 87.7 versus 2, and so on. The first argument is plotted as the *x*-variable (more formally known as the "abscissa"), the second argument as the *y*-variable (known as

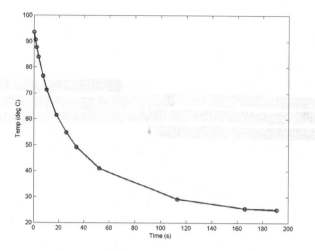

Figure 3.4.   A plot of the coffee temperature data made with `plot(time,temp,'-o')`.

the "ordinate"). The third argument is a character string that describes how to draw the points being plotted. (See Discussion 3.3 in the Exercises.)

By itself, `plot` doesn't add labels to the plot. This can be done with the graphical plot-editing tool, or with the `xlabel`, `ylabel`, `title`, and `text` commands.

Constructing Sequences of Numbers

A commonly encountered task is to plot a mathematical function $f(x)$. For example, Figure 3.5 shows $\sin 3x$ for x over the domain 0 to 5. One possible algorithm for making such a plot is simple:

Repeat the following for each value of x in the domain 0 to 5: Compute $f(x)$, then plot a point at coordinates $(x, f(x))$.

For example, let

```
≫ x = 0.01
```

Then $\sin 3x$ is

```
≫ sin( 3.*x )
  ans: 0.029996
```

Plot a point at $(0.01, 0.02996)$. Repeat this for the other values of x in the interval 0 to 5.

A moment's reflection reveals a difficulty with this algorithm: There is an infinite number of different values of x in the domain. Plotting all of them would take an infinite amount of time.

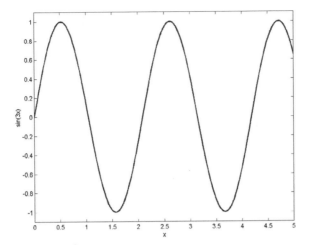

Figure 3.5. A plot of sin(3*x*).

This difficulty is easily dealt with simply by taking a finite number of different *x* values spaced closely enough to fool the eye into seeing a continuum. So we need only create a vector of closely spaced x ranging from 0 to 5, then compute sin(3.*x), then plot one versus the other. For example,

```
≫ x = [0, 0.01, 0.02, ··· and so on up to 5];
≫ y = sin(3.*x);
≫ plot(x,y)
```

Unfortunately, the computer isn't programmed to interpret "and so on" in the manner that a moderately educated human would do. Instead, we can use the colon operator:

```
≫ x = colon(0, 0.01, 5)
```

Colon takes three arguments: the starting number in the sequence, the step size, and the ending number, and returns as output a vector containing the sequence. For example,

```
≫ colon(0, .5, 2)
  ans: 0 .5 1 1.5 2
```

The ending number may not be in the returned sequence; for example,

```
≫ colon(.1, .5, 2)
  ans: .1 .6 1.1 1.6
```

Admittedly, colon is a pretty strange name for an operator that produces sequences: seq or sequence or makeseq would have been better. The funny name derives from the special syntax that can be used in place of the `colon` operator. This syntax involves separating the arguments to the operator by colons. For example,

```
≫  0:.5:2
   ans: 0 .5 1 1.5 2
≫  .1:.5:2
   ans: .1 .6 1.1 1.6
```

If the middle argument isn't given, it is taken by default to be 1:

```
≫  0:5
   ans: 0 1 2 3 4 5
```

The functional syntax, using `colon`, has the advantage that the grouping of operations is obvious to the reader. For example,

```
≫  colon(1, 5) + 1
   ans: 2 3 4 5 6
```

is obviously quite different from

```
≫  colon(1, 5+1)
   ans: 1 2 3 4 5 6
```

But what is `1:5+1`? When using the special syntax, remember to use parentheses to avoid any ambiguity. Note the difference between the following two expressions:

```
≫  1:(5+1)
   ans: 1 2 3 4 5 6
≫  (1:5)+1
   ans: 2 3 4 5 6
```

The collection bracket operator will work on vectors as well as numbers:

```
≫  [1:3, 2:5]
   ans: 1 2 3 2 3 4 5
```

Using the special colon syntax, our plot can be made with this sequence of commands:

```
≫  x = 0:.01:5;
≫  y = sin( 3.*x );
≫  plot(x,y)
```

A Matter of Style: _____

In the preceding example, we could have made the entire plot with a single command

```
≫  plot( 0:.01:5, sin(3.*(0:.01:5)) )
```

The problem here is redundancy: The same information is being given twice. It's too easy to make a mistake when repeating information. The following command has a "small" mistake and generates an error message.

```
≫  plot( 0:.01:5, sin(3.*(0:.1:5)) )
   ans: ??? Error using ==> plot
        Vectors must be the same lengths.
```

A good rule of thumb is to use assignment statements to avoid having to type exact duplicates. Using assignment to x allows us to guarantee that exactly the same values are used in all places where duplication is intended, and makes it clear to the human reader what is intended.

The Empty Vector

A surprisingly useful vector is the *empty vector*, written []. The uses for this will become apparent in Chapters 8 and 12. The empty vector plays the same role in collecting as the number 0 plays in addition or 1 plays in multiplication:

```
≫  [ [2, 5, 1], [] ]
   ans: 2 5 1
≫  [ [], [2, 5, 1] ]
   ans: 2 5 1
```

Scalars

MATLAB treats a single number as a vector, but with only one element.

```
≫  length(7.3)
   ans: 1
≫  length([7.3])
   ans: 1
≫  [7.3] == 7.3
   ans: 1
```

The term *scalar* is used to refer to such single numbers.

3.4 Booleans: True or False

What use is a single bit? The types we have encountered so far are made up of collections of bits: 8 bits for a conventionally packaged ASCII character, and so on. Yet a single bit is capacious enough to hold the answer to an important class of questions: yes or no. Are you 21 or older? It takes only 1 bit to hold the answer.

Key Term

A 1-bit quantity is termed a *boolean* value, named after an nineteenth-century mathematician, George Boole, who developed the logic of reasoning with true and false.

In English, we have a specialized syntax for asking questions. "Are you 21 or over?" really is a question about numbers:

Is   your age   greater than or equal to   21   ?
      quantity 1   relationship   quantity 2

The verb *is* and the question mark indicate the beginning and end of the phrase but don't otherwise govern its meaning. The question is really about the truth or falsity of a relationship between two numbers: your age and 21.

Such a question, when part of an exchange between humans, might have several types of answers:

- "I'm 18" gives the questioner enough information to answer the question herself—it's really the answer to another question, "What is your age?"
- "I knew this would happen if I forgot my ID" Dissimulation and retreat, not quite a lie.
- "Here's my identification card" anticipates the next question that might be asked.
- "Of course" is a way of saying that the questioner should have been able to figure out the answer herself without asking, or perhaps is an acknowledgement that the question was a formality whose answer is obvious.
- "No" is one of the two boolean answers. The other is "yes." People often think it is impolite to answer questions with a simple yes or no and so we are not practiced in doing so.

In a court room, a lawyer will sometimes direct the witness to answer yes or no. Such instructions are an attempt to prevent the witness from introducing additional information such as "It was an accident" or "I didn't mean to." The lawyer wants to restrict the information to 1 bit, a boolean quantity.

Questions with boolean answers involve a computation: a translation from inputs to a boolean output. Seen this way, there is no reason not to treat boolean questions as an ordinary operator and to use the ordinary operator sequence. The infix operator for "greater than or equal to" is `>=`. It takes two arguments and returns a boolean that indicates whether the first argument is greater than or equal to the second. Thus, `yourage >= 21` is the computer equivalent of "Is your age 21 or older?" (assuming that the variable `yourage` has a value giving your age in years).

MATLAB prints out the 1-bit boolean answer to the question as a 0 or a 1. Zero means "no" or "false"; one means "yes" or "true."

```
>> 18 >= 21
ans: 0
```

```
≫ 27 >= 21
  ans: 1
```

There are a variety of operators for comparing two numbers:

Numerical Relational Operators

Traditional Notation	Infix Operator	Meaning	Functional Operator
=	yourage == 21	equality	eq
≠	yourage ~= 21	inequality	ne
<	yourage < 21	less than	lt
>	yourage > 21	greater than	gt
≥	yourage >= 21	greater than or equal	ge
≤	yourage <= 21	less than or equal	le

Note that the equality operator's infix symbol is a token composed of two equal signs (==). Remember that a single equal sign (=) is part of the assignment command and has a completely different meaning.

The numerical relational operators take two numbers as arguments. There are other boolean operators that take only one input. For example, isprime answers whether the input is a prime number:

```
≫ isprime(7)
  ans: 1
```

Boolean Operators on Vectors and Strings

Consider the following two vectors x, and y:

```
≫ x = [5 6 7];
≫ y = [6 5 7];
```

Is x the same as y? The answer depends on what you mean by "same." x and y contain the same numbers, but they are in a different order. So they are the same in one sense but different in another. A boolean equality statement like x == y asks a definite question: Are the corresponding members of x and of y the same?

```
≫ x == y
  ans: 0 0 1
```

Key Term

The answer is that the last number in x is the same as the last number in y, but the other corresponding numbers differ. The output from the == operation is a *boolean vector*, a vector composed of boolean elements. A synonym used in MATLAB documentation is *logical vector.*

The equality operator == always compares two numbers at a time, as do the other numerical relational operators. When there are two vectors involved, the operators compare the corresponding numbers; correspondence always refers to position and not to value. When the two vectors have the same length, as is the case with x and y, then the operators compare the first number on one vector to the first number in the other, and so on. When one of the vectors is a scalar, the operators compare that scalar to each of the numbers in the vector, producing a boolean answer for each number in the vector:

```
≫ 5 == x
   ans: 1 0 0
```

Otherwise, if the vectors differ in length, then the operators cannot return a sensible output and are set up to generate an error message instead:

```
≫ [3 4] == x
   ans: ??? Error using ==> ==
```

When comparing character strings, our notion of "sameness" is based on the entire string, and not just the individual members: what we require for two strings to be the same is an exact match between each pair of corresponding characters. Despite the fact that character strings are represented as vectors of ASCII numbers, the numerical comparison operators do not work in a way suitable for string comparisions. For instance,

```
≫ greeting = 'Hello.';
≫ greeting2 = 'Hello!!';
```

The two greetings cannot be compared effectively with == since they are not the same length.

```
≫ greeting == greeting2

??? Error using ==> ==
Array dimensions must match for binary array op.
```

We need an operator that doesn't generate an error in such a situation but that reports that the two strings are different, if only because they are different lengths. The strcmp operator implements this requirement: It takes two character strings and answers whether they are identical:

```
≫ strcmp(greeting, greeting2)
   ans: 0
```

"Hello!!" and "Hello." are not the same strings, even if they are practically the same sentiment.

Only when the string match is exact does strcmp give a positive answer:

```
≫ strcmp(greeting, 'Hello.')
   ans: 1
```

Even capitalization makes a difference to the `strcmp` operator:

```
≫ strcmp(greeting, 'hello.')
   ans: 0
```

There is another operator, `strcmpi`, that ignores capitalization:

```
≫ strcmpi(greeting, 'hello.')
   ans: 1
```

3.5 Logical Operators

Key Term

We often need to combine simple boolean expressions to construct more complicated conditions. In everyday language, we do this using words like *and, or, but not, unless,* and so on. The corresponding computer operators, called *logical operators*, listed in Table 3.1, allow us to combine boolean values. For example, "Are you between 25 and 30?" can be written

```
≫ (yourage >= 25) & (yourage <= 30)
```

As an example, consider a situation where your birthday is represented with two variables, `m` for the birth month and `d` for the day of the month. A birthday of February 17th would be

```
≫ m=2;
≫ d=17;
```

Answering the question of whether a given birthday is in the winter involves a compound expression. Let's define winter to be days between December 22nd and March 21st. Then a birthday is in the winter if the month is either January or February, **or** if it is in December but after the 22nd, **or** if it is in March but before the 21st. In terms of the logical operators, we have

```
≫ (m==1) | (m==2) | (m==12 & d>=22) | (m==3 & d<=21)
```

Another example: A leap year (that is, one with 366 days) is defined to be a year evenly divisible by 4, except for years evenly divisible by 100, except for years evenly divisible by 400. If the year number is stored in variable `y`, the corresponding statement expressed as logical operators is

```
≫ rem(y,4)==0 & ~(rem(y,100)==0 & ~rem(y,400)==0)
```

There is always more than one way to write a logical statement. For example, `a & b` is completely equivalent to `~( (~a) | (~b) )`.

Table 3.1   Some Common Logical Operators

Operator	Infix Notation	Meaning
not(a)	~a	Returns 0 if a is 1, and 1 if a is 0.
or(a,b)	a | b	Returns 0 if both a and b are 0, otherwise returns 1.
and(a,b)	a & b	Returns 1 if both a and b are 1, otherwise returns 0.
xor(a,b)	none	Returns 1 if both a and b are different, 0 otherwise.
any(a)	none	Returns 1 if any of the elements in the Boolean vector a are 1, otherwise 0.
all(a)	none	Returns 1 if each and every one of the elements in the Boolean vector a are 1.

The any operator is a generalization of or to vectors or booleans: It returns 1 if any of the elements in the vector are 1, and 0 if they are all zero. For instance,

```
≫ any( 10 < [4 2 52])
  ans: 1
```

Similarly, all is a generalization of and to vectors.

```
≫ all( 10 < [4 2 52])
  ans: 0
```

The xor operator, called "exclusive or," is not used much in everyday language. It has the interesting property that it is self-inverting. To illustrate, take two random boolean vectors:

```
≫ a = (rand(1,10)>.5)
  a: 1 0 1 0 1 1 0 0 1 0
≫ b = (rand(1,10)>.5)
  b: 1 1 1 1 0 0 1 1 0 1
```

Combining a and b using xor

```
≫ c = xor(a,b)
  c: 0 1 0 1 1 1 1 1 1 1
```

By self-inverting, we mean that c combined with b gives back a, and c combined with b gives back a:

```
≫ xor(b,c)
  ans: 1 0 1 0 1 1 0 0 1 0
```

```
≫  xor(a,c)
   ans: 1 1 1 1 0 0 1 1 0 1
```

or, demonstrating the equivalence more compactly,

```
≫  all(a == xor(b,c))
   ans: 1
≫  all(b == xor(a,c))
   ans: 1
```

The `xor` operator is important in computer graphics, where it provides a way to draw overlays, such as a mouse cursor, in a way that that can be erased simply by drawing the overlay again.

3.6 Exercises

Discussion 3.1:
Explain the following commands and why the returned values are what they are:

```
≫  '3+2' + 7
≫  'hello' + 'world'
≫  'hello' + 'fred'
```

Discussion 3.2:
Explain how the following expressions are broken down into tokens and whether the expressions are well formed. If not, explain why.

```
≫  quote = 'Now is the time ... newline
for all good men'
≫  'a bird' = 'cardinal';
≫  answer = ''Who can say?'';
≫  seven = 'three' + 4;
≫  b = 7(3+4);
≫  3e1.72;
≫  3 e1
≫  z = 4.5e1i;
```

Discussion 3.3:
Try the following as the third arguments to `plot`:

```
'.'   '+'   'g*'   'ro'   '-'   'k^'
```

What is the default third argument (that is, the value of the third argument that is implied if only two arguments are given)?

Discussion 3.4:
The ASCII code involves an identifiable case bit that indicates whether a letter

is upper- or lowercase. Which bit is it? Does 1 or 0 mean lower case? (*Hint:* You can figure this out by looking at the numerical ASCII codes.)

Describe (in words) an algorithm for doing what the `toupper` does.

Discussion 3.5:

The `num2str` operator takes a number as an argument and returns that number formatted as a character string.

```
≫  num2str(38)
   ans: 38
```

This seems quite confusing, because the character string ′38′ prints out in a way that looks just like the number 38. Explain what's happening in the following:

```
≫  length(38)
   ans: 1
≫  length(num2str(38))
   ans: 2
≫  double(num2str(38))
   ans: 51 56
```

Discussion 3.6:

Perhaps 10^{11} people have lived on earth in all history since writing began. Estimate how many characters (that is, the total amount of text) all of these people could possibly have written. Give a reasonable upper bound (a number that you are confident exceeds the unknown actual number of characters written). Say how many bits would be needed to give each of these characters a unique identifier.

Discussion 3.7:

Explain what

```
≫  - sort( - x)
```

does when x is a numerical vector.

Discussion 3.8:

Explain each of the following statements and its output:

```
≫  sort('The quick brown fox jumps over the lazy dog.')
   ans: .Tabcdeeefghhijklmnoooopqrrstuuvwxyz
≫  lower(sort('The quick brown fox jumps over the lazy dog.'))
   ans: .tabcdeeefghhijklmnoooopqrrstuuvwxyz
≫  (sort(lower('The quick brown fox jumps over the lazy dog.'))
   ans: .abcdeeefghhijklmnoooopqrrsttuuvwxyz
≫  unique(sort(lower('The quick brown fox jumps over the lazy dog.')))
   ans: .abcdefghijklmnopqrstuvwxyz
```

Discussion 3.9:
Consider the function $f(x) = 3e^{-x/4}\sin(2x)$. This is an oscillatory function whose amplitude decreases exponentially with x. The *envelope* of the function is $E(x) = \pm 3e^{-x/4}$.

Make a solid-line plot of $f(x)$ for x ranging from 0 to 10. Overlay the envelope—both top and bottom—as a dotted line.

Ordinarily, each `plot` command erases any previous plot. The command `hold on` allows you to overlay new plots on top of an existing plot, without erasure. To go back to the erasure mode, use `hold off`.

Exercise 3.1:
Write a one-line expression that computes $\sum_{j=0}^{4} 2^j$. (*Hint:* Construct a vector and use `sum`.)

Exercise 3.2:
Define a variable n to have an integer value: for example,

```
≫   n '= 25;
```

Write a one-line expression that computes $\sum_{j=0}^{n} 2^{-j}$ and that works for any $n \geq 0$.

Exercise 3.3:
Make plots of the following functions over the indicated domains. Try to find the largest step size that gives visually reasonable results and explain what "visually reasonable" might mean.

Function	Domain	
	Start	End
$\log x$	.1	10
e^x	0	5
e^{-x}	0	5
$e^{-x}\sin 3x$	0	5
$1/x$	0	20
$\sin \frac{1}{x}$	0	100

Exercise 3.4:
The *mean* of a vector of numbers v is given by

$$\frac{1}{N}\sum_{i=1}^{N} v_i$$

where N is the length of the vector and v_i is the ith value in the vector. The *standard deviation* of v is defined when $N \geq 2$ and is

$$\sqrt{\frac{1}{N-1} \sum_{i=1}^{N} (v_i - \bar{v})^2}$$

where $\bar{v}$ is the mean of the values in the vector.

Without using the built-in `mean` or `std` operators,

1. Write a one-line statement to compute the mean of the numbers in a vector named v. Assign the variable v a value before you execute your statement.
2. Write a one-line statement to compute the standard deviation of a numerical vector named v. Assign the variable v a value before you execute your statement.

Exercise 3.5:

Write a sequence of statements that will plot a circle of radius r at location x, y. Assign the variables r, x, and y to specific values before executing your statements. (*Hint:* A circle of radius r is a set of points at $x = r\cos\theta$, $y = r\sin\theta$ for θ in $[0, 2\pi]$.) Create a vector of x-coordinate positions and a corresponding vector of y-coordinate positions. Both of these are generated from an array of angles that ranges from 0 to 2π in small steps. You can use `linspace(0, 2*pi, 100)` to generate the angles and the `sin` and `cos` operators to convert these into x and y-coordinates.

Note that depending on how the plotting axes are set, your circle may look quite elliptical. Make sure to set the plotting axes appropriately to make your circle appear circular. You can use the commands `xlim` and `ylim` for this. (Alternatively, the command

```
>> axis('equal')
```

will do the job.)

Exercise 3.6:

Write a statement that will draw a square whose lower left corner is at x1, yb and top right corner is at xr, yt. Assign the variables xr, x1, yb, yt to specific values before executing your statements. Use the `xlim` and `ylim` commands to set the coordinate axes to be bigger than the drawn square. For instance, `xlim([-10,10])` will set the x-axis to range from -10 to 10.

Exercise 3.7:

Construct the following text as character strings, including the quotation marks

- I said, "John can't do this!"
- He said, "The apostrophe's meaning is possessive."
- Joe said, "Sarah said, 'Don't quote me.' "

Exercise 3.8:

Write a MATLAB expression that evaluates to boolean 1 if the sum of a and b is greater than N, unless b is less than 0. Otherwise the statement should evaluate to boolean 0.

Exercise 3.9:

Write the MATLAB version of the mathematical notation boolean statement $a \leq b \leq c$ involving three variables a, b, c.

Exercise 3.10:

Let m hold a single integer that gives the number of a month. Write an expression that tells whether the English word for that month has the letter "r" in it. For example, m=1, for January, has an "r" in it but m=6, for June, does not. (*Hint:* create a vector rmonths that lists the numbers of the months with "r". Your expression should be written in terms of m and rmonths.)

Exercise 3.11:

Let d be the day of the month of your birthday and m be the month. For example, for February 21st d = 21; m = 2;. Let df and mf be the same birthday information for a friend. Write boolean expressions to answer whether:

1. Your birthday is earlier in the calendar year than your friend's.
2. Your birthdays are closer together than 2 months.

Exercise 3.12:

How many leap years are there in the interval 1780 to 2832, inclusive? In the Gregorian calendar, a year is a leap year if it is evenly divided by 4, unless it is evenly divided by 100, so long as it is not evenly divided by 400.

CHAPTER 4

Collections
and Indexing

In Chapter 3, we saw how to store collections of numbers and of characters and some of the ways to operate on these collections. In this chapter, we focus on two main issues: how to access individual elements of a collection and how to group related elements together, even when they are of different types. We also encounter the "matrix" type that is of great utility in many scientific computations.

4.1   Indexing

Consider the following vector:

```
≫ elmstreet = [3 5 2 0 4 5 1];
```

Let us imagine that elmstreet was produced by a census worker counting the number of residents in each house on Elm Street. In this format, we can easily compute several statistics:

- the total number of Elm Street residents:

```
≫ sum(elmstreet)
  ans: 20
```

- the mean household size:

```
≫ mean(elmstreet)
  ans: 2.8571
```

- the largest household size:

```
≫ max(elmstreet)
  ans: 5
```

- the smallest household size:

>> `min(elmstreet)`
  ans: *0*

This last result is suspect: Is the house really unlived in or perhaps was there no one home when the census worker called? The census bureau needs to send someone back to Elm Street to check things out.

It would be wasteful to recollect all of the data for the street just to verify one house's data; the census worker need only go back to the one house in question. But which house?

Looking at the `elmstreet` vector, we can count down the vector to find that the problem is with the fourth house. This illustrates an important point about vectors:

Each element of a vector has two attributes:

Key Term

- **The *value* of the element**

Key Term

- **A position, or *index*, of the element**

The index is a kind of address. It describes where in the vector the element is to be found. The index is roughly analogous to the number on a house, which tells where on the street the house is to be found.

Many of the operations we perform on vectors involve finding the indices of certain elements or finding the value of an element at a particular index, or assigning a new value to an element.

The `find` operator enables us to compute the index of those elements of a vector whose values satisfy a specified criterion. `find` takes as its single argument a boolean vector that specifies for each element whether the criterion is satisfied. For instance, to find the empty houses on Elm Street the criterion is described using the boolean comparison `== 0`:

>> `elmstreet == 0`
  ans: *0 0 0 1 0 0 0*

This result, like any other, can be assigned to a variable:

>> `emptyhouses = (elmstreet == 0);`
  emptyhouses: *0 0 0 1 0 0 0*

The preceding statement no doubt looks a bit odd, combining as it does the = and == tokens. But it is an ordinary assignment statement

$$\underset{\text{name}}{\underline{\qquad\qquad}} = \underset{\text{value}}{\underline{\qquad\qquad}}$$

where the right-hand side value is computed using the boolean == operator.

Although it's easy enough to find the 1s in `emptyhouses` by visual inspection, `find` does it automatically:

>> `find(emptyhouses)`
  ans: *4*

Or, avoiding the step of assigning the boolean value to a variable,

```
≫ find(elmstreet==0)
   ans: 4
```

The returned value is the index of the element that satisfies the criterion.

Sometimes more than one element meets the criterion. In this case, find returns a vector of indices:

```
≫ find(elmstreet>2)
   ans: 1 2 5 6
```

In other cases, no element satisfies the criterion:

```
≫ find(elmstreet>100)
   ans: []
```

Something needs to be returned from this unsuccessful search; the empty vector serves the role well.

Key Term

Given an index, we can find out the value of the corresponding element by *referencing*, or, synonomously, *indexing*. This is done with a referencing operator that is invoked with a special syntax:

```
≫ elmstreet(4)
   ans: 0
```

There is also a version of the assignment operator that takes an index as an argument and allows us to make an assignment to an individual element. For instance, once the Census Bureau has valid data on the fourth house, the record can be updated.

```
≫ elmstreet(4) = 2
   elmstreet: 3 5 2 2 4 5 1
```

Although a statement like elmstreet(4) looks like the application of an operator called elmstreet to the argument 4, it is not. It is the application of an index referencing operator to the two arguments elmstreet and 4.

The First and Last Elements of a Vector

The first element of a vector is referenced with index 1, that is,

```
≫ elmstreet(1)
   ans: 3
```

This may appear obvious, but note that in some computer languages, such as C, C++, and Java, the first element is numbered 0. MATLAB follows the tradition set by FORTRAN in indexing the first element with 1.

Our `elmstreet` vector has seven elements, so the last element will be index 7:

```
≫ elmstreet(7)
  ans: 1
```

Key Term   The number of elements in a vector is called the *length* of the vector. The operator `length` returns the length of any vector:

```
≫ length(elmstreet)
  ans: 7
```

So a general way of referring to the last element of a vector is to use the output of `length`, as in

```
≫ elmstreet( length(elmstreet) )
  ans: 1
```

This sort of construction is so common that there is a special syntax for it involving the keyword `end`, which when used inside the indexing parentheses has the meaning "the last element." For instance,

```
≫ elmstreet( end )
  ans: 1
```

Erroneous Indices

An index must be an integer in the range 1 up to the vector's length. Statements involving numerical indices smaller than 1 will generate error messages. References involving indices that are larger than the length of the vector similarly will generate an error message. (But see "Extending a Vector" on page 70.) Indices that have a fractional component will return a sensible result (rounding the index down to the closest integer value) but will generate a warning message:

```
≫ elmstreet(3.4)
   Warning: Subscript indices must be integer values.
          ans = 2
```

Subsets of a Vector and the Empty Vector

By using a vector of indices, we can extract more than one element from a vector. Some examples should suffice to illustrate how this works. Remember that

```
≫ elmstreet = [3 5 2 0 4 5 1];
≫ elmstreet([1 2 3])
  ans: 3 5 2
≫ elmstreet( find(elmstreet >= 4) )
  ans: 5 4 5
```

```
≫ elmstreet( find(elmstreet > median(elmstreet)))
   ans: 3 5 4 5
```

The assignment operator works with subsets of a vector as well. For example, let's set up a vector

```
≫ a = [10 20 30 40 50];
```

and change some of its elements, those in the subset with indices 1, 3, and 5:

```
≫ a([1 3 5]) = [6 7 8]
   a: 6 20 7 40 8
```

For subset assignment to work, the subset referenced on the left-hand side of the assignment operator must be the same size as the subset on the right-hand side. There is an exception to this; there can always be single number, a scalar, on the right-hand side. The scalar value will be assigned to each of the members of the indexed subset:

```
≫ a([1 2 4]) = 0
   a: 0 0 7 0 8
```

A statement that might initially seem odd is

```
≫ elmstreet( [] )
   ans: []
```

Key Term
The vector [] is the *empty vector*, the vector with no elements. Far from being a useless thing, the empty vector allows an indexing statement to return a value that indicates that no elements were available. Indexing with the empty vector is like asking for nothing. When you ask for nothing, you get nothing back. [] stands for "nothing" but is packaged in the form of a vector. Although there's no point in explicitly indexing with [], this type of indexing happens implicitly when you specify a criterion that none of the elements in the vector happen to satisfy. For instance, there are no households with 1.5 members:

```
≫ elmstreet( find(elmstreet == 1.5))
   ans: []
```

The empty vector also provides a convenient notation for indicating that an element is to be deleted from a vector.

```
≫ v = [10, 20, 30, 40, 50];
≫ v(2) = []
   v: 10 30 40 50
```

More than one element can be deleted at a time:

```
≫ v([1,3]) = []
   v: 30 50
```

Boolean Vectors and Indices

The find operator takes a boolean vector as an argument; it returns the indices of the nonzero elements. It's possible to index directly with booleans without using find as an intermediary. For instance,

```
>>  a = elmstreet >= 4
    ans: 0 1 0 0 1 1 0
>>  elmstreet(a)
    ans: 5 4 5
```

When using booleans as indices, the length of the boolean index vector must match that of the vector being indexed. That is, there must be a boolean element for each and every element in the vector being indexed. It's also important that the index be a genuine boolean and not just zeros and ones that look like a boolean. This distinction becomes important when the elements are collected by hand:

```
>>  a = [0 1 0 0 1 1 0];
```

Although this looks like a is a boolean vector, it is not. Instead, a is just a numerical vector that consists of the numbers 0 and 1. Indexing elmstreet with a generates an error message:

```
>>  elmstreet(a)
    ???: Index into matrix is negative or zero.
```

The problem is that using numerical 0 as an index is not allowed; there is no zeroth element of a vector.

Booleans are created using the boolean operators such as the numerical relational operators. To turn the numerical vector a into a proper boolean vector, we can apply a relational operator:

```
>>  a = (a ~= 0);
    ans: 0 1 0 0 1 1 0
```

Now a is a boolean vector. Alternatively, the logical operator can be used to turn a vector of zeros and ones into a boolean vector:

```
>>  elmstreet(logical([0 1 0 0 1 1 0]))
    ans: 4 5 4
```

Extending a Vector

It can be helpful to think of a vector as a sort of organized box: something we can put numbers into and access the individual numbers using indices. The square brackets, [and], create a new box.

As we have seen, we can create a box and specify its contents at the same time:

```
>> boxOne = [1, 2, 3]
   boxOne: 1 2 3
```

Or we can make an empty box:

```
>> boxTwo = []
   boxTwo: []
```

In either case, we can add new contents to the box at any time. There are two main ways to do this: by using indexing and by creating a new box with the contents of an old box.

To use the index operator to add into a box, we simply specify where we want the new element to go:

```
>> boxOne(4) = 7
   boxOne: 1 2 3 7
```

We can even add elements past the end of the box,

```
>> boxOne(6) = 10
   boxOne: 1 2 3 7 0 10
```

Since we never said what we want in position 5, MATLAB has placed a zero there.

To extend a vector by one element, we need to compute the what the index of that slot will be and assign a value to that index. Since end resolves to the index of the last slot, end+1 is the first slot after the current end:

```
>> boxOne(end+1)= 9
   boxOne: 1 2 3 7 0 10 9
```

Remember that boxOne(end+1) is short for

```
boxOne(length(boxOne)+1)
```

Another distinct style for adding new elements to a vector uses the collection brackets. The main idea is to create a new box and put the contents of the old box into the new one. To illustrate, we start with the empty boxTwo and add something to it:

```
>> boxTwo = [boxTwo, 9, 8, 7]
   boxTwo: 9 8 7
>> boxTwo = [boxTwo, 6, 5, 4]
   boxTwo: 9 8 7 6 5 4
```

This style can also be used to prepend new elements to the start of the vector:

```
>> boxTwo = [3, 2, 1, boxTwo]
   boxTwo: 3 2 1 9 8 7 6 5 4
```

Notice that this is not a box within a box; the contents of the old box have been copied into the newly created one.

It's important to remember that the collection brackets can never change the value of an element in an existing vector. Only the assignment operator can change values. The collection brackets create a new vector. For instance,

```
≫   [boxTwo, 99]
  ans:  3 2 1 9 8 7 6 5 4 99
≫   boxTwo
  ans:  3 2 1 9 8 7 6 5 4
```

This important point is worth emphasizing:

The only way to change a variable is by using an assignment operator.

▶

Example: The Fibonacci Numbers, Indexing Style

In Section 2.2, we computed a few of the Fibonacci numbers: the sequence $0, 1, 1, 2, 3, 5, 8, 13, 21, \ldots$. Denoting the nth number in the sequence by x_n, the numbers follow the recursive relationship

$$x_{n+1} = x_n + x_{n-1} \tag{4.1}$$

To create a vector containing the Fibonacci numbers, we can start with the first two and construct the next one using Eq. (4.1). This process can be repeated to generate as many numbers in the sequence as desired:

```
≫   x = [0,1];
≫   x(end+1) = x(end) + x(end-1)
  x:  0 1 1
```

In the preceding statement, end plays a role analogous to n in Eq. (4.1). The relationship of Eq. (4.1) is both described and implemented by the command expression. The command can be repeated, verbatim, to extend the array:

```
≫   x(end+1) = x(end) + x(end-1)
  x:  0 1 1 2
≫   x(end+1) = x(end) + x(end-1)
  x:  0 1 1 2 3
≫   x(end+1) = x(end) + x(end-1)
  x:  0 1 1 2 3 5
```

In Chapter 8, we will see how to arrange to have the interpreter repeat the command automatically.

One of the interesting properties of the Fibonacci numbers is that the ratio of successive numbers approaches the golden ratio as n becomes large.

That is,

$$\lim_{n \to \infty} \frac{x_{n+1}}{x_n} = \frac{1 + \sqrt{5}}{2} \approx 1.6180 \qquad (4.2)$$

To verify Eq. (4.2), we can compute

```
≫  x(end)./x(end-1)
   ans: 1.6667
```

The value will change as we add more terms to the sequence, but with enough terms the value will approach 1.6180.

◀

A Matter of Style:

The use of parentheses for indexing, as well as the application of operators to arguments, raises the possibility of a name conflict: You might create a vector with the same name as an operator. In this case, the operator is hidden. This can have unanticipated results. For example, consider the following sequence of statements:

```
≫  sqrt(0)
   ans: 0
≫  sqrt(4)
   ans: 2
≫  sqrt = [1 2 3 4 5 6 7 8]
   sqrt: 1 2 3 4 5 6 7 8
```

The preceding statement defines a variable called sqrt, and now the operator sqrt cannot be accessed. The same statements now give different results:

```
≫  sqrt(4)
   ans: 4
```

This answer actually is correct; it's just that the question being asked is, "What is the fourth element of the vector sqrt?" and not "What is the square root of 4?"

It can be bewildering to get an error message such as the following:

```
≫  sqrt(0)
   ???: Index into matrix is negative or zero
```

This error message only makes sense when we remember that sqrt(0) is trying to access the vector sqrt.

In Chapter 6, we'll see how to package commands in a way that makes it easy to avoid problems with such vector-operator name conflicts.

4.2 Matrices

Some data come to us in tabular form. For instance, the census collects data that goes beyond the number of residents of each house:

Elm Street Census

Household Index	# of Residents	# of Adults	# of Children	Household Income
1	3	2	1	35,000
2	5	2	3	41,000
3	2	1	1	25,000
4	2	2	0	56,000
5	4	2	2	62,000
6	5	3	2	83,000
7	1	1	0	52,000

The question we face is how to represent this table of data to the computer. We could collect the numbers in this table into one long vector, such as

```
1 3 2 1 35000 2 5 2 3 410000 3 2 ···
```

but doing so would hide the fact that the table's columns stand for different things.

Key Term

To preserve the distinct meaning of each column and of each row, we can store the data as a *matrix*. A matrix is a rectangular array of numbers. The numerical part of the preceding table is an example. The matrix data type provides an organization that allows us to keep track of the columns individually while collecting the data into one variable.

A matrix can be constructed using the same collection brackets as for a vector, with the rows separated by a semicolon (;) or by typing a new line. For example,

```
≫  elmstreet = [3, 2, 1, 35000;
5, 2, 3, 41000;
2, 1, 1, 25000;
2, 2, 0, 56000;
4, 2, 2, 62000;
5, 3, 2, 83000;
1, 1, 0, 52000]
```

This is a matrix with seven rows and four columns. (A mnemonic: The columns run vertically, analogous to columns in a classical building.) We didn't bother to enter the "household index" in the matrix; it's really not data about the household but simply a way to make it easier for humans to find and reference specific rows of the table. Instead, we'll use the MATLAB indexing operations to find the rows we need. Later, in Chapter 11, we'll come back to the idea of using explicit ID numbers for each case.

The `size` operator returns a vector of two values: the number of rows and the number of columns, in that order:

```
≫  size(elmstreet)
   ans: 7 4
```

The indexing operations work much as for vectors, but with a matrix we need to specify both the row index and the column index:

```
≫  elmstreet(3,4)
   ans: 25000
≫  elmstreet(4,3)
   ans: 0
```

The row and column indices are separated by a comma. This seems natural enough if you remember that the row and column indices are two distinct arguments to the matrix referencing operator that is being invoked with a special syntax; we can imagine that `elmstreet(4,3)` is short for the functional style invocation `matrixref(elmstreet,4,3)`.[1]

As with vectors, we can pull out a subset by using a vector of indices. For example, the incomes of the families in houses 4, 5, and 7 are the matrix elements in rows 4, 5, and 7, in column 4:

```
≫  elmstreet([4, 5, 7], 4)
   ans: 56000 62000 52000
```

Often, we want to pull out all of the data in a given row or a column. There is a special syntax for this involving the colon (:). To pull out the first row, use a 1 to index the row and a colon (:) to index the column, meaning "first row, all columns":

```
≫  elmstreet(1, :)
   ans: 3 2 1 35000
```

To pull out a specific column, use : to index the row and the desired column number to index the column:

```
≫  elmstreet(:, 1)
   3
   5
   2
   2
   4
   5
   1
```

---

[1]MATLAB doesn't have a `matrixref` operator, but it does have a similar functional-style operator called `subsref` that works in a more flexible way.

Key Term

The subsets that we pull out of a matrix must always have the form of a matrix; that is, a rectangular array of numbers. When this array has only one row, we call it a *row vector*. When it has only one column, we call it a *column vector*. The term *vector* can refer either to a row or to a column vector. In fact, the basic numerical type in MATLAB is really a matrix. A single number—called a *scalar* in mathematics—is really just a very small matrix. For example,

```
>> size(7)
   ans: 1 1
```

MATLAB is strict in maintaining rules about combining row vectors with column vectors. In general, operations that involve two matrices (or vectors) on an element-by-element basis require that the two matrices have the same size. An important exception is scalars; that is, matrices with one column and one row. An example should make the matter clear. We define three different shape matrices:

```
>> scalar = 10;
>> rowvector = [1, 2, 3]
   rowvector: 1 2 3
>> columnvector = [1; 2; 3]
   columnvector: 1
                 2
                 3
```

Addition of a matrix with a scalar always works regardless of the shape of the matrix:

```
>> rowvector + scalar
   ans: 11 12 13
>> columnvector + scalar
   ans: 11
        12
        13
```

Also valid is to add two matrices of the same shape:

```
>> rowvector + rowvector
   ans: 2 4 6
>> columnvector+columnvector
   ans: 2
        4
        6
```

But addition isn't defined for two matrices of different shapes:

```
>> columnvector+rowvector
???: Error using ==> +
     Matrix dimensions must agree.
```

The matrix indexing operations provide considerable power to access our data. For example, if we want to count all of the adults on Elm Street,

```
≫  sum( elmstreet(:, 2) )
   ans: 13
```

or the children,

```
≫  sum( elmstreet(:, 3) )
   ans: 9
```

One inconvenient aspect of using matrices to store data is that it is up to us to remember which column index refers to which census variable; for instance, that column 2 is the number of adults and column 3 is the number of children. In Section 4.3, we will see another data type that overcomes this inconvenience.

Sometimes it's useful to copy out the data from a row or column in order to handle it more conveniently. For instance, suppose we wanted to calculate the mean income for families with children and for families without children:

```
≫  children = elmstreet(:,3);
≫  nokidshouses = find( children==0 )
   nokidshouses: 4 7
≫  incomenokids = elmstreet(nokidshouses, 4)
   incomenokids: 56000
                 52000
≫  mean(incomenokids)
   ans: 55000
≫  withkidshouses = find(children > 0)
   withkidshouses: 1
                   2
                   3
                   5
                   6
≫  incomewithkids = elmstreet(withkidshouses,4)
   incomewithkids: 35000
                   41000
                   25000
                   62000
                   83000
≫  mean(incomewithkids)
   ans: 49200
```

On Elm Street, the average income of households without children is some-what higher than the average household income with children.[2]

Boolean operators can be used with matrices in much the same way as with vectors. For example,

```
≫ elmstreet > 3 & elmstreet <=5
  ans: 0 0 0 0
       1 0 0 0
       0 0 0 0
       0 0 0 0
       1 0 0 0
       1 0 0 0
       0 0 0 0
```

The resulting boolean matrix can be used as an index:

```
≫ elmstreet(elmstreet > 3 & elmstreet <=5)
  ans: 5
       4
       5
```

The find operator returns the indices of the elements that are 1 in the boolean argument. This is most useful when two return values are called for, in which case the corresponding row and column indices are returned:

```
≫ [r,c] = find(elmstreet > 3 & elmstreet <= 5);
```

Since there are three elements that satisfy the boolean criterion, each of r and c will contain three indices. r gives the row index of each of the three elements, and c gives the column index:

```
≫ r
  ans: 2
       5
       6
≫ c
  ans: 1
       1
       1
```

Matrices are more than a convenient way of organizing data. Matrix computations (matrix multiplication, decompositions, etc.) play a funda-mental role in advanced scientific computation. Some of the applications of matrix computations are presented in Chapter 17.

---

[2]But a statistical analysis, using the *t*-test, indicates that there is not enough data to establish the statistical significance of the difference in average incomes.

4.3 Mixed Data Types

Matrices and vectors provide a mechanism for storing lists of numbers on the computer. But not all data is a list of numbers. Consider the situation of the census bureau, which wants to record the name, age, sex, and profession of each member of the Smith family, who live at 11 Elm Street:

The Smith Family

	First Name	Middle Names	Last Name	Age	Sex	Profession
(1)	Emily	Lisa	Smith	38	F	Mathematician
(2)	George	Robert	Smith	27	M	Primary School Teacher
(3)	Felicia	Grace	Smith	11	F	Student
(4)	Nancy	Emily Grace	Smith	7	F	Student

Address: 11 Elm St., Covington, NY 10483

Household income: 62,000

We could, of course, contrive to represent all of this information as numbers. Indeed, ultimately the computer will store it all as a collection of bits. For instance, the names will be stored as character strings: vectors of numbers interpreted as ASCII. The sex can be stored as, say, 1 for female, 0 for male; the profession as a character string or possibly as a numerical code keyed to a list of professions.

Here's the problem. Although the preceding table looks something like a matrix, the elements of the table are not single numbers. To illustrate, suppose we created a four-row, six-column matrix to hold the six pieces of information for each of the four family members. As soon as we attempt to fill the matrix, we run into a problem:

```
>> smithhousehold(1,1) = 'Emily'
   ???: Subscripted assignment dimension mismatch.
```

This admittedly obscure error message is pointing out the fact that 'Emily' is a vector of five numbers to be interpreted as ASCII.

```
>> length('Emily')
   ans: 5
```

We cannot shoehorn five numbers into one element of a numerical matrix as the statement smithhousehold(1,1) = 'Emily' is trying to do.

An old-fashioned approach to storing information like that of the Smith family involves formatting the data as character strings so that each row has exactly the same length. I mention this method not to encourage you to use it but merely because it is used commonly and you will, no doubt, encounter it. To use this method, we might specify that each person's data will be stored

Key Term

as one long character string. We need to set a *format* for the string so that we know how to interpret it. For example, we might specify the following format:

Character Positions	Contents
1 to 10	First Name
11 to 20	Middle Names
21 to 30	Last Name
31 to 32	Age
33	Sex
34–43	Profession

In this format, the data are a matrix of ASCII characters:

```
smithhousehold = ['Emily      Lisa      Smith      38FMathematic';
                  'George     Robert    Smith      37FPrimary Sc';
                  'Felicia    Grace     Smith      11FStudent    ';
                  'Nancy      Emily GracSmith       7FStudent    ']
```

Now the data are stored as a matrix of numbers that can sensibly be printed with an ASCII interpretation.

Data stored in this format are hard to manipulate, and the format itself imposes limitations. What happens if someone's name is greater than 10 characters? (Notice that Nancy's middle name was truncated to 'Emily Grac'.) How can we handle people whose age is 100 or greater given just two characters?

We could have addressed these problems by more careful planning of the format, leaving more space for the names and profession and adding a third character for the age. Failure to do such planning led to the famous Y2K problem of the transition to the twenty-first century. But even adding a third digit to the age wouldn't help us if it were decided to record children's ages with a fraction (e.g., 4.5). And where do we store the address and household income data in such a matrix? And how will we store such data for the many different households involved in a census?

Let us remember the saying:

If the only tool you have is a hammer, everything looks like a nail.

What we need is a new tool; matrices, however useful they may be for other purposes, aren't doing the job here. A matrix lacks two important qualities needed for storing data such as these: It provides no good mechanism for heirarchical or structured data and it can store only character or numerical data. Ultimately, for the purposes of storing large amounts of structured data, one wants to use the rather sophisticated tools of modern databases. (Some database concepts are introduced in Chapter 11.) But many of the small and midsized problems encountered in practice can be solved with two rather simple new tools embodied by two new data types: *cell arrays* and *structures*.

Names as Indices: Structures

Key Term

A *structure* is a type of variable with one or more components, each of which can hold any type of data. But instead of indexing the elements with a number, the elements are referred to by a character name. Each element is called a *field* and has a *fieldname*, which follows the same construction rules as variable names.

Key Term

Construction of a field can be accomplished by assignment. For example,

```
≫  pt.x = 3
  pt: x: 3
```

creates a variable named pt with a field named x and assigns the value 3 to that field. New fields can be constructed at any time; for instance,

```
≫  pt.name = 'Henry'
  pt: x: 3
        name: Henry
```

Fields can have values that are matrices, character strings, booleans, and even other structures.

Structures are particularly useful for making variables self-documenting. Each field, via its fieldname, gives a little description of its purpose (if the fieldnames are chosen appropriately).

For example, to store information about a person we might create a structure:

```
≫  people.firstname = 'Emily';
≫  people.sex = 'F';
≫  people.yob = 1966;
```

Key Term

When there is more than one item to be represented, we can create a *structure array*. Each component of the array is indexed numerically, like a matrix, and has the same fieldnames, but, of course, the value of the fields can be different.

```
≫  people(2).firstname = 'Nancy';
≫  people(2).sex = 'F';
≫  people.yob = 1997;
```

The length and size operators give information about the number of components in the structure array, not the number of fields. The fieldnames operator returns the names of the fields.

A new field can be created at any time by making an assignment to a new field name. For instance,

```
≫  people(2).profession = 'Student'
   people: 1x2 struct array with fields:
           firstname
           sex
           yob
           profession
```

The new field is created for all of the elements of the structure array. The value of the field for those elements where assignment has not yet been performed is [].

Storing Sets: Cell Arrays

Key Term   A *cell array* is a type that is much like a vector or matrix, but each element can hold any type of data. Like a vector or matrix, it has a length and a size, and the individual components are indexed with integers. Here's a cell array holding the first names of the Smith family:

```
≫  firstnames = {'Emily', 'George', 'Felicia', 'Nancy'}
```

Note that no extra spaces are needed to pad the names: in a cell array, the character string components don't need to be the same length.

Each of the elements of a cell array can be any data at all: character strings, numbers, matrices, and so on. In contrast, each element of a matrix must be a number, or, for a character matrix, an ASCII character.

For a cell array, we use the curly braces ({ and }) to collect the elements rather than the square collection brackets ([and]) used for vectors and matrices. Referencing of single elements of a cell array is also done using curly braces:

```
≫  firstnames{2}
   ans: George
```

If we want to reference multiple elements, then we have to think about what sort of thing is going to be returned: that is, how the information will be packaged. Consider the expression

```
≫  firstnames{2:4}
   ans: ans = George
        ans = Felicia
        ans = Nancy
```

This has returned three values, the values of firstnames{2}, firstnames{3}, and firstnames{4}. But these three values are not packaged into a cell array, and so a perfectly sensible-looking assignment statement returns an error:

```
≫ a = firstnames{2:4}
  ???: Illegal right hand side in assignment.
       Too many elements.
```

What is needed is a way to instruct the interpreter to package the returned subset as a cell array. Perhaps this operator would be called getCellArraySubset. This operation is distinct from getCellArrayElementContents. It therefore has a distinct special syntax. This syntax uses the ordinary parentheses ((and)) rather than the curly braces. So the statement to extract the second through fourth names from the firstnames cell array and return them packaged as a cell array is

```
≫ firstnames(2:4)
  ans: 'George' 'Felicia' 'Nancy'
```

Use parentheses with cell arrays to refer to subsets packaged as cell arrays themselves. Use curly brackets to refer to the contents of individual cells.

Recall that when concatenating two vectors, the individual identity of the two groupings is lost. For instance,

```
≫ a = [1,2];
≫ b = [3,4,5];
≫ c = [a,b]
  c: 1 2 3 4 5
≫ c(2)
  ans: 2
```

For cell arrays, the grouping is maintained, allowing a form of heirarchy:

```
≫ a = {1,2};
≫ b = {3,4,5};
≫ c = {a,b}
  c: {1x2 cell} {1x3 cell}
≫ c2
  ans: [3] [4] [5]
```

This behavior seems so much like "putting b in c" that it's easy to forget that only a copy of b is in c; changing b doesn't change c.

```
≫ b = 99;
≫ c{2}
  ans: [3] [4] [5]
```

4.4 Exercises

Discussion 4.1:
Create a structure suitable for storing information about a music collection (CDs or MP3 files).

Discussion 4.2:
For the data in the `elmstreet` matrix on page 74, write sets of commands to answer the following questions. It's important that your commands be written generally—imagine a matrix with hundreds of thousands of entries where you don't have the ability to scan manually through the rows.

1. How many residents are there altogether?
2. How many children?
3. What is the average number of children per household?
4. What is the average number of children per household in households with children?
5. What is the average income per household?
6. What it the average income per adult?
7. What is the average income per adult in households with and without children?
8. Are there any cases where the number of adults and children isn't equal to the total number of residents?

Discussion 4.3:
Create a matrix m with the same contents as the `elmstreet` matrix on page 74. Write commands to modify the contents of m as follows:

1. Change the income of household 3 to 28,000.
2. Add 500× the number of children to each household's income.
3. Add 1000 to the household income of a single parent household.

Discussion 4.4:
Design a structure for holding information about intervals of dates in a standard historical format; for instance, 300 B.C. to 120 A.D.

Discussion 4.5:
Suppose we are storing information about people in a structure array, as on page 82. We wish extend the structure to create a family tree. To do so, we need to create a field that says who each person's parents are. Let's call these fields `father` and `mother`. Explain what information you would put into these fields to represent the father and mother, who may or may not have information stored about them.

Exercise 4.1:
Write statements to

- Find the mean income per adult for each of the families on Elm Street.

- Find the fraction of families with children that have a single adult.

Exercise 4.2:
Write statements to do the following operations on a vector x:

1. Return the odd indexed elements.
2. Return the first half of x. We specify, somewhat arbitrarily, that if `length(x)` is odd, then the first half will include the middle element.
3. Return the second half of x. For odd length vectors, this is also somewhat arbitrary, but for consistency this should be all of the elements that were not in the first half of x.
4. Return the vector in reverse order; that is, with the last element in x coming first and the first element in x coming last in the returned vector.

Exercise 4.3:
Suppose we have a vector

```
>> vec = [2,8,5,6,3,4,7,4,3,6,9,0];
```

and we wish to exclude the elements at indices

```
>> k=[2,4,6,7]
```

Write a one- or two-line expression that returns the nonexcluded elements of vec.

Exercise 4.4:
Consider the two statements

```
>> 6 ~= (1:10)
```

and

```
>> (6 ~= 1):10
```

Which of them will give a boolean vector telling which of the numbers in 1:10 is not equal to 6? What does the other one do?

Without parentheses, the statements would be somewhat ambiguous to a human reader. Note that

```
>> 6 ~= 1:10
```

means one thing in newer versions of MATLAB and another thing in older versions. Using parentheses, in addition to helping a human reader, can avoid problems in such cases.

Exercise 4.5:
Write statements to return the elements of a vector x in random order by indexing the vector with a set of nonrepeating random indices. (*Hint:* The difficulty here is in producing nonrepeating random indices.) Try

`rand(size(x))` to generate random numbers, and `sort` to produce indices. Note that `sort` can return two values, the second of which is an index.

Exercise 4.6:

Write statements to do the following operations on a vector x:

1. Return a vector that is the odd indexed elements of x followed by the even indexed elements of x.
2. Return a vector that is the result of a perfect, nonrandom shuffle of x; that is, the first element in the first half of x followed by the first element in the second half, followed by the second element in the first half, then the second element in the second half, and so on.

Exercise 4.7:

The `reshape` operator rearranges a vector into a matrix shape. The operator takes three arguments: the vector to be rearranged, the number of rows, and the number of columns. For instance,

```
≫ x = reshape(1:10, 2, 5)
  ans: 1 3 5 7 9
       2 4 6 8 10
```

Use `reshape` to create the matrix x as indicated, and find the row and column numbers that index the number given. For example, the number 8 in the preceding x can be indexed with `x(2,4)`:

1. `x = reshape(1:100, 25, 4)`. Find the indices of 17, 29, 55, and 87.
2. `x = reshape(1:100, 5, 20)`. Find the indices of 10, 20, 35, and 99.
3. `x = reshape(1:36, 2, 18)`. Find the indices of 2, 19, and 25.
4. `x = reshape(0:2:71, 3, 12)`. Find the indices of 0, 14, 52, and 68. Also, explain why 71 doesn't appear in the matrix.

Exercise 4.8:

The type of recurrence relationship seen in Eq. (4.1) can be generalized to

$$x_{n+1} = ax_n + bx_{n-1} \qquad (4.3)$$

Depending on the values of a and b, the sequence will follow different patterns. The Fibonacci sequence, with $a = b = 1$, generates a sequence that grows geometrically.

Try Eq. (4.3) with $a = 1.3$ and $b = -0.9$, starting at $x_1 = x_2 = 1$ and iterating for about 50 times. Plot x_n versus n and describe the shape of the plot.

CHAPTER 5

Files
and Scripts

Anyone who has used a word processor or spreadsheet is familiar with the idea of a "file," a place for storing information on a computer disk. We tend to think of a file as being a document, a picture, a musical recording, or so on. But a file is merely a collection of bytes. The meaning given to those bytes depends on the software used to examine them. The term ASCII "file" is used to denote files intended to be interpreted as characters stored in the ASCII format. Other files are generically called "binary files".

Ironically, although the files produced by most word processors contain text, they generally are stored as binary files. This is because, in addition to the text, information is stored about font types and sizes and the layout of the page; the codes used to convey this information are not text based. Such files look like gibberish when interpreted as ASCII characters but appear sensible when given their intended binary interpretation by your word processor.

The contents of a binary file make sense only when interpreted in terms of a particular format or "filetype." There are many such formats. Some of them have been standardized and are in general use (such as the JPEG or RTF or MP3 used to store images, documents, or music) and others are specialized and idiosyncratic to a particular piece of software. When you try to open a file with the wrong software, you generally get a message such as "Unknown File Type."

This chapter is about how files are stored and how they are accessed by software. We will encounter several different kinds of file operators, particularly those that are used in MATLAB programming. Of special importance will be the "m-file," which allows us to package sets of commands for repeated use.

Key Term

5.1   Filenames

We refer to computer files by using names. This is a clever bit of engineering intended to make it easier for people to deal with data. Ultimately, the computer needs to refer to the data via its physical location on a disk drive or other storage device.

The situation is similar to that of a library. In order to find a book, you need the book's library reference number. This can be found in the library's catalog. The catalog itself can be searched by topic, book title, or author. The reference number tells what shelf in the library the book is kept on; the shelves have labels to make them easy to locate and the books on the shelf (hopefully!) are sorted, so that it is a simple matter to find a specific book on a shelf.

The people who designed operating systems realized that it would be convenient to have the computer keep track of the catalog and to hide completely the steps of looking up the reference number and translating this into a physical location. All that was needed was a way for the human user to search the catalog; the computer could do the rest. By analogy to the way a library works, a title was given to each data set. But, breaking the library analogy, the collection of data encompassed by the title was not called a "book" or a "document" or an "item" or an "object." An individual object

Key Term

on disk is called a *file*. Since early computers had very limited storage space, the referencing data that we call a title in a library was limited to only one short word and was called the file's "name": the *filename*.

A filename may or may not be an ordinary word, but it usually is not just arbitrary characters. It is conventional with many computer operating systems for the filename to end with a short suffix (for example, .doc) that indicates the filetype; that is, how the bytes in the file are intended to be

Key Term

interpreted. This suffix is called a *filetype extension*. The filetype extension is an integral part of the filename[1] and should always be used in the filename. Some examples of commonly encountered filetype extensions are given in Table 5.1.

Organizing Files in Directories

It becomes hard to keep track of files when there are a lot of them. On the computer I am using to write this book, there are at present 29,134 files. This is about as many as the number of books in a small public library. Libraries

---

[1]In the various versions of the Windows operating system, the default setup is not to display the filetype extension. Instead, the filetype is displayed using an icon. This can be confusing because the filetype extension is actually part of the filename. You can change this user-friendly omission of critical information. To see how, search the Start/Help facility under "file extensions."

Table 5.1   Some Filetype Extensions in Common Use

Extension	Filetype
.html	WWW files in the hypertext markup language.
.mp3	A compressed format for storing music.
.jpg	A compressed format for storing photographs.
.xls	Spreadsheet files from the Excel program.
.csv	Comma Separated Values: another spreadsheet format.
.doc	A word processing format.
.rtf	Rich Text Format: another word processing format often used to communicate between different word processing programs.
.pdf	The Portable Document Format from Adobe often used for Web-based documents.
MATLAB-Related File Suffixes	
.m	A file containing commands.
.mat	Binary file containing MATLAB variables.
.dat	ASCII file containing numerical or character data.

manage the trick of organizing by assigning a unique reference number to each physical item and keeping the items sorted according to this number. This means that every time a new item is acquired, it is necessary to look through the existing catalog to find a reference number that is not already in use.

Another strategy for organizing large numbers of things is "divide-and-conquer." We split the collection into two or more smaller collections and organize these separately. To some extent, libraries use this strategy as well; they have reference sections, general circulation sections, and perhaps special collections for maps and pamphlets.

On the computer, the divide-and-conquer strategy is taken to the extreme. In addition to dividing the entire collection of files into several subcollections, each of the subcollections can be divided further into subcollections, and so on. The result is that files are organized like the leaves on a tree: Each leaf is on a twig that comes off of a branch, but the branch itself might be an offshoot of another branch, which might emerge from a larger limb, which itself joins the trunk. One such tree is shown in Figure 5.1.

To locate a file, we have to be able to say what collection it belongs to. This means that collections themselves must have names. Of course, a collection might itself be a subcollection in a more comprehensive collection,

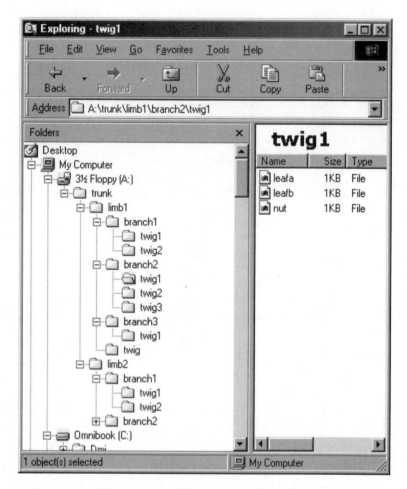

Figure 5.1.   A file tree on a Windows computer displayed using the "Windows Explorer" program. The left pane in the window shows only directories. The right pane shows the files or subdirectories within a single directory: in this case, the three files in the directory with pathname a:/trunk/limb1/branch2/twig1. (Note that the corresponding folder in the left pane is represented with an open-folder icon.) Dotted lines are used to show which directories share a relative pathname.

and so on. Returning to the tree analogy, we can simplify the task of finding a particular leaf by describing the path to take up the tree: Start at the trunk, take this limb, then that limb, then that branch, then this offshoot, then that twig.

The same strategy is followed on the computer or a computer network. Window-based operating systems such as Windows or Mac OS, which treat people like squirrels, allow them to navigate the file system by making a sequence of one-step-at-a-time decisions about which branch they are going to go up from their current position or whether they want to go back down to another level.

Since we focus in this book on writing computer programs, we cannot rely on the slow, point-and-click, one-at-a-time decision-making method of specifying the path to a file. Instead, we will give the complete itinerary for the path all at once. Such an itinerary is called a *pathname*. On a Windows system, it might look like `c:\myfiles\compsci\`, on a Unix system `/usr/cathy/compsci/`, and on a Macintosh like `harddisk:compsci:`.

Key Term

Although the term *branch* would serve perfectly well to describe the bifurcations of a file storage tree, the conventional terms are directory or *folder*. It's confusing that there are two different terms for the identical concept; only the term *directory* is used in this book.

Just as branches can come off of other branches, directories can be within other directories, and these are sometimes referred to as "subdirectories." It's important to keep in mind that a subdirectory is simply a directory that happens to be located in another directory.

A path name consists of a sequence of directories. The directory names are separated by some punctuation. The most common punctuation in the computer world is the forward slash (`/`). Although Windows uses the backslash (`\`) as the directory-separation punctuation and Mac OS uses the colon (`:`), we will always use the forward slash in this book and MATLAB can use this regardless of the platform on which you are working.

Key Term

The *complete filename* is the concatenation of the pathname and filename, (e.g., `/usr/cathy/compsci/project2/sunspots.dat`).

All files on a computer must have unique complete filenames; there can be no repeats. But this doesn't mean that two files can't have the same name, just that two such files must be in different directories so that their complete filenames are different. Similarly, two different directories can have the same name, but only if the path that leads to the two directories is different. But within a single directory, all of the files must have unique filenames since they all share the same path.

One uses a similar system of pathnames to locate files on other computers on a network. On the Internet, this might look like

```
www.fuzzy.org/shareware/science/sunspots.html
```

Key Term

For historical reasons, the term *uniform resource locator (URL)* rather than *complete pathname* is used to denote such a construction. This example specifies a file called `sunspots.html` in the directory `science`, which is in the directory `shareware` on the server `www.fuzzy.org`. Even the server name takes the form of a path. This server is the one called `www` found in the domain `org`, subdomain `fuzzy`.

The Inside Story: _____

What's at the start of a path? The pathname header!

The terms "directory" and "folder" describe the same thing: a branch of the file-storage tree. Many more terms are used to describe the start of a path, and these multitudinous terms reflect the diversity of hardware and file-storage techniques. Given that we analogize the file-storage system to a tree, Unix has the simplest and most intuitive term for the start of a path: **Key Term** *root*. The word *trunk* might also have been used, but this has an earlier, different meaning in telecommunications. In Unix, every computer has one directory that is called root and from this directory all other directories on the computer's file system can be reached even if these directories physically are stored on different disk drives. Root is denoted by the first / on the pathname.

Key Term In Windows, the term *drive* is used (short for the term *disk drive*). Dating back more than 20 years, the denotation of a disk drive is a letter followed by a colon (e.g., `c:`). On a Windows computer, there can be several disk drives; the floppy disk drive is usually `a:`. This system has now been augmented by a Unix-like one where the root file directory is called the "Desktop," which contains subdirectories like "My Computer" with the various disk drives and the "Network Neighborhood" by which you can reach other computers.

For resources stored on networked computers, the path has to include some indication of which computer to look on. A common convention is to start a network pathname with a double backslash, followed by a name. On a local network, the double backslash might be followed by a computer name; on the Internet, it might be followed by a "server" name in the format of a network name such as `www.macalester.edu`. Internet names like this actually are a pathname in their own right with the different components being separated by dots. The order of the pathname components is reversed from that used in filesystem pathnames: The last component (e.g., `.edu`, `.com`, `.uk`) gives the first branch in the path. Some network pathnames, called uniform resource locators (URLs), start with a component like `http:` or `ftp:`. This leading component describes the protocol to be used in transmitting and interpreting the file; it is analogous to the use of filetype extensions to describe the format of files.

Specifying Files in MATLAB

In MATLAB, a file's name is represented, naturally, as a character string (for instance, `'sunspots.dat'`). For a pathname, one uses a single long character string (for example, `'/usr/cathy/compsci/project2'`). A complete filename is the concatenation of a filename and a pathname, for instance,

```
'/usr/cathy/compsci/project2/sunspots.dat'
```

A complete filename gives an unambiguous specification of a file: There can be no other file with the same complete filename. The concept here is simple—we can just think in terms of a complete filename as a long filename that is unique for every file—but it poses both an inconvenience and some genuine difficulties. The inconvenience is having to use the long complete filename rather than just the filename. The difficulty comes about when we move files around on our computer or copy the files to another computer. The moved file or the copied file will have a different complete filename than the original, therefore any commands that use the complete filename will have to be changed. If we always had to use complete filenames, it would be difficult to share, install, or update software.

Key Term

Key Term

The simple solution to this difficulty involves *relative pathnames*. A relative pathname is the last parts of a complete path. The first parts are called the *path stem*. For instance, the path `'/usr/cathy/compsci/project2'` can also be specified as the relative pathname `'compsci/project2'` whose stem is `'/usr/cathy'`. Or it can be the relative pathname `'project2'` with a stem of `'/usr/cathy/compsci'`.

A *relative filename* is the concatenation of a relative pathname and a filename. Relative pathnames and filenames can be ambiguous; there can be more than one file with a given relative filename. In order to eliminate the ambiguity, we need to specify the information that would turn a relative pathname into a complete pathname. That is, a relative pathname is useful only in conjunction with a path stem. There must always be a path stem involved in any access of a file by a relative filename.

The Working Directory

Key Term

In MATLAB, the path stem is set to be a particular directory, called the *working directory*. The `cd` operator is used to change the working directory; the `pwd` command returns the name of the current working directory.

```
≫ pwd
  ans: /usr/local/MATLAB11/work
```

Depending on how the MATLAB software was installed on your computer and what type of computer it is, the output you get might look different. In any event, the output of `pwd` is a complete pathname.

It's usually a good idea to change the working directory to one you create just for your own work. For instance, to change the working directory to `/usr/cathy/compsci`, if there happened to be such a directory, one would use the command

```
≫ cd('/usr/cathy/compsci')
```

We can also use relative pathnames to set the working directory. These relative pathnames will be relative to the current working directory. For example, in place of the above `cd` command, we might have used a sequence of commands:

```
>> cd('/usr')
>> cd('cathy')
>> cd('compsci')
```

or, instead of the preceding but with equivalent effect:

```
>> cd('/usr')
>> cd('cathy/compsci')
```

In order to use relative pathnames successfully, it is imperative that it always be possible to decide whether a given pathname is relative or complete. To provide this information, complete pathnames must always start with a pathname header (for instance, / in Unix or c:\ in Windows). Relative pathnames must never have this header.

A convenient shorthand for the relative directory above the current directory in the pathname is the double-dot (..) Suppose we are in the directory /usr/cathy/compsci; then .. refers to /usr/cathy.

```
>> pwd
   ans: /usr/cathy/compsi
>> cd('..')
>> pwd
   ans: /usr/cathy
```

The Search Path

Key Term

When a filename is used without a pathname, the interpreter looks first for the file in the working directory. If the file is not found there, the interpreter consults a list of directories called the *search path*. Each directory in the search path is examined, in turn, until a file with the specified filename is found. If there is no such file in any of the directories in the search path, the interpreter prints an error message: "file does not exist." Of course, the file may well exist somewhere on the computer or the network; it just doesn't exist in the current directory or in the directories in the search path.

The `path` function can be used to examine or set the search path. There is also a SET PATH menu item in the FILE menu that provides a graphical means to add or delete directories from the search path.

5.2 File Operators

File reading operators do the work of reading the file's contents from the storage medium (typically a disk drive) and interpreting the contents according to the rules of a given format. *File writing operators* do the reverse job: formatting data in the appropriate way and writing the formatted data to the storage medium. There are many operators simply because there are many different formats.

In subsequent chapters, we will encounter file operators for formats relating to images, sounds, graphics, video, and so on. But for now, we will focus on operators and file formats for storing MATLAB variables.

When a MATLAB session ends, all of the variables cease to exist. When a new MATLAB session is started, no user-defined variables exist. In order to maintain the values of variables between sessions, it's necessary to save them from the earlier session and load them into the later session. This idea will be familiar to users of word processors who need to save and open documents to transfer them between sessions.

In MATLAB, there are several ways to save the values of variables. The way that works most like a word processor is to use the MATLAB graphical user interface. Under the FILE menu there is an item SAVE WORKSPACE AS . . . . This will save all of the variables in the workspace to a file whose filetype extension is .mat in just the same way that a word processor will save a document to a file with a filetype extension of .rtf, .doc, and so on.

To load the variables back into MATLAB, the OPEN item in the FILE menu can be used in just the same way as a word processing document is opened.

These menu-based commands are convenient when stopping a MATLAB session with an expectation that it will be continued at another time. But for programming purposes, the important operators are save and load. Save allows individual variables or groups of variables to be stored in files. Load enables variables or data from other sources to be read into MATLAB.

To illustrate, suppose we have created three variables:

```
>> a = [3 4 2];
>> b.field1 = 8;
>> b.field2 = 'welcome';
>> c = {3, 'gtacctaggact'};
```

The save operator will create a binary file that holds variables and their names:

```
>> save('myvariables', 'a', 'b', 'c');
```

The first argument is the name of the file in which to store the variables. The remaining arguments are the names, as quoted strings, of the variables to store in the file.

The filename is relative to the working directory. By default, save will attach a filetype extension of .mat to identify the file as using the conventional MATLAB binary format for storing variables. We could have given this explicitly:

```
>> save('myvariables.mat', 'a', 'b', 'c');
```

We could even use some other filetype extension (for instance, `'myvariables.foo'`), but this usually just causes confusion.

The `load` operator is used to copy the variables from the file into the workspace. By default, all of the variables are loaded:

```
>> load myvariables.mat
```

but it is also possible to load in just selected variables:

```
>> load myvariables.mat a c
```

Key Term

The preceding two expressions have a syntax that we have not yet introduced called the *command syntax*. The command syntax doesn't employ parentheses in organizing arguments. An expression using the command syntax is parsed and executed in a special way: Each of the tokens—characters separated by spaces—is treated as a character string. The operator named by the first token is then invoked with the character strings as arguments. So

```
>> load myvariables.mat a c
```

is more or less equivalent to

```
>> load('myvariables.mat', 'a', 'c')
```

The difference, however, is that the no-parentheses–no-quotes use of `load` will create the loaded variables in the workspace. That is, it will do assignment without the assignment operator = having been invoked. In contrast, the ordinary functional syntax will not do assignment unless = is used. The value returned by the parentheses–quotes invocation of `load` will be a structure whose fields are named as the loaded variables.

```
>> foo= load('myvariables.mat', 'a', 'c')
   foo: a: [3 4 2]
        c: {[3] 'gtacctaggact'}
```

5.3 Importing and Exporting Data

Using `save` in the way described previously creates files in a special binary format. These files are not readable by general-purpose software outside of MATLAB.

By "exporting data" we mean the process of transferring data from MATLAB to some other program; "importing data" is the reverse process. Successfully importing or exporting data will depend inevitably on the other program. To simplify things, since we cannot possibly anticipate all of the programs that might be involved in an import or output transfer, we'll consider only reading from and writing to standard file formats.

There was a time when a programmer would design his or her own file formats particularly suited to the specific data to be stored. Such specialized formats could be extremely efficient if properly designed and implemented. The problem was that special software was required to read and write these formats. Getting this software often involved contacting the programmer, and there was no guarantee that the software would work on another computer.

For many types of data where efficiency is important (i.e., images, sounds, video, and documents), there are now standardized formats that have been carefully designed. The software for reading and writing these formats is distributed widely so that the files are highly portable.

A common type of format for general-purpose data transfer use is the spreadsheet table. A spreadsheet table is in many ways like a cell array; the spreadsheet is a rectangular array of cells in which information can be sorted: numbers, characters, dates, and so on. Spreadsheet software is used to edit this information. The software can also be used to perform calculations by entering formulas into the cells, and the software can format the information for printing, display, and even making graphs. We will limit ourselves, however, to spreadsheets without formulas, formatting, or graphics. We'll consider simple arrays of character and numerical information.

Several file formats are in widespread use for simple spreadsheet data. Among these are the following:

.csv Stores data as ASCII, indicating the boundaries between cells using commas and newline characters. The data within the cells cannot contain commas.

.dlm Similar to .csv, but using some other delimiter for marking boundaries between cells. The tab, space, and semicolon are commonly used. The data within the cells cannot contain this delimiter.

.xls A proprietary binary format that can also store formula and formatting data.

ASCII **Flat Numerical Files**

Key Term

Spreadsheet layouts can be quite elaborate, particularly if the sheet is intended for a human reader. We shall emphasize *flat layouts* that have a particularly simple structure:

1. The data are stored in a rectangular array of spreadsheet cells.
2. Each row, except possibly the first, has an identical format.
3. Each column, again possibly excluding the first cell in the column, has all of its cells containing the same sort of data.
4. The first row may consist of text headers that identify each column.

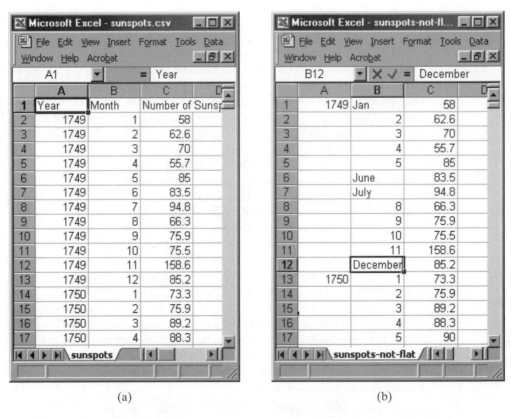

(a)   (b)

Figure 5.2. Two spreadsheets containing data about the number of sunspots each year. (a) A flat spreadsheet with a header row. (b) A nonflat spreadsheet. Note that the first column contains two types of data: numbers and blanks. Similarly, the second column contains two types of data: numbers and character strings.

Figure 5.2 shows two spreadsheets containing historical data about the number of sunspots each year. One of the spreadsheets is flat; the other is not. Although the flat format may seem limiting, it greatly simplifies the handling of data—a subject we'll return to in Chapter 11.

Reading and writing flat spreadsheets containing numerical data is extremely simple because the contents of the spreadsheet can be represented naturally as a matrix within MATLAB. The main issue is whether there is a header on the spreadsheet. To illustrate, we'll use two files with identical numerical data: sunspots.csv has a header; sunspots-no-header.csv does not. For the file without a header, the csvread operator simply translates the numbers inside the spreadsheet into a MATLAB matrix:

```
≫  d = csvread('sunspots-no-header.csv')
   d: 1749 1 58
      1749 2 62.6
      1749 3 70
      1749 4 55.7
      1749 5 85
      and so on for 2820 rows altogether
```

Somewhat more generally, the `load` or `importdata` operators can be used in exactly the same way—these operators can also read files that are delimited in other ways.

When there is a header, things are a bit more complicated. Neither `csvread` nor `load` has been arranged to deal with mixed numerical and text data:

```
≫  d = csvread('sunspots.csv')
   ???: Trouble reading number from file (row 1, field 1)
```

A simple way to avoid this is to instruct `csvread` to skip the header row:

```
≫  d = csvread('sunspots.csv',1)
   d: 1749 1 58
      and so on
```

The `dlmread` operator works in a similar way, but `load` will not work with spreadsheet data.

The `importdata` operator is more intelligent; it attempts to deduce the structure of the file both from the filename extension and the file contents. `Importdata` does a sensible thing with numerical flat files with headers, returning a structure with the numerical matrix in a field named `data` and the text headers in a field named `colheaders`.

```
≫  d = importdata('sunspots.csv')
   d: data: [2820x3 double]
      textdata: {'Year' 'Month' 'Number of Sunspots'}
      colheaders: {'Year' 'Month' 'Number of Sunspots'}
```

Writing flat numerical files is extremely straightforward. The file writing operator is called with a matrix and a filename. For example,

```
≫  mat = [1, 2, 3; 5, 4, 3];
≫  csvwrite('mydata.csv', mat)
```

This will not put a header row on the file.

Remember that a spreadsheet stores the equivalent of only one MATLAB variable. A separate file is needed for each variable.

Historically, scientists and engineers faced with the need to store large amounts of numerical data have, for reasons of efficiency, used binary formats in which the bits representing numbers are written directly, without special formatting. Appendix C gives some of the commands and techniques for reading and writing such files.

Nonnumerical ASCII Files

An ASCII file is any file that contains bytes intended to be interpreted as ASCII characters. As we've seen, ASCII files can be used to store numerical data, but they are also used for more general purposes than this (e.g., storing formatted nonnumerical data but also simple text).

Consider the file mobydick-exerpt.txt, the first part of which looks like this:[2]

```
    Call me Ishmael.  Some years ago--never mind how
long precisely--having little or no money in my purse, and nothing particular
to interest me on shore, I thought I would sail about a little and see the
watery part of the world.  It is a way I have of driving off the spleen, and
regulating the circulation.  Whenever I find myself growing grim about the
mouth; whenever it is a damp, drizzly November in my soul; whenever I find
myself involuntarily pausing before coffin warehouses, and bringing up the
rear of every funeral I meet; and especially whenever my hypos get such an
upper hand of me, that it requires a strong moral principle to prevent me
from deliberately stepping into the street, and methodically knocking
people's hats off--then, I account it high time to get to sea as soon as I can.
```

The filetype extension, .txt, is used to indicate plain text data; this is the extension that will be generally assigned by word processors when saving a document in ASCII form.

To treat the ASCII file as a long string of characters, we can read in the file as a character string. The fread operator, described in Appendix C, allows us to do this:

```
≫  fid = fopen('mobydick.txt', 'r');
≫  s = fread(fid, Inf, 'uchar');
≫  fclose(fid);
≫  size(s)
   ans: 101412 1
```

We used Inf to read in all the 101,412 characters in the file, but we might have chosen to read in a smaller number.

Note that s is not a character array; fopen returns a numerical array. Here are the first several values with the transpose operator ' used to turn the column vector into a row vector for ease of printing:

```
≫  s(1:15)'
  ans: 32 32 32 32 32 67 97 108 108 32 109 101 32 73 115
```

---

[2]The text file of *Moby Dick,* or *The Whale,* by Herman Melville, is file moby10b.txt from Project Gutenberg, www.gutenberg.org. The file mobydick-exerpt.txt contains the first chapter of the book.

To interpret these values as ASCII, we can use the char operator:

```
≫   char(s(1:15)')
    ans:  ____Call me Is
```

Note the five spaces at the beginning of the string, corresponding to the five 32 values in s.

Reading in a text file as one long character string is not always the most appropriate way to process text. For example, we often consider text one line at a time, but the fread operator does not have a concept of lines. It reads in the newline character (ASCII value 10) as just another character.

The textread operator can be used to read in the file one line at a time.

```
≫   lines = textread('mobydick.txt', '%s', 'delimiter', '\n');
```

The second argument, the string '%s', is a format string that directs textread not to perform any conversions of numerals to numbers. The last two arguments designate the newline character ('\n') as dividing up the file into chunks.

The returned value is a cell array, with one line of the text file in each cell:

```
≫   lines{1}
  ans: 'Call me Ishmael. Some years ago--never mind how'
≫   lines{2}
  ans: 'long precisely --having little or no money in my purse,'
```

Many of the characters in the mobydick.txt file were not written by the author, Herman Melville. For instance, the breaks between lines were inserted by some software designed to make the text easier to print. Perhaps we would like to have one character string, without line breaks, for each paragraph. Or perhaps we would like a list of words, each word being stored as a character string. By setting the delimiter argument in textread appropriately, we can control how the text is divided up into units. For example, setting the delimiter argument to '\n ,.?!' as delimiters will break up the text into words. (See Exercise 5.3.)

The Inside Story: _____

In order to save some typing, MATLAB can use the command syntax for operators like cd that take character strings as arguments. In the unified syntax used previously, the arguments are enclosed in parentheses and the quotation mark is used to delimit the character string. In the functional syntax, the quotation mark is not used and a space is used to separate between the operator name and the arguments. For example,

```
≫  cd /usr/cathy/compsi
```

is equivalent to

```
≫  cd('/usr/cathy/compsi')
```

Although the command syntax is often convenient, you should use it sparingly since it can give confusing results. For example, the command

```
≫  plus(3,2)
   ans: 5
```

can give a surprising result when invoked with the command syntax

```
≫  plus 3 2
   ans: 101
```

This odd-looking output stems from the fact that the character strings '3' and '2' are ASCII numbers whose values are 51 and 50, respectively.

The command syntax can be used with load as a form of assignment. When used with a .mat file, load will copy over the variables included in the file into the workspace. With a .dat file, load produces a single variable. For example,

```
≫  load sunspots.dat
```

will create a variable sunspots with the contents of sunspots.dat.

The command syntax cannot be used with the ordinary assignment operator. For example,

```
≫  a = plus 3 2
```

gives an error message.

5.4 Scripts

In MATLAB, commands are ordinary text, and a command can be represented by a text string. For instance, suppose we define the string mystr:

```
≫  mystr = 'sqrt(9)'
```

As expected, using mystr as a expression simply returns its value, the string:

```
≫  mystr
   mystr: sqrt(9)
```

However, passing the string to the `eval` operator causes the string to be evaluated just as if it had been typed directly at the command prompt:

```
≫ eval(mystr)
  ans: 3
```

This ability to execute commands stored in strings means that it is possible to store commands in a text file and execute them automatically. Such a file is called a *script*.

Key Term

A script file typically is created in a text editor or word processor and stored in a file whose filetype extension is `.m`. This filetype extension gives the name to script files, as well as the similar function files that will be introduced in the next chapter, collectively as m-files.

The m-file can be created with any text editor. The MATLAB package provides a convenient one. But whatever editor is used, the file must be stored as ASCII and have a filetype extension of `.m`.

Figure 5.3 shows an m-file script created using the MATLAB editor. The m-file has been saved under the name `plotsunspots.m` and contains the commands for drawing Figure 5.4.

```
c:\kaplan\book\progsExamples\files\plotsunspots.m

File  Edit  View  Text  Debug  Breakpoints  Web  Window  Help

 1    % load in the data and plot them
 2    %! A comment line
 3    d = importdata('sunspots.csv'); %! load the spreadshee
 4    x = d.data; % just the numerical part of sunspots.csv
 5    % The first column is the year
 6    year = x(:,1);
 7    % The second column is the months
 8    month = x(:,2);
 9    % The third is the number of spots
10    spots = x(:,3);
11    % Compute the date as year + fraction
12    time = year + (month-.5)./12;
13    plot(time,spots); %! Plotting the data
14    title('Sunspot Counts'); %! Add a title
15    xlabel('Year');  %! and labels
16    ylabel('# of Sunspots');
17

Ready
```

Figure 5.3. M-files are created in a text editor and saved with a filetype extension of `.m`.

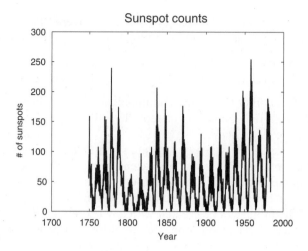

Figure 5.4. The plot generated by the plotsunspots command implemented using the m-file described in the text.

We can execute the commands in the m-file just by invoking the script using its name. For example, to execute the commands in plotsunspots.m we simply give the command:

```
>> plotsunspots
```

Note that the .m part of the m-file name isn't included in the command.

Hidden behind the surface simplicity of the execution of scripts is a lot of activity. Note that plotsunspots is a perfectly valid variable name. When the MATLAB interpreter encounters a token that is a variable name, the interpreter checks whether any variable of that name has been defined. If so, the value of that variable is returned. But if there is no such variable, the interpreter starts looking for alternative possibilities. In particular, the interpreter looks for a file whose name matches the token and whose filetype extension is .m. The interpreter looks in several places for such a file—first the working directory and, if the file isn't found there, then the directories in the search path.

When creating a script file, these points should be kept in mind:

- The m-file name should be a valid variable name followed by the .m filetype extension.
- The file should be saved in the current working directory. Or, conversely, the current working directory should be set to the directory containing the file. (Actually, the file can be anywhere in the "search path," as described in Chapter 9.)

- Variables with the same name as the script should not be used. If they are used, the MATLAB interpreter will treat any such token as referring to the variable rather than the m-file.

The contents of the m-file are ordinary commands of the sort we have been using all along. M-files can also include comments. The % sign indicates that the rest of the line is a comment intended to be ignored by MATLAB. Such comments are extremely useful for documenting scripts so that other users—and yourself!—can better understand what the script is intended to do.

We will be using m-files extensively throughout the rest of this book. In order to make it easy to explicate individual lines and refer to them, we will display the m-file in an annotated format, including line numbers, thus:

A comment line	[1]	`% load in the data and plot them`
Load the spreadsheet w/header	[2]	`d = importdata('sunspots.csv');`
	[3]	`x = d.data; % just the numerical part`
	[4]	`% The first column is the year`
	[5]	`year = x(:,1);`
	[6]	`% The second column is the months`
	[7]	`month = x(:,2);`
	[8]	`% The third is the number of spots`
	[9]	`spots = x(:,3);`
	[10]	`% Compute the date as year + fraction`
	[11]	`time = year + (month-.5)./12;`
Plotting the data	[12]	`plot(time,spots);`
Add a title	[13]	`title('Sunspot counts');`
and labels	[14]	`xlabel('Year');`
	[15]	`ylabel('# of sunspots');`

Keep in mind that the line numbers, annotations, and the filename are not an actual part of the contents of the m-file. The contents are just that part displayed in `typewriter font` on the right-hand side of the display.

A Matter of Style:

It's not uncommon to see files with names like `Draft 1 of my biology report.doc` Such verbose and multiword names may be helpful for the human reader, but they can cause difficulties. MATLAB has a particular difficulty in using multiword names stemming from the syntax of the language and the way that new operators are given names.

In this book, we will use only one-word filenames with a filetype extension indicating the filetype (for instance, `average.m`). You should do the same when programming. One way to get the benefits of multiword names without the difficulties is to use a "caps" style (e.g., `Draft1OfMyBiologyReport.rtf`) or a "bumpy" style (e.g., `draft_1_of_my_biology_report.rtf`).

5.5 Scripts as Computations

We defined a computation as a transformation of inputs into an output. Scripts are obviously performing a computation, but there doesn't seem to be an input or output involved. Indeed, attempting to use a script name on the right-hand side of the assignment operator will generate an error, regardless of the contents of the script.

```
≫  x = plotsunspots
  ???: Attempt to execute SCRIPT plotsunspots as a
       function.
```

Trying to pass arguments to a script produces the same error.

```
≫  plotsunspots(3)
  ???: Attempt to execute SCRIPT plotsunspots as a
       function.
```

We'll see in Chapter 6 how to create operators called functions that can accept arguments and return values. To illustrate a pitfall of using scripts, suppose you are graphing some mathematical functions in order to compare them to the sunspot data.

```
≫  x = 1:100;
≫  y = sqrt(x);
≫  plot(x,y)
```

Seized by an astrophysical curiosity, you decide to look at the sunspot data once again.

```
≫  plotsunspots
```

And, back again to your work:

```
≫  plot(x,y)
  ???: Error using ==> plot
       Vectors must be the same lengths.
```

Something strange has happened: A command that was working fine has stopped working. Take a minute and think about it before reading on.

As you may or may not recall, the script file `plotsunspots.m` contains a line that makes an assignment to the variable x. This has overwritten your previous value for x. In this case, the error is easily detected because `plot` produced an error message. Worse would be the case where there is no error message, because there would still be an error but you might not know it.

Whenever assignment is done in a script, any data stored previously in that variable are lost. You might think this is not such a big deal; you just need to be careful not to use variables that have the same names as those in scripts. But this is not a real solution because there is no way that you can know what is going on in all of the commands that you give. For example, the `plot` command is doing something complicated and might be making assignments to many variables as it computes things like the appropriate scales for axes. It would be unreasonable to expect you to memorize the names of all the variables in `plot` so that you could avoid creating variables with the same names in your own scripts. Even if you could do this, adding any new software to your system would be a nightmare since you would have to check it against all of the existing software for name conflicts.

A much better solution to the problem posed by name conflicts is to limit the assignments done inside scripts so that they cannot have effects outside of the script. We shall see how this is done in the next chapter.

The Inside Story:

In interpreted languages such as MATLAB, commands are ASCII text, and script files are text files. In compiled languages such as C, C++, and FORTRAN, a human programmer[3] writes a text file that contains instructions, but these files are not directly executed. Instead, the text files are translated by a special program called a *compiler* into a binary format that is usually completely unintelligible to a human but is efficient for a computer to process. The text files that are the input to the compiler (usually with filetype extensions like `.c`, `.cpp`, or `.f` for C, C++, and FORTRAN, respectively) are called *source files* and the binary output of the compiler is called an *object file* (conventionally with a filetype extension of `.o`). Another step, called *linking*, performed by a program called a *linker*, combines the object files together and arranges things as a single file that can be executed by the computer. This file is called an *executable* file. In Windows, such files have the filetype extension `.exe`, but on Unix they need not.

Key Term

Sometimes the compiled but unlinked files are collected together into a library so that other programs can use the facilities provided by the compiled files. In Windows, such a library file is called a dynamically linked library with filetype extension `.dll`. In Unix, it typically is called a `shared library`.

The advantage of interpreted languages like MATLAB is that it is relatively easy to add new commands and change old ones. All you need to do is

---

[3]Actually, not all such programs are written by humans. Many are written automatically by computers.

edit the command file and save it. No compilation or linking is required. In addition, the interpreters often provide standard commands for things like graphics and file access.

There are two main advantages of compiled and linked languages: They typically perform computations faster because the work of interpreting the source files has already been completed by the compiler, and they can produce stand-alone applications that do not require any software other than the operating system to work.

There has been a debate of long-standing between the proponents of interpreted languages and those of compiled languages. The debate generally comes down to speed of execution versus ease of use. Which of these is more important depends strongly on the task at hand. For beginning programmers, or advanced programmers designing prototype systems, ease of use can be paramount. But for those with very large computations (e.g., searching a gene database or running a detailed model of the weather) speed of execution can be the most important issue.

The MATLAB system itself, considered as a piece of software rather than as a language, is a mixture of interpreted, compiled, and executable software that reflects the relative advantages of compiled and interpreted software for different uses. The MATLAB system contains a stand-alone executable that provides the functionality of the interpreter and some of the basic numerical and graphical capabilities and libraries of object files that provide additional features. It also contains many hundreds of dot-m files that extend the capabilities. Although this book deals only with dot-m files, you should know that it also is possible dynamically to link object files to MATLAB, so that c, c++, or FORTRAN programs can run within MATLAB.

5.6 Exercises

Discussion 5.1:
Find the characters that are illegal in file names on your computer. (*Hint:* Use your computer's help facility. Also, try renaming a file [whose contents you don't mind losing] with a questionable character and examine any error message that results.)

Exercise 5.1:
Make sure you have some coins in your pocket. Now, use an ordinary word processor or spreadsheet program to create a flat ASCII file containing a matrix with two columns. The first column is the denominations of the coins. The second column is the dates of the coins. There will be one row for each coin. Make sure to save the file with an appropriate extension (e.g., .txt or .csv). Most word processors have an ASCII option (often found in the SAVE AS dialog box). Spreadsheets can generally write .csv files.

Read the file you created into MATLAB and compute the total face value of those coins, the mean age of the coins, and the mean age for different denominations of coins. Package the file reading and computation into a script. Your script should work on any set of data in the right format.

Exercise 5.2:

Write a script, `swap.m`, that swaps the values of the variables a and b. For example,

```
>>  a=1;
>>  b=2;
>>  swap
>>  a
  ans: 2
>>  b
  ans: 1
```

Exercise 5.3:

Find a chapter of a novel in ASCII form on the World Wide Web and save it as a text file on your computer. The file `mobydick.txt` is such a file, but you might want to pick your own favorite. Project Gutenberg (`www.gutenberg.org`) is a good source of such materials. The goal of this exercise is to produce from that text a list of words in alphabetical order with duplicates removed.

Read the text file into MATLAB using the `textread` operator so that each word is one element of a cell array. (You will want to think carefully about the values of the `'delimiter'` and `'whitespace'` arguments.) The result should be a cell array with one word in each element. Process the cell array to produce the sorted list of words. (*Hint:* Look at the `unique` operator.) Make sure that you are not fooled by the uppercase/lowercase distinction: Words are the same regardless of capitalization for the purposes of this exercise.

You will undoubtedly make mistakes and find punctuation characters in some of your words. If so, fix the mistaken arguments to the `textread` operator.

Package your entire set of commands into a shell script called `makeWordList.m` that reads from a file whose name is stored in the variable `textname` and that puts the sorted list into a variable called `wordlist`.

Exercise 5.4:

Read the file `counting.bin` using the binary flat file operator `fread` but treating each bit as a single item. (*Hint:* The format string `'ubit1'` will read individual bits.) Write an m-file that contains the commands for opening the file, reading the first 16 bits and assigning them to a variable called `one`, then the second 16 bits assigned to `two`, and so on. Finally, the m-file should close the file.

5.7 Project: Time for a Cool Cup of Coffee

Key Term

Newton's law of cooling says that the rate at which an object cools is proportional to the difference between the object's temperature and the temperature of the environment. The bigger the temperature difference, the faster the cooling. The result is that, as a function of time, temperature $T(t)$ tends exponentially toward the temperature of the surrounding environment T_{env}. The function is

$$T(t) = T_{\text{env}} + (T(0) - T_{\text{env}}) \exp(-kt) \tag{5.1}$$

The *time constant k* that governs how fast the cooling occurs depends on the properties of the object.

The coffee cooling data on page 47 does not follow Newton's law of cooling exactly. For the first few minutes, the temperature follows Newton's law well, with a time constant of $k = 0.0426$ with units of min^{-1}. But at longer times, the coffee temperature no longer follows Newton's law, or, perhaps, the "constant" k is changing.

The ambient temperature was measured to be $T_{\text{env}} = 24.5\,°\text{C}$, and the initial temperature of the coffee was $T(0) = 93.5\,°\text{C}$.

Using this information, write a script to produce the graph shown in Figure 5.5, which compares the measured data with the simple Newton's law theory.

Here are some suggestions:

- You can read in the data, which is stored as a comma-delimited spreadsheet file, using

  ```
  ≫ data = csvread('coffeecooling.csv');
  ```

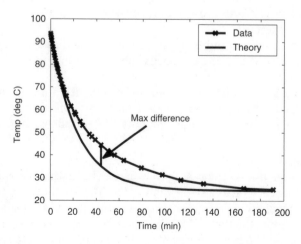

Figure 5.5. Comparing theory with the cooling data in Section 3.3.

- Do things one at a time. You might start by constructing the *x*- and *y*-coordinates of one of the curves and then plot out that curve. Then construct the coordinates of the other curve and plot that out.
- Once you have the basic outline of the plot, put all of the necessary commands for making the data and the plot into a script file, and refine that file gradually.
- Note that the units on the graph are not the same as the units in the data.
- Use the MATLAB help features to find information about the functions you will need to use: `plot`, `xlabel`, `ylabel`, `text`, `legend`, and `max`. You will find that there is much more information about these functions than you need. An important skill to develop is quickly winnowing out unnecessary information.
- You can use the plot editing and annotation capabilities in the figure window to draw arrows and add text. For instance, the mouse can move the legend box to an appropriate position.

Exercise 5.5:

Make a printed copy of your graph and the script file that you used to produce it. The script file should be able to produce the graph starting from a fresh session of MATLAB. Make sure to test it this way. Don't be surprised if a script that seems to work fine doesn't work so well from a fresh session of MATLAB. This may be because the script uses a variable that was not defined within the script. In the next chapter we'll see how the use of "functions" can avoid this difficulty.

CHAPTER **6**

Functions

As we saw in the previous chapter, the use of script files to implement new computations allows us to automate repetitive tasks and avoid and fix typographical errors, but it involves a serious and substantial complication: Seemingly irrelevant details of the existing computations—such as variable names—profoundly can influence the outcome of the overall computation. This means that to write a successful program using scripts you must either be lucky enough to avoid all name conflicts or you must be fully knowledgeable about each of the existing computations you are using. This is an unreasonably burdensome responsibility for a programmer.

Taking the point of view that a computation is a transformation from inputs to an output, it seems reasonable to expect that the output of a computation should depend only on its inputs and not on internal details such as variable names. The knowledge we should require of the user of a computation should be quite limited: what are the inputs and what are the outputs. Knowing exactly how the transformation from input to output is accomplished (what algorithm is being used, how the algorithm is implemented, what variables are named, etc.) should not be a prerequisite to using the computation.

6.1 Computations without Effects

In order to avoid any possibility of a computation having an undesired effect, many computer languages, including MATLAB, adopt a simple but highly effective strategy: Computations can be packaged so that they have absolutely no effect at all. For example, the logarithm operator works this way.

```
≫  log10(1)
 ans:  0
≫  log10(20)
 ans:  1.30102999566398
```

Although these two commands have produced outputs, no variables have been changed from the values they had before the computation.

Of course, we might decide to use the output of the logarithm computation in an assignment

```
≫  a = log10(20);
```

Now something has changed, namely the value of the variable a. But the logarithm computation hasn't brought about the change; rather, the assignment statement has. Log10 is an innocent bystander to the assignment.

One consequence of the no-effect nature of computations is that an operator serves the same purpose as a lookup table. Younger readers may have limited experience with such tables because, for mathematical purposes, they have been almost completely replaced by calculators. Figure 6.1, for example, shows part of a table of logarithms. To use it to compute the logarithm of a number, you look up the number in the left-hand column labeled "Num.": The logarithm is the number in the column labelled "Logarithmi." (This table can also be used in reverse, to compute the antilogarithm of a number.)

An obvious feature of such tables is that they are printed and therefore cannot change as a result of your using them. The table in Figure 6.1 is almost 400 years old but gives the same values that we compute today. Every time you look up a given number, you will find the same answer; the table describes a static, unchanging relationship between quantities. In the case of Figure 6.1, this is the relationship between numbers and their logarithms.

Key Term

Essentially, all modern computer languages provide ways to package sequences of commands in a way that implements a computation as an ideal transformation of inputs to outputs, one that has no effect on the world other than providing the output. A computation packaged in this way is called a *function*. (The term is drawn from the mathematical term; in mathematics, "function" refers to a *single-valued relationship* between an input and an output. "Single-valued" means that for any given input there is one, and only one, output.)

Functions can implement a no-effect computation: a static, unchanging description of the relationship between quantities. Every time you provide a given input, you will get exactly the same output.

Chilias prima.

Num. abfolu.	Logarithmi.	Num. abfolu.	Logarithmi.	Num. abfolu.	Logarithmi.
1	0,00000,00000,0000	34	1,53147,89170,4226	67	1,82607,48027,0082
			1258,91573,0801		643,41100,0242
2	0,30102,99956,6398	35	1,54406,80443,5028	68	1,83250,89127,0624
	17609,12590,5568		1223,45564,1701		634,01780,3102
3	0,47712,12547,1966	36	1,55630,25007,6719	69	1,83884,90907,3726
	11291,87366,0830		1189,92123,9971		614,89492,7700
4	0,60205,99913,1796	37	1,56820,17140,6700	70	1,84509,80400,1426
	9691,00130,0806		1158,18725,4982		616,03087,0481
5	0,69897,00043,3602	38	1,57978,35966,1682	71	1,85125,83487,1908
	7918,12460,4761		1128,19104,0968		607,41477,1219
6	0,77815,12503,8364	39	1,59106,46070,2650	72	1,85733,24964,3127
	6694,67896,3061		1099,53843,0147		599,03636,8919
7	0,84509,80400,1426	40	1,60205,99913,2797	73	1,86332,28601,2046
	5799,19469,7768		1072,38653,9177		590,88596,1052
8	0,90308,99869,9194	41	1,61278,38567,1974	74	1,86923,17197,3098
	5115,25224,473?		1046,54336,7816		582,95436,6721
9	0,95424,25094,3932	42	1,62324,92903,9790	75	1,87506,12633,9170
	4575,74905,606b		1021,91651,8163		575,23288,8905
10	1,00000,00000,0000	43	1,63346,84555,7952	76	1,88081,35922,8079
	4139,26851,5823		998,42209,0660		567,71728,9169
11	1,04139,26851,5823	44	1,64345,26764,8619	77	1,88649,07251,7248
	3778,85008,8939		975,96371,8915		560,38775,1800
12	1,07918,12460,4762	45	1,65321,25139,7534	78	1,89209,46026,9048
	3476,21062,5922		954,53179,0623		553,24885,9996
13	1,11394,33523,0684	46	1,66275,78316,8157	79	1,89762,70912,9044
	3218,46833,7140		934,00262,5416		546,28957,0150
14	1,14612,80356,7824	47	1,67209,78579,3573	80	1,90308,93869,9194
	2996,32233,7744		914,33794,3986		539,50318,8671
15	1,17609,12590,5568	48	1,68124,12373,7559	81	1,90848,50188,7865
	2802,87236,0024		895,48416,5292		532,88335,0507
16	1,20411,99826,5591	49	1,69019,60800,2851	82	1,91381,38523,8372
	2632,89387,2235		877,39143,0751		526,42399,9235
17	1,23044,89213,7827	50	1,69897,00043,3602	83	1,91907,80923,7607
	2482,35837,2504		860,01717,6192		520,11936,8518
18	1,25527,25051,0331	51	1,70757,01760,9794	84	1,92427,92860,6188
	2348,10958,2495		843,31675,3686		513,96396,5241
19	1,27875,36009,5283	52	1,71600,33436,3480	85	1,92941,89257,1429
	2227,63947,1115		827,25259,6599		507,95255,2918
20	1,30102,99956,6398	53	1,72427,58696,0079	86	1,93449,84512,4357
	2118,92990,6994		811,78901,2218		502,08013,7605
21	1,32221,92947,3392	54	1,73239,37598,2297	87	1,93951,92526,1862
	2020,33860,8819		796,89296,7127		496,34195,2155
22	1,34242,26808,2211	55	1,74036,26894,9424	88	1,94448,26721,5017
	1930,51551,9538		782,53375,1196		490,73344,9474
23	1,36172,78360,1759	56	1,74818,80270,0620	89	1,94939,00066,4491
	1848,34056,9401		768,68286,6629		485,25027,9441
24	1,38021,12417,1161	57	1,75587,48556,7249	90	1,95424,25094,3932
	1772,87669,6043		755,31378,9045		479,88828,8177
25	1,39794,00086,7204	58	1,76342,79935,6294	91	1,95904,13923,2109
	1703,33391,9878		742,40180,7920		474,64350,8447
26	1,41497,33479,7082	59	1,77085,20116,4214	92	1,96378,78273,4556
	1639,04161,8817		729,92587,4150		469,51211,0838
27	1,43136,37641,5899	60	1,77815,12503,8364	93	1,96848,29485,5394
	1579,42671,8323		717,85846,2713		464,49050,4576
28	1,44715,80313,4222	61	1,78532,98350,1077	94	1,97312,78535,9970
	1523,99665,5674		706,18544,8748		459,57516,8915
29	1,46239,79978,9896	62	1,79239,18350,9825	95	1,97772,36052,8885
	1471,32568,2070		694,88599,5133		454,76377,5072
30	1,47712,12547,1966	63	1,79934,05494,5358	96	1,98227,12330,3957
	1414,04391,1461		683,94245,3031		450,05012,1667
31	1,49136,16938,3427	64	1,80617,99739,8389	97	1,98677,17341,6624
	1378,82844,8564		673,33826,5897		445,43414,2621
32	1,50514,99783,1991	65	1,81291,33566,4286	98	1,99122,60756,9249
	1336,39615,5798		663,05788,9901		440,91189,0506
33	1,51851,39398,7789	66	1,81954,39359,4187	99	1,99563,51945,9755
	1296,39771,6437		653,08671,5895		436,48054,0245
34	1,53147,89170,4226	67	1,82607,48027,0082	100	2,00000,00000,0000
	1258,91573,0801		643,41100,0242		432,12727,8268

A   101

Figure 6.1. The logarithms (base 10) of the integers 1 to 100 computed by Henry Briggs in 1624. The small numbers between entries are the differences between the successive entries. *Source:* Used with permission, Linda Hall Library of Science, Engineering & Technology.

On the computer, the logarithm table is implemented as the function named log10. Using the table is a matter of invoking the function:

```
>> y = log10(x);
```

x represents the quantity in the "Num." column of the table, y is the corresponding quantity from the "Logarithmi." column of the table. The computerized version of the table has the advantage of ease of use and the ability to hold a great many more numbers than can be printed even in a book-sized table. For instance, we can check the values in the table

```
>> log10(9)
   ans: 0.95424250943932
```

but also compute logarithms not in the table

```
>> y = log10(9.5);
   ans: 0.97772360528885
```

Such between-entry problems would have been solved in earlier generations by interpolation, an operation that is still useful.

6.2   Creating Functions

In MATLAB, a function can be created in two distinct ways. The first and most common way is the subject of the rest of this section. It involves creating an m-file and writing commands in the same way a script is written (see Section 5.4) but with one more step: adding a declaration line that identifies the inputs and outputs. The second way to create a function, to be covered in Section 6.5, involves using a special-purpose function-creation operator named inline.

When creating a function using an m-file, there is a distinctive first line that is used to indicate that the commands that follow are part of a function and to identify explicitly the inputs and outputs. Here's an example that computes the length of the hypotenuse of a right triangle given the lengths of the two sides:

Function declaration line

```
[1]   function c = hypotenuse(a,b)
[2]   c = sqrt( a.^2 + b.^2 );
```

Putting the preceding two lines[1] into an m-file named hypotenuse.m gives us a new operator called hypotenuse that takes two inputs. We can

---

[1]Remember that the m-file contains only the text on the right side, displayed here in monospaced typewriter font. The line numbers and comments on the left side are added in this book only to help explain the commands.

invoke this operator (we will sometimes use the terminology "evaluate the function") using the standard operator-calling syntax:

```
≫ hypotenuse(3,4)
   ans: 5
```

or, using the explicit function evaluation syntax:

```
≫ feval('hypotenuse', 3, 4)
   ans: 5
```

Key Term

When the statements in the function have been executed, we say that the function "returns." The value of the variable identified as the output in the m-file, c in this case, is called the *return value*.
If we like, we can assign the return value to a variable:

```
≫ a = hypotenuse(3,4);
≫ a
   ans: 5
```

Note that although we identified the output name as c within the function, we are free to call the output by any valid variable name *outside* the function's m-file. The assignment on line 2 of hypotenuse.m to a variable named c has absolutely no effect outside of the function. To see this, let's define a variable c and check what effect evaluation of the function has on it:

```
≫ c = 0;
```

Evaluate the function

```
≫ d = hypotenuse(3,4);
```

and check to see if the value of c has changed:

```
≫ c
   ans: 0
```

It hasn't.
Line 1 of the m-file, beginning with the word function, indicates to the interpreter that there will be two arguments to the function. When the function is evaluated, the value of the first argument will be given the name a and the value of the second argument will be given the name b. The line also indicates the name of the variable whose value will become the output of the function. Although c = hypotenuse(a,b) looks like an expression involving assignment to a variable c, it is not; it is a special syntax that says which variables hold input values when the function is invoked and which variables hold output values when the function returns.

Hiding Variables

Functions rigorously sequester their internal variables from the rest of the computational world; the internal variables are completely hidden to any outside command.

Here's the scene metaphorically: Imagine yourself in an office containing many pieces of paper with important information. Executing a script is like leaving your office and inviting someone else, a stranger, to come in to perform an operation. When the person is done, he or she leaves your office and you return. Unfortunately, in the course of the operation, the stranger might have done just about anything: change the information on a paper, delete some papers, or create new pieces of paper.

Evaluating a function is different. It is as if you write a memorandum on a fresh piece of paper saying what operation you want performed and giving the value of each input to the operation. You mail this memorandum to a distant office where it is processed. By return mail, you receive an envelope containing the output of the operation given the inputs you specified. You can do whatever you want with the output contained in the envelope, but you can be assured that performing the operation has changed nothing else in your office; there can be no surprises.

How does the interpreter prevent the statements inside a function from having an effect on variables outside the function? In Chapter 9, we'll answer this question more completely, but for now the following picture will suffice. Whenever a function is invoked, the interpreter goes through the following steps:

Key Term

1. Start up a new *environment*. An environment is, essentially, a fresh session of MATLAB with no variables defined.
2. In the new environment, create the variables identified as inputs on the `function` line. Assign these variables the values of the arguments given in the function evaluation. For instance, in `hypotenuse(3,4)` the variable a is assigned the value 3 and the variable b is assigned the value 4.
3. Execute all of the function's statements inside the new environment.
4. When the function's statements have been executed, take the value of the variable identified as the output return that has the value of the evaluated function.
5. Eliminate the environment created for the function invocation.

Multiple Outputs

Sometimes a computation involves more than one output. We have seen this with the max function, which returns the largest value in a numeric vector as well as the index of that value.

```
≫  [bval, ind] = max([5 4 2 7 3 2]);
≫  bval
   ans: 7
≫  ind
   ans: 4
```

The number 7 is the largest value in the vector, and it's in position 4.

This type of behavior works only when the assignment operator is used to handle the returned values from a function. If there is no assignment operator involved, then only one value, the first, is returned by a function:

```
≫  max([5 4 2 7 3 2])
   ans: 7
```

Setting up a function to return multiple values is simple and requires only specifying which of the variables internal to the function is to be returned as the first value, which as the second, and so on. This is done in the function declaration line. To illustrate, here's a simple function that takes two numbers and returns their arithmetic and geometrical means:

Arithmetic mean is $\frac{a+b}{2}$

Geometric mean is $\sqrt{ab}$

```
[1]  function [arith, geom] = twomeans(a,b)
[2]  arith = (a + b)./2;
[3]  geom  = sqrt( a.*b );
```

The function declaration line simply lists the returned values one at a time, collected between square brackets and separated by commas. Although this looks like the special syntax for the numerical vector collection operator, it has the meaning of "return multiple values" when it is on the function-definition line of an m-file.

Functions with multiple returned values work the same way as functions with a single returned value; except that when the function returns, the assignment operator, if there is one, is handed the multiple values rather than just the single one.

```
≫  [am,gm] = twomeans(1,4)
   ans: am = 2.5000
        gm = 2
```

This named, multiple assignment is a special bit of syntax that only works with assignments from functions. For example, the statement

```
≫  [a,b] = [1,2];
```

appears to make perfect sense. It is intended to set a to 1 and b to 2. But the assignment operator is not handling the returned values of a function, and the statement results in an error:

```
≫  [a,b] = [1,2]
   ???: Improper matrix assignment. A variable cannot
        return multiple arguments.
```

A function, however, is allowed to return multiple values to the assignment operator.

The assignment operator is able to sort out how many variables are on its left-hand side and how many values are on its right-hand side and do the appropriate thing. If there are fewer left-hand variables than right-hand values, only the first values are used; the rest are ignored. If there are more left-hand variables than values, an error statement results:

```
>> [a,b,c] = max([5 4 2 7 3 2])
   ???: Error using ==> max
        Too many output arguments.
```

Most computer languages permit a function to return only one value— but this value can be an array or a structure that contains multiple pieces of information. MATLAB is unusual in permitting multiple named outputs.

Comments and Help

Key Term

The % character denotes that the rest of the line is to be ignored by the interpreter. This is useful for communicating with a human reader, and therefore % is called the *comment sign* or *comment character*.

Although comments have no effect on the way the interpreter handles the file, they can have a critical effect on the usability of the functions you create. It's a good idea to get into the habit of putting comments into your functions. These comments can be oriented toward several ends:

- Giving information to the user about what are the arguments to a function and what are the outputs. It's also useful to give some examples of how the function can be used.
- Describing the algorithm being used. Such descriptions are useful both for those who want to understand a function and those who want to modify it.
- Describing the role of variables. Although a variable name provides some description, in many cases it's helpful to supplement this with a comment.
- Providing information about the author or authors of a function, contact information (for instance, an e-mail address or telephone number), copyright information, and often revision information. (For example, if you look at the function m-files that come with this book, you will see comments where the first character is an explanation mark. We use these comment lines to provide information about how to format the m-files for printing in the book.)

MATLAB has a built-in help system that uses comments in a function to provide help to users. This works through the help command. For instance,

```
>> help max
MAX     Largest component.
    For vectors, MAX(X) is the largest element in X.
    For matrices, MAX(X) is a row vector containing
    the maximum element from each column.

    [Y,I] = MAX(X) returns the indices of the maximum
    values in vector I. If the values along the first
    non-singleton dimension contain more than one
    maximal element, the index of the first one is
    returned.
```

Help works through a very simple method that makes it extremely easy for you to include your own functions in the help system. help prints out the first contiguous comment lines in a function. These comments are placed at the start of the function's m-file. Here's an example revising the twomeans m-file to include help:

```
[1] %   [arith,geom] = twomeans(a,b)
[2] %   Computes the arithmetical and geometrical means
[3] %   of the two scalar, numerical arguments.
[4] %   Example: [a,g] = twomeans(1,4) gives a->2.5, g->2
[5] function [arith, geom] = twomeans(a,b)
[6] arith = (a + b)./2;
[7] geom  = sqrt( a.*b );
```

Arithmetic mean is $\frac{a+b}{2}$ [6]
Geometric mean is $\sqrt{ab}$ [7]

The first three lines are comments. help scans the m-file, identifies these three lines as the first contiguous comment lines, and displays them.

```
>> help twomeans
[arith,geom] = twomeans(a,b)
    Computes the arithmetical and geometrical means
    of the two scalar, numerical arguments.
    Example: [a,g] = twomeans(1,4) gives a->2.5, g->2
```

You could put any comments you like in these help-comment lines: poetry, love notes, or speculation on the meaning of life. Possibly the most useful thing to put in these lines (there can be as many as you care to have) is a short statement of the syntax of the function and what the function does. For example, the help-comment lines for twomeans shows that there are two values returned and indicates the meaning of these two values. It's also important to include information about the meaning of the input arguments. An example, when appropriate, can be extremely helpful to the reader.

The first of the help lines is special. The lookfor command takes a character string as an argument and reads through all of the m-files accessible to MATLAB and returns the first help-comment line from any function where that line contains the character string. For example,

```
≫ lookfor('mean')
```

finds all of the functions where the string 'mean' appears in the first help
line. This includes the twomeans function as well as some others.

```
twomeans.m: % [arith,geom] = twomeans(a,b)
exponential.m: % EXPONENTIAL(n,mean)---probability distribution
MEAN    Average or mean value.
MEAN2 Compute mean of matrix elements.
```

lookfor is helpful for finding functions when you don't remember the
function name, but only if the author of the function has put helpful information in the first help line.

One trick for keeping track of the m-files you write is to put your name
or initials in the first help-comment line. Then by using the lookfor command you can easily get a listing of all of the functions you have written.

6.3 Functions as Arguments

In Section 5.7, we encountered the task of graphing a function of time, $\theta(t)$
from Eq. (5.1). This is a common sort of task in scientific graphics, and there
is a simple procedure for making such a graph. To illustrate, we'll graph the
function $\sqrt{x}\sin(2x)$ for x running from 0 to 20.

First, we make a finely spaced vector of points, say 100 points, to cover
the domain of x:

```
≫ xpts = linspace(0,20,100)
  xpts: 0 0.2020 0.4040 0.6061 and so on
```

Next, evaluate the mathematical function at these points.

```
≫ ypts = sqrt(xpts).*sin(2.*x)
  ypts: 0 0.1767 0.4595 0.7290 and so on
```

Finally, plot the graph:

```
≫ plot(xpts, ypts)
```

It is convenient to package these steps as an m-file function so that they
can be repeated for other mathematical functions and other endpoints. The
first step in doing so is to identify the inputs. These are the left and right
endpoints of the domain of x (0 and 20 in the preceding example) and the
mathematical function itself. Since we can assemble the two endpoints into a
vector, the syntax funplot(f, limits) seems natural enough (the function implementing this is given on page 124).

Before we consider how to write funplot, let's examine the problem of invoking it. It might seem obvious to set the first argument to be
sqrt(x).*sin(2.*x), but this generates an error:

🐞 ≫ `funplot(sqrt(x).*sin(2.*x), [0 20])`
    ???: *Undefined function or variable 'x'.*

The problem here is that `sqrt(x).*sin(2.*x)` is not a function, it is an expression involving a variable named x.

The way we have seen to create a function is to write an m-file. Here is one:

```
[1]  function res = mathfun(x)
[2]  res = sqrt(x).*sin(2.*x);
```

Mathfun isn't a very descriptive name; it would be nice if we didn't need to think up a name just to graph a mathematical function. The function itself works just fine:

≫ `mathfun(xpts)`
    ans: *0 0.1767 0.4595 0.7290 and so on*

but we still can't use it as an argument to `funplot`:

🐞 ≫ `funplot(mathfun, [0 20])`
    ???: *Input argument 'x' is undefined.*

The problem now is that the interpreter attempts to evaluate the arguments to `funplot` before invoking the function itself. The first argument is the expression `mathfun`; this expression invokes `mathfun` without any arguments, just as if we had given the command

🐞 ≫ `mathfun`
    ???: *Input argument 'x' is undefined.*

What is needed is a way of treating the function as a value rather than as an expression to be evaluated. MATLAB provides several ways of doing this, but we will consider just one here. When a function name is enclosed in quotes, it is just an ordinary character string. The string can be stored in a variable and passed as an argument to a function in the ordinary way. The `feval` operator can take that character string, and execute it as a function with arguments. For example,

≫ `feval('mathfun',10)`
    ans: *2.8870*

This provides the basic capability needed to pass a function as an argument.

The function that receives the character-string argument can use `feval` to evaluate the function: Here is the `funplot` program that uses `feval` for the critical step:

```
[1] function funplot(f, lims)
[2] %  funplot(f, [left,right]) graph a function f(x)
[3] %  in the domain left <= x <= right
[4] %  A poor substitute for the built-in FPLOT
[5] xpts = linspace(min(lims), max(lims), 100);
[6] ypts = feval(f,xpts);
[7] plot(xpts, ypts);
```

Here is `funplot` at work:

```
>> funplot('mathfun', [0 20])
```

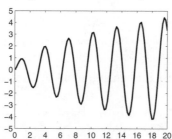

Although `funplot` does create a plot, it contains some substantial mathematical shortcomings. `Funplot` uses a fixed number of plotting points, a simple but unsophisticated method that isn't a reliable way of graphing mathematical functions. The built-in function `fplot` does a much better job and has the same argument interface as `funplot`. Use `fplot` instead.

In later chapters we will see many examples of operators that act on functions and that allow us to express ideas and algorithms for tasks such as iteration, equation solving, and minimization. The style of passing functions as arguments allows us to write such operators in a general way, applicable to any function in the correct format.

6.4 Wrapper Functions

When a function is passed as an argument, it generally has to be in an agreed-upon format in order to work with the function that is taking it as an argument. For example, `fplot` takes a function of a single variable as an argument. This sometimes requires reformatting the argument function.

Consider the m-file `newtoncooling` that implements the coffee-cooling relationship given in Eq. (5.1) on page 110.

```
[1]  function T = newtonCooling(T0, Tenv, k, time)
[2]  %   newtonCooling(T0, Tenv, k, time)
[3]  %   Temperature of a cooling object at any given time
[4]  %   T0 --- object's temperature at time 0
[5]  %   Tenv --- environment's temperature
[6]  %   k --- time constant of exponential cooling
[7]  %   time --- the time
[8]  T = Tenv + (T0 - Tenv).*exp(-k.*time);
```

From Eq. (5.1) — line [8]

This function takes several arguments, but `fplot` wants a function of just a single argument.

We would like to make a plot of temperature versus time. If we know the values of `T0`, `Tenv`, and `k`, then we can compute the temperature at any time. For example, at time 10 minutes, using the values from Section 5.7, we can compute

```
>> newtonCooling(93.5, 24.5, .0426, 10)
   ans: 69.5650
```

To use `newtonCooling` in `fplot`, we need to repackage it so that it takes a single argument. Of course, we could edit the file `newtonCooling.m` to change the arguments, but this might interfere with other uses of `newtonCooling`.

Key Term   Instead, we can write another function, called a *wrapper function*, that provides an interface to `newtonCooling` that is in the right format for `fplot`. Here is one:

```
[1]  function res = newtonWrapper(t)
[2]  % a wrapper for newtonCooling to make it a function of t
[3]  res = newtonCooling(93.5, 24.5, 0.0426, t);
```

`NewtonWrapper` can be passed directly to `fplot`:

```
>> fplot('newtonWrapper', [0, 100]);
```

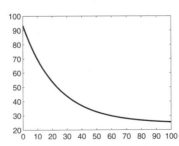

A wrapper function is a perfectly ordinary sort of function: It transforms an input to an output. But the body of the wrapper function is usually very simple, as in this case.

6.5 Returning Functions as Values

Although the trick of quoting a function name and using `feval` to evaluate it handles the problem of passing functions as arguments, it can be tedious to have to create and edit an m-file just for the purposes of writing a wrapper function. MATLAB provides an operator that can create a function "on the fly" without having to create an m-file.

Notice that m-file functions have a simple template: We can write one just by filling in the blanks:

```
function res = _____ ( _____ )
                  name of function      argument name
res = _____ ;
         expression using the argument name
```

Key Term

The `inline` operator takes a set of character strings as arguments and returns an *executable value*: a value that can be assigned to a variable or passed to a function as an argument (in this sense it is like any other value such as a numerical vector or a character string). But unlike vectors or character strings, an executable value can be used in function invocation.

To illustrate, we'll make an executable value for a mathematical function we want to graph:

```
≫ f = inline('sqrt(x).*sin(2.*x)', 'x')
  f: Inline function:
     f(x) = sqrt(x).*sin(2.*x)
```

The first argument to `inline` is a character string containing the MATLAB expression implementing the body of the function. The second argument gives the name of the argument to the functions. These arguments to `inline` provide all of the information for the fill-in-the-blanks template, except for the function name. `Inline` returns an executable value that can be assigned to a variable; that variable gives the name to the function.

Now f can be used just like the m-file `myfun.m`:

```
≫ f(xpts)
  ans: 0 0.1767 0.4595 0.7290 and so on
```

Executable values can be copied so the value can be given new names. For instance,

```
≫ aNewName = f
  aNewName: Inline function:
            aNewName(x) = sqrt(x).*sin(2.*x)
≫ aNewName(xpts)
  ans: 0 0.1767 0.4595 0.7290 and so on
```

As always, there is no need to name a value in order to pass it as an argument to a function, so we can plot out the mathematical function without naming it:

```
≫ funplot(inline('sqrt(x).*sin(2.*x)', 'x'), [0 20])
```

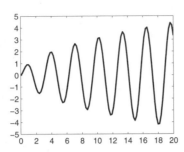

Here is an `inline` wrapper function for plotting `newtonCooling` as a function of time:

```
≫ newtonWrapper = inline('newtonCooling(93.5, 24.5, .0426, ...
       t)', 't');
```

Since `newtonWrapper` is now an executable value and not a character string, it should not be quoted when passed as an argument to another function:

```
≫ fplot(newtonWrapper, [0, 100])
```

A Matter of Style: _____

When is it best to create a wrapper m-file and when an inline value? Since any function that can be created with `inline` can also be created as an m-file, and since `feval` will handle executable values or strings naming m-files, there is always a choice.

Use `inline` in situations where you need to pass a function as an argument, when you have no other use for that function, and when your use of that function will be so fleeting that it is not worthwhile giving a name to the function.

6.6 Exercises

Discussion 6.1:

Imagine a function `arrange2(a,b)` that takes two variables as arguments and puts the smaller value in the first variable and the larger value in the second variable. For instance, we'd like `arrangeInOrder` to work like this:

```
≫ high = 6; low = 9;
≫ arrangeInOrder(low,high);
≫ low
  ans: 6
```

```
≫  high
   ans:  9
```

Explain why this behavior is not possible using functions or scripts. (The function is implemented with a different invocation syntax in Section 7.1.)

Discussion 6.2:
Consider the following sequence of statements:

```
≫  a=7;
≫  f = inline('a*x');
≫  f(5)
```

It looks like the answer should be 35, but instead there is an error message. Explain why.

Discussion 6.3:
inline functions can take more than one argument. Read the documentation and figure out how to write a function of three arguments, a, b, and c defined as $f : (a,b,c) \rightarrow a+b*c$.

Exercise 6.1:
Write a function, powersum, that computes $\sum_{j=0}^{n} z^{-j}$ for any specified value of z and n.

Exercise 6.2:
Create a directory called MyFunctions and create m-files in that directory for the following functions. In each case, make sure that there is an appropriate help line or lines. Also, in the first help line, add your initials.

1. A function add that takes two variables as arguments and returns their sum.
2. A function square that takes one argument and returns its square.
3. A function pow that takes two arguments, x and y, and returns the value x^y.
4. A function eliminate that takes a character string as an argument and returns a character string of the same length, but where the letters e and E have been replaced with *. (*Hint:* Use find and the index operator.)
5. A function HMStoDays that takes three arguments, hours, minutes and seconds, and adds them up to give the sum in days (which might have a fractional part).
6. A function mean that takes a vector v of numbers as an argument and returns the arithmetic mean

$$\frac{v_1 + v_2 + \cdots + v_N}{N}$$

7. A function geomean that takes a vector v of numbers as an argument and returns the geometric mean

$$(v_1 \times v_2 \times \cdots \times v_N)^{1/N}$$

Add the MyFunctions directory to the front of the search path, and show that MATLAB can find the functions even when the working directory is not MyFunctions. Show that the function mean that you wrote, rather than the built-in one, is being used in this case. (*Hint:* You can try help mean or which mean to see which one is being used.) Now delete MyFunctions from the front of the search path and add it to the end of the search path instead. Show that the functions you wrote are still accessible even when the working directory is not MyFunctions but that the built-in mean is being used instead of your version.

Show that lookfor, when given your initials as an argument, finds the functions you wrote.

Exercise 6.3:

Write an addToEnd function that takes two arguments: a row vector and a scalar or row vector and concatenates the two.

```
≫ addToEnd([3 4], 5)
  ans: 3 4 5
```

Suppose you have a vector a = [1 2 3] and you concatenate 5 to it

```
≫ addToEnd(a,5)
  ans: 1 2 3 5
```

Has the value of a changed? Give a statement using addToEnd that modifies a to be [1 2 3 5].

Exercise 6.4:

Write a function numToBoolean that converts numbers to booleans according to the rule that any nonzero number corresponds to true and zero corresponds to false. The function should have help lines that explain clearly what it does.

Exercise 6.5:

What's wrong with the following function? It is intended to return a boolean that indicates whether the argument is numerically bigger than 3.

```
function res = biggerThan3(x)
x > 3;
```

Exercise 6.6:

Write a function that translates between a boolean vector and a TF character string, and another that translates the other way.

```
≫ BooleanToTF([1 0 0 1 1 1 0])
  ans: 'TFFTTTF'
≫ TFToBoolean('TFFTFFFT')
  ans: [1 0 0 1 0 0 0 1]
```

(*Hint:* Use indexing on the vectors 'TF' and [0 1]==1.)

Exercise 6.7:

A perfect square is the square of an integer, for instance, 9, 16, 25, and so on. Write a function implementing a boolean operator that says whether a number is a perfect square. (*Hint:* Use sqrt, floor, and ==.)

Exercise 6.8:

Write a function sumdiff that returns two values: the sum and the difference of two variables, assumed to be numbers. It should work like this:

```
≫ [s,d] = sumdiff(4,2);
≫ s
  ans: 6
≫ d
  ans: 2
```

Exercise 6.9:

Write a function wholeAndPart that takes a number and returns the integer part and the fractional part. For example,

```
≫ [int,fract] = wholeAndPart(24.3);
≫ int
  ans: 24
≫ fract
  ans: 0.3
```

(*Hint:* The rem function can be used to compute the fractional part of a number.)

```
≫ rem(24.3,1)
  ans: 0.3
```

Check your function against negative numbers and make sure that it produces sensible results. In particular, you want the sum of the returned integer part and the fractional part to equal the argument to the function.

Exercise 6.10:

Write a function daysToHMS that takes a time duration in days (perhaps fractional) and returns three values, the number of hours, minutes, and seconds in that number of days.

```
≫ [h,m,s] = daysToHMS(1)
  ans: h=24
       m=0; s=0
```

(*Hint:* The rem, floor, and ceil functions may be useful.)

Exercise 6.11:

Write a function polarToRect that takes two arguments, the radius r and angle θ (in radians) of a position in polar coordinates, and returns two values, the corresponding x- and y-positions in rectangular (Cartesian) coordinates. (Given polar r and θ, the corresponding rectangular x and y are $x = r\cos\theta$, $y = r\sin\theta$. Given x and y, the equivalent polar coordinates are $r = \sqrt{x^2 + y^2}$ and θ such that $\tan\theta = \frac{y}{x}$. Solving for θ is not so simple, because the tangent function is not uniquely invertible. However, computer systems such as MATLAB provide a related function that performs the desired inversion. In MATLAB this is `atan2(y,x)`.)

Write another function, `rectToPolar`, that converts the other way. Give some test cases to show that the functions work. Further test the function by writing another function, `testPolarToRect`, that takes two numbers as arguments, converts them using `rectToPolar`, and then converts back using `polarToRect` and returns the relative error of the back-and-forth converted arguments compared to the originals.

Exercise 6.12:
Some statements are genuinely ambiguous but do not violate the grammatical rules of the language. For example, before typing anything, think about the statement

```
>> [a,a] = max([ 5 4 8 2 7 ])
```

and try to figure out what the value of a will be. Then try it on the computer and see if the result matches your prediction. Try to explain why the computer answered as it did.

It is bad style to use such a statement, and there is no reason to do so unless your intention is to confuse a human reader or entertain the possibility of a program working differently on different computers. If your goal is to throw away the first returned argument from max, then it's better to signal this explicitly to the reader:

```
>> [trash, a] = max([5 4 8 2 7]);
```

Exercise 6.13:
Without writing any m-files, use `fplot` and `inline` to draw graphs of the following mathematical functions, over a range from 0 to 100:

1. $e^{-x/10}$
2. $\cos x/5$
3. $\frac{x^{10}}{50^{10} + x^{10}}$

Hand in the statements that you wrote to make the plots, but you don't need to hand in the plots themselves.

Exercise 6.14:
The function `strrep` takes three arguments: a string, a string to look for, and a string to replace any instances of the second string in the first. For example,

```
≫ strrep('She sells sea shells by the sea shore.',...
    'sea', 'ocean')
ans: She sells ocean shells by the ocean shore.
```

Write an inline function that takes one string as an argument and replaces any instances of the substring 'if' with 'when'. (*Hint:* When you want to have a quotation mark within a quoted string, you use two single quotes in a row. For example, to produce *can't* as a quoted string, you would write 'can''t'.)

Exercise 6.15:
A palindrome is a sentence or phrase that spells the same backward and forward. For example,

Madam, I'm Adam.

Able was I ere I saw Elba.

Write a function palindrome(str) that takes a character string as an argument and returns a boolean result indicating if the string is a palindrome. Only the letters themselves, and not the capitalization or punctuation, should be used in deciding whether the string is a palindrome. (*Hint:* You may want to use the lower and isletter functions.)

CHAPTER 7

Conditionals

Annual income twenty pounds, annual expenditure nineteen nineteen six, result happiness. Annual income twenty pounds, annual expenditure twenty pounds ought and six, result misery.
— Charles Dickens, *David Copperfield*, Chap. xii.

Consider how to write an algorithm for the absolute value of a number, traditionally denoted $|x|$. This computation involves one input and produces one output, suggesting the following start to a function m-file:

Function declaration   [1]   `function res = absval(x)`
The help line   [2]   `%  absval(x) --- computes |x|`

To implement the body of the function, we need to express the $|x|$ computation in terms of other computations that we already know how to perform.[1] One possible algorithm uses the - operator: Negate the argument if it is negative; otherwise leave it alone. This chapter is about how to express the "if" and "otherwise" parts of the previous statement.

---

[1]There is a built-in MATLAB function abs that does exactly the computation that we want. Although there's little practical point in reimplementing functions that already exist—indeed, having duplicates can make software hard to manage—we do so here since we want to study how such functions can be written.

7.1 The if Statement

The natural way to express an algorithm for $|x|$ using arithmetic operators involves a conditional: If x is less than zero, then $|x|$ is $-x$. Otherwise, $|x|$ is simply x.

The MATLAB-language syntax for expressing such a conditional is quite similar to English. Here is the continuation of the function absval declared previously:

The test condition	[3]	`if x < 0`
The YES block	[4]	`    res = -x;`
	[5]	`else`
The NO block	[6]	`    res = x;`
	[7]	`end`

Key Term

This type of expression is called an *if statement*. The words if, else, and end are just punctuation. Such punctuation words (we've already seen another one, function) are called *keywords* or *reserved words*; they provide special formatting information to the parser and therefore shouldn't be used as the names of variables. In an if statement, the keywords if, else, and end identify the three different parts of the expression:

The conditional: A boolean expression that will always be evaluated and will return a scalar boolean 0 or 1

The YES block: An expression or set of expressions that will be evaluated only if the conditional expression returned 1

The NO block: Another expression or set of expressions that will be evaluated only if the conditional expression returned 0

In no case will both the YES and the NO blocks be evaluated; it's either one or the other. But the conditional is always evaluated.

A Matter of Style:

The use of new lines to break up the if statement across several lines is a format strongly to be preferred for human readers. The indentation of statements within the YES block and the NO block marks to the human reader the extent of those blocks. Semicolons within the YES or NO block are used to suppress printing of the values of the individual expressions and to separate expressions from one another. This becomes important as the blocks become more complicated, as we will see later.

The editor that is bundled with the MATLAB program will automatically attempt to indent lines in a useful way. This indentation not only makes it easier to read the function once it is finished, but it can help to spot mistakes when writing a function. For example, an accidental omission of the end statement at the end of a block of statements becomes apparent when the MATLAB editor continues to indent the statements after the intended end of the block.

When editing and revising programs, the indentation sometimes becomes confused. The MATLAB editor has a "reindent" command, invoked by highlighting the text and using the CTRL-I keystroke.

Still, the parser does not need either the indentation or the new lines since the YES and NO blocks are delimited by the else and end keywords. So it is technically possible to write an entire if statement on one line; for example,

```
if x<0; res= -x; else res = x; end
```

This compact style is to be discouraged except, perhaps, for very short statements.

Any number of expressions can go in the YES and NO blocks. Here is a short function that takes two numbers as input and returns the same two numbers but with one identified as the bigger and the other as the smaller of the two.

```
[1]   function [smaller, bigger] = arrange2(a,b)
[2]   %  which of a and b is smaller and bigger
[3]   if a < b
[4]       smaller = a;
[5]       bigger = b;
[6]   else
[7]       smaller = b;
[8]       bigger = a;
[9]   end
```

The YES block is these two lines — lines [4] and [5].
The NO block is these two lines — lines [7] and [8].

Such a function might be useful to put the two endpoints of a numerical interval in proper order. In testing this function, we should remember that we need to name the two returned values if we want both returned:

```
≫  [s,b] = arrange2(7, 5);
   ans: s = 5
        b = 7
```

7.2 More Than Two Cases

The structure of an `if` statement suggests that we can test only for two possibilities: Either the conditional is true or it is false. For example, consider the problem of finding the largest of three numbers, a, b, and c. Of course, there are three possible outcomes: a is the biggest, b is the biggest, or c is the biggest. It's helpful to make a table of the possible outcomes and the conditions under which each one is right.

Outcome	Condition
a	$a \geq b$ and $a \geq c$
b	$b \geq a$ and $b \geq c$
c	$c \geq a$ and $c \geq b$

A situation with three possibilities does not lend itself well to the either-or framework imposed by the `if-else-end` syntax.

The `elseif` keyword allows us to insert additional cases:

```
if       boolean test for first case
         block for first case
elseif   boolean test for second case
         block for second case
elseif   boolean test for third case
         block for third case
         (remaining cases using additional elseif lines)
else
         block for the last case
end
```

Using the `if-elseif-else-end` construction, here's a function for finding the largest of three numbers:

```
                        [1]  function res = max3V1(a,b,c)
a is the biggest        [2]  if a >= b & a >= c
                        [3]      res = a;
b is the biggest        [4]  elseif b >= a & b >= c
                        [5]      res = b;
c is the biggest        [6]  else
                        [7]      res = c;
                        [8]  end
```

7.3 Completeness and Exclusivity

It's possible to leave out the `else` component of an `if` statement. For example, the absolute value function might have been written as follows:

Version 2	[1]	`function res = absvalV2(x)`
This assignment will always be done, but it will be wrong if $x < 0$	[2]	`res = x;`
	[3]	`if x<0`
Fix things when $x < 0$	[4]	`    res = -x;`
	[5]	`end`

If the conditional expression (x < 0) is false, then the YES block will not be evaluated and, since there is no NO block, no expression at all will be evaluated. This works here because nothing needs to be done when the conditional expression is false; `res` has already been set to the correct value in line 2.

An `if` statement is "complete" if it is arranged so that one of the conditional blocks will always be evaluated. All `if-else-end` statements are complete because either the YES block will be evaluated or the NO block will be evaluated; it can never happen that neither block is evaluated. Similarly, an `if-elseif-····-else-end` statement is complete; even if none of the booleans on the `if` or `elseif` lines evaluate to 1, the block following the `else` keyword will be evaluated.

The `if` statement in `absvalV2` is incomplete. This is not in itself a problem; the function works perfectly well. It's pretty easy to get such a simple function to work, but it can be much more difficult to write and debug a more complicated function and to ensure that it works.

One effective strategy for doing this is to use complete `if` statements, explicitly identifying each possible case. Reserve the `else` block for a warning or error about impossible cases.

Impossible cases? Why worry about things that are impossible? The answer is that we don't need to worry about cases that are truly impossible but often something that we think is impossible is not. Murphy's law says,

Anything that can go wrong, will.

An extension is this:

When you think nothing can possibly go wrong, the impossible is bound to happen.

To illustrate, here's an example in a version of `max3` where there's been a slight mistake, but the mistake is somewhat hard to find.

```
[1] function res = max3V2(a,b,c)   %  V2, broken
[2] if a > b &  a > c
[3]     res = a;
[4] elseif b > a &  b > c
[5]     res = b;
[6] else
[7]     res = c;
[8] end
```

Test if a is the biggest — [2]

Test if b is the biggest — [4]

Neither a nor b is the biggest, therefore (?) c is the biggest — [6]

We try this out on a few cases representing each possible outcome, and it seems to work:

```
≫  max3V2(1,2,3)
  ans: 3
≫  max3V2(2,3,1)
  ans: 3
≫  max3V2(3,2,1)
  ans: 3
```

At this point, we might well decide that max3 is correct and ready to be used as part of a larger computation.

But max3 doesn't work reliably. Let's try a case with two identical arguments:

```
≫  max3(3,3,1)
  ans: 1
```

The answer is wrong. The problem is that we assumed that the conditions on the three outcomes were complete: Either a>b & a>c or b>a & b>c or c>a & c>b. Since we thought that these three cases cover all possibilities, we felt safe in using else instead of a conditional to cover the last case. This was a mistake, since the conditions actually are incomplete; they ignore the possibility that two of the arguments are equal. The result is that the else block was evaluated in cases where it shouldn't have been.

Even worse than giving the wrong answer, the function doesn't tell us it's broken: we have to figure this out ourselves. That's easy here, since it's easy to figure out that the largest of the set $\{3,3,1\}$ is 3. But imagine that we didn't know the correct answer to max3(3,3,1), as might be the case if the operation were more complicated. How would we see that max3 is broken? If max3 were being used in a larger computation, we might never be aware of an error and the output of the larger computation might be completely wrong.

By using an impossible-else construction, we make it possible to identify situations that violate our assumptions. We put each possible outcome into it's own `if` or `elseif` block and put an error statement into the `else` block.

	[1]	`function res = max3V3(a,b,c)`
	[2]	`% Version 3, still broken`
Test if a is the biggest	[3]	`if a > b & a > c`
	[4]	`    res = a;`
Test if b is the biggest	[5]	`elseif b > a & b > c`
	[6]	`    res = b;`
Test if c is the biggest	[7]	`elseif c > a & c > b`
	[8]	`    res = c;`
	[9]	`else`
We should never get here if the if and elseif cases are complete. If we do, we know something's wrong.	[10]	`    error('Impossible case.');`
	[11]	`end`

The `error` operator sends a message that something is wrong: the argument to the `error` operator is the message to be sent. We'll see more about `error` statements in Chapter 8.

When the computation is attempted with inputs that violate our assumptions, we get an error message.

```
≫ max3(3,3,1)
  ans: ??? Error using ==> max3
       Impossible case.
```

Although we still do not know what the problem is, at least we know that something is wrong in `max3`. This is the first, critical step in finding and fixing errors. We can get more detail by using a debugger set to stop on the `error` statement, as we shall see in Chapter 9.

Exclusivity

Key Term

The completeness of an `if-else-end` or `if-elseif-...-else-end` construction is guaranteed by the `else` block. Another feature of these constructions is that the blocks are *exclusive*: Only one of the blocks will be executed.

Exclusivity is also guaranteed by the construction, even if the statements aren't logically completely exclusive. For example, the two conditions in `max3elseif` (on page 136) are `a >= b & a >= c` and `b >= a & b >= c`. These aren't actually completely exclusive: When a, b, and c are all equal, both of the conditions are true. In this case, it seems that both of the corresponding blocks might be executed, but this would be a violation of exclusivity. In fact, only the block corresponding to the first true conditional will be executed; any others are ignored even if their conditionals are true.

To illustrate how the enforced exclusivity of an `if-elseif-else-end` construction can help, consider the problem of translating a numerical test score into a letter grade. Although grading policies differ from school to school (and even from class to class), one common policy is that a grade above 90% becomes an A, a grade between 80% and 90% becomes a B, and so on. One way to write a function that uses this policy is illustrated by the following:

```
[1]   function res = convertGradeV1( score )
[2]   if score >= 90
[3]       res = 'A';
[4]   elseif score >= 80 &  score <= 89
[5]       res = 'B';
[6]   elseif score >= 70 &  score <= 79
[7]       res = 'C';
[8]   elseif score >= 60 &  score <= 69
[9]       res = 'D';
[10]  else
[11]      res = 'Fail';
[12]  end
```

An error! (marginal note at line [4])

The `else` clause guarantees that the conditional is complete. The clauses are also exclusive because the number ranges in the `elseif` conditions do not overlap. Yet there is a serious mistake. Imagine the unhappy student who scored 89.5 in the course:

```
≫  convertGradeV1(89.5)
   ans: Fail
```

The mistake comes from the gap between `>= 90` and `<= 89` in the first two conditions; the case of 89.5 is therefore handled by the `else` clause.

In an abstract sense, the mistake stems from an inconsistency between redundant numbers in the program. In each of the `elseif` conditions there are two numbers, a maximum and a minimum value of the range for that grade. But the minimum score for one grade is the same as the maximum score for the next lower grade. Or, at least this is intended to be the case. Whenever there is redundancy of this sort, it's easy to introduce an inconsistency.

This problem of redundancy leading to contradiction is quite important. Due to human forgetfulness and the imprecision of language, humans appear to benefit from being told the same thing in several different ways. But this is not true for computers.

Another style for writing the `convertGrade` function uses the cascading property of the `elseif` to ensure exclusivity without needing redundancy:

```
[1]   function res = convertGradeV2 ( score )
[2]   if score >= 90
[3]       res = 'A';
[4]   elseif score >= 80
[5]       res = 'B';
[6]   elseif score >= 70
[7]       res = 'C';
[8]   elseif score >= 60
[9]       res = 'D';
[10]  else
[11]      res = 'Fail';
[12]  end
```

Here there certainly is not exclusivity between the conditions; it may well happen that more than one of the conditions is satisfied. For instance, a grade of 85 will satisfy the second, third, and fourth conditions. But the interpreter guarantees that only one of the conditional blocks will be executed, the first one whose condition is satisfied. In order for this style to work, however, the conditions must be placed carefully in order. For example, the following conditional would have no effect if placed just on line 4 or after, since all cases with `score >= 90` are handled by lines 2 and 3.

```
elseif score >= 97
  res = 'A+';
```

7.4 Switch/Case

Often, a conditional is used to pick one of several named cases. For example, in performing the addition of numbers with units, we need to convert the two numbers to a common unit. We might decide to convert all lengths to units of meters for the purposes of adding them. To do this, we need to multiply by a conversion factor that depends on the name of the units being used. For example, to convert inches to meters, we multiply by 0.02540. To convert yards to meters, multiply by 0.9144.

Here's how we might do this using an `if-elseif-else` statement:

<table>
<tr><td></td><td>[1]</td><td><code>function res=conversionToMetersV1(unitname)</code></td></tr>
<tr><td></td><td>[2]</td><td><code>% Return a conversion factor to meters</code></td></tr>
<tr><td></td><td>[3]</td><td><code>% for the given character-string unitname</code></td></tr>
<tr><td></td><td>[4]</td><td></td></tr>
<tr><td></td><td>[5]</td><td><code>if strcmp(unitname, 'yards') | ...</code></td></tr>
<tr><td>Check for match to "yards"</td><td>[6]</td><td><code>   strcmp(unitname, 'yd')</code></td></tr>
<tr><td></td><td>[7]</td><td><code>   res = 0.9144;</code></td></tr>
<tr><td></td><td>[8]</td><td><code>elseif strcmp(unitname, 'inches') | ...</code></td></tr>
<tr><td>Similarly for "inches"</td><td>[9]</td><td><code>   strcmp(unitname, 'in')</code></td></tr>
<tr><td></td><td>[10]</td><td><code>   res = 0.0254;</code></td></tr>
<tr><td></td><td>[11]</td><td><code>elseif strcmp(unitname, 'miles') | ...</code></td></tr>
<tr><td></td><td>[12]</td><td><code>   strcmp(unitname, 'mi')</code></td></tr>
<tr><td></td><td>[13]</td><td><code>   res = 1609.0;</code></td></tr>
<tr><td></td><td>[14]</td><td><code>elseif strcmp(unitname, 'furlongs')</code></td></tr>
<tr><td></td><td>[15]</td><td><code>   res = 201.125;</code></td></tr>
<tr><td></td><td>[16]</td><td><code>elseif strcmp(unitname, 'light-years')</code></td></tr>
<tr><td></td><td>[17]</td><td><code>   res = 9.460913e15;</code></td></tr>
<tr><td></td><td>[18]</td><td><code>else</code></td></tr>
<tr><td></td><td>[19]</td><td><code>   error('No conversion to meters known');</code></td></tr>
<tr><td></td><td>[20]</td><td><code>end</code></td></tr>
</table>

The `else` block handles cases where the unit name is unknown to the function, for instance, "cubits" or "hands" or the printers' units "points" or "ems."

The function takes a unit name as an argument and returns a conversion factor from the named unit to meters:

```
>> conversionToMetersV1('inches')
  ans: 0.0254
```

What's distinctive about this function is that all of the boolean conditions are tests for sameness. Since `unitname` is a character string, the appropriate boolean comparison operator is `strcmp`.

The `switch/case` construction provides a special syntax for situations where every boolean condition checks for sameness.

```
[1]  function res=conversionToMetersV2(unitname)
[2]  switch lower(unitname)
```

Does unitname match "yards" [3] `case { 'yards', 'yd'}`
[4] `    res = 0.9144;`
Similarly for "inches" [5] `case { 'inches', 'in'}`
[6] `    res = 0.0254;`
and so on [7] `case { 'miles', 'mi'}`
[8] `    res = 1609.0;`
[9] `case 'furlongs'`
[10] `    res = 201.125;`
[11] `case 'light-years'`
[12] `    res = 9.460913e15;`
[13] `otherwise`
[14] `    error('No conversion to meters known');`
[15] `end`

Version 2 of this function is considerably easier to read and revise than version 1.

Key Term

The statement after the `switch` keyword is called the *switch expression*. It must be evaluated as either a character string or a number. (In `conversionToMetersV2` the switch expression involves the `lower` operator so that we can write all of the comparisons with lowercase letters and still cover uppercase inputs.) After each `case` keyword is a statement whose value is tested for sameness with the value of the switch expression. Alternatively, a cell array can follow the `case` keyword as on lines 3, 5, and 7, in which case each element in the cell array is tested for sameness with the value of the switch expression. The statements in the block after the first matching `case` statement are evaluated; none of the other blocks are evaluated. If none of the `case` statements match, then the `otherwise` block is evaluated.

If the switch expression is a number, then the `==` operator is used to test for sameness with the `case` expressions. If the switch expression is a character string, then `strcmp` is used to test for sameness.

A Matter of Style: _____

Whenever possible, a function whose return value is set within a conditional statement should return the same kind of thing in all of the cases. For example, `conversionToMetersV2` always returns a number when it returns successfully. The `otherwise` block handles situations where no number can sensibly be returned. Although it might seem "friendly" to return a character string in such situations, say, `'No conversion known.'`, this would be bad style; it would force the user of `conversionToMetersV2` to check the returned value to see if it is a number or a string. Instead, `conversionToMetersV2` generates an error if there is no conversion. `Error` prevents the function from returning any value whatsoever.

7.5 Exercises

Discussion 7.1:

What's wrong with this program?

Either a or c is the biggest

```
[1]   function res=max3wrong(a,b,c)
[2]   %  An incorrect implementation of max3
[3]   if a >= b
[4]       if a >= c
[5]           res = a;
[6]       else
[7]           res = c;
[8]       end
[9]   else
[10]      %  So b must be the biggest
[11]      res = b;
[12]  end
```

Exercise 7.1:
Without using sort, write a function med(a,b,c) that returns the median of the numbers a, b, and c. The median of three numbers is the number that has a value in between the other two. If there is a tie, then the value of the tying numbers is the median. (*Hint:* If $a \leq b$ and $b \leq c$, then b is the median.)

Exercise 7.2:
Write a function max4 to find the largest of four numerical arguments: max4(a,b,c,d). Because this is intended to exercise your skills with compound if statements, don't use the built-in max function, which would allow you to produce the answer with just one line: max([a b c d]).

Exercise 7.3:
Write a function max5 that works like max4 but handles five inputs.

Exercise 7.4:
If you were unusually clever in doing Exercise 7.3, you might have written max5 using max4 and a simple if-else-end construction that needs to look at two cases. Otherwise, write max5 this way now. (In Chapter 12, we'll explore the general principle at work here, recursion.) Then write max6 using max5. You might also rewrite max4 and max3 in this style. Explain why it doesn't make sense to rewrite max2. (In Section 9.8, we'll see a better way to handle families of functions for different numbers of arguments.)

Exercise 7.5:
Modify the functions you wrote in Exercise 7.4 to return two arguments: the

maximum value and an index indicating which argument was the maximum. (*Hint:* You can break ties arbitrarily.)

Exercise 7.6:

Write a function `absolutevalue` that works for both real and complex numbers. For a complex number $z = x + iy$, the absolute value is $|z| = \sqrt{x^2 + y^2}$.

Exercise 7.7:

Write a function `isvowel` that takes a single letter as an argument. It returns a boolean 1 when a letter is a vowel and 0 when it is a consonant. Such a boolean return value has no place for the idea of "sometimes y." You will have to decide whether or not y is a vowel.

Exercise 7.8:

Write a function to return the number of protons in an atom taking as input the element symbol as a character string (e.g., `'H'`, `'Ag'`). (*Hint:* There are a lot of elements. For this exercise, cover at least the lightest 10 in order to show how the function would be written.)

Exercise 7.9:

Write a function `signum` that takes one number as an argument and returns 1 or -1 depending on whether the number is positive or negative. If you use an `if-else-end` construction, the function will not work for vectors or matrices. Figure out an arithmetic/boolean expression to replace `if-else-end` that will work for vectors and matrices.

Exercise 7.10:

The roots of a polynomial (for instance, the quadratic $x^2 + bx + c$) are those values of x for which the polynomial is zero. For a quadratic, there are two roots and they can be found with the famous quadratic formula:

$$\lambda = \frac{-b \pm \sqrt{b^2 - 4ac}}{2a}$$

Write a function that takes the three numbers a, b, and c and returns one of the strings `'real'`, `'degenerate'`, or `'complex'` depending on whether the roots are two different real numbers or both are the same real number (degenerate), or both are complex numbers. (*Hint:* Some relevant operators for testing the results of the calculation are `real`, `imag`, and `conj`.)

Exercise 7.11:

Write a function `month2num` to convert a month name to a number. It should work with both the full name (e.g., `'January'`) and abbreviations (e.g., `'Jan'`).

Exercise 7.12:

Write a function to convert three numerical arguments standing for year, month, and day to a decimal year.

7.6 Project: The Morse Code

Samuel Finley Breese Morse (1791–1872) was an American painter and an early daguerreotypist. His daguerreotypes of his Yale class reunion are the first known photographic group portraits. He was also the inventor of the eponymous Morse code, a system of dots, dashes, and spaces used in telegraphy to represent the letters of the alphabet and numerals. The following table gives the "international" code, which is a modification of Morse's original code necessitated by distortions of the signals in undersea cables. Morse's original code continued to be used in the United States for land communication.

The International Morse Code

A	•—	B	—•••	C	—•—•
D	—••	E	•	F	••—•
G	——•	H	••••	I	••
J	•———	K	—•—	L	•—••
M	——	N	—•	O	———
P	•——•	Q	——•—	R	•—•
S	•••	T	—	U	••—
V	•••—	W	•——	X	—••—
Y	—•——	Z	——••	1	•————
2	••———	3	•••——	4	••••—
5	•••••	6	—••••	7	——•••
8	———••	9	————•	0	—————
period	•—•—•—	comma	——••——	colon	———•••
?	••——••	apostr.	•————•	hyphen	—••••—
÷	—••—•	( )	—•——•—	quote	•—••—•

A • is one time unit long.

A — is three time units long.

Space between components of one character is one time unit.

Space between characters is three time units.

Space between words is seven time units.

•••••••• means delete the last word.

The telegrapher's idea of using sequences of on/off pulses to transmit information prefigures the way that digital computers would be designed more than a century later. Inside the computer, communication is generally accomplished across multiple wires, but for many external purposes just one

Key Term

wire contains the signal. These are called *serial lines*. Morse code, which involves short sequences of different duration pulses for each letter, has been replaced with the ASCII code of fixed-length sequences (seven or eight pulses, typically) in which each on/off pulse lasts for a standard length of time.

Exercise 7.13:

Write a function `morsechar` that translates a single ASCII character into its Morse code equivalent. Handle properly those cases where there is no exact equivalent, perhaps substituting one character for another. For instance, the Morse code doesn't include lowercase characters, but it is reasonable to treat all letters as if they were uppercase. Save your program for later use in Exercise 8.22

The first step in developing an algorithm is to think of what the type and information content of the inputs and outputs should be. For now, the input will be a single character, for instance, `'d'`. In Chapter 8, we'll encounter the tools we need to extend the function to translate entire text messages (which we'll represent as character strings) such as `'What hath God wrought'`, the demonstration message sent by Morse on May 24, 1844.

The output is trickier. The specification of the problem doesn't indicate what the output is going to be used for. We might want one form of output if the output was going to be used as the input to an actual telegraph or HAM radio, and another if we wanted to produce the sound that a telegraph makes. In Exercise 8.22, we'll see how to generate an output that when played on a speaker sounds like Morse code.

The problem of incomplete specification is common when writing computer programs. To avoid restricting our future possibilities, let's make the output a sequence of numbers with 1 standing for a •, 2 for a −, 3 for the short interval between components of one character, 4 for the short space between characters, and 5 for the long space between words. Later on, when we know what the output really should look like, we can write another program to translate from our intermediate form of output to the final form. To illustrate, the output for D would be the vector of numbers 2 3 1 3 1; that is, $\begin{array}{ccccc} 2 & 3 & 1 & 3 & 1 \\ - & & \bullet & & \bullet \end{array}$ The output for a white-space character would be a single 5.

8

Loops

Now, in English, the letter which most frequently occurs is e.
Afterward, the succession runs thus: a o i d h n r s t u y c f g l m w b k p q x z

— Edgar Allan Poe, *The Gold Bug*[1]

Many algorithms involve repeating the same steps over and over. Here is a simple algorithm for computing the square root of a number: Start by bounding the prospective output between two values: an upper value that is too high and a lower one that is too low. Then, narrow the bounds by replacing one of them with the midpoint of the two bounds. Repeat the process.

Here is a possible implementation.

Set up the interval

```
[1]  function res = squarerootV1(x)  %Version 1
[2]  if x > 1
[3]      lower = 1;
[4]      upper = x;
[5]  else
[6]      lower = x;
[7]      upper = 1;
[8]  end
```

---

[1] In Section 8.9, we'll see that Poe's description is only partially correct.

The first part of the program sets up the upper and lower bounds. We know that $\sqrt{x}$ is always between 1 and x. The rest of the program narrows these bounds three times:

Narrow the bounds

```
[9]   middle = (lower+upper)./2;
[10]  if middle.*middle > x
[11]      upper = middle;
[12]  else
[13]      lower = middle
[14]  end
[15]
```

Narrow again

```
[16]  middle = (lower+upper)./2;
[17]  if middle.*middle > x
[18]      upper = middle;
[19]  else
[20]      lower = middle;
[21]  end
[22]
```

And again

```
[23]  middle = (lower+upper)./2;
[24]  if middle.*middle > x
[25]      upper = middle;
[26]  else
[27]      lower = middle;
[28]  end
[29]
```

The output

```
[30]  res = (lower+upper)./2;
```

Let's try it:

```
≫  squarerootV1(9)
   ans: 3.5
≫  squarerootV1(0.16)
   ans: 0.4225
```

The answers are off by a little. Only the first digit is right. This is because doing just three narrowings doesn't give a very precise answer. Since each narrowing reduces the interval by half, the final interval is only one-eighth [that is, $(\frac{1}{2})^3$] as long as the original, almost one digit of a base-10 number. To have an output that is precise to even four or five digits requires that the final interval be $\frac{1}{1000}$ to $\frac{1}{10,000}$ the size of the original.

We can improve the precision by adding more narrowings; just insert more copies of the block of the function that performs the narrowing (lines 10 to 15) before the final output statement. With 10 copies of this block, the final interval will be $\frac{1}{2}^{10} = \frac{1}{1024}$ the size of the original. To gain the 52 bits of precision of the mantissa held in a double-precision floating point number, we would, in principle, require 52 copies of the interval-narrowing block.[2]

The style of `squarerootV1` is awful: Repeating the steps in an algorithm by copying sets of commands over and over makes the program hard to write, hard to read, hard to modify, and hard to check for errors. Here is a dictum to remember when communicating with the computer:

Never repeat yourself.

When you repeat yourself in instructions to the computer, you introduce the possibility of inconsistencies and errors. This is not just a matter of typing mistakes; the same problems can arise when you use a text editor to cut and paste copies. Ambrose Bierce (1842–1914) anticipated some of the problems in a definition in *The Devil's Dictionary*:

Quotation, *n*: The act of repeating erroneously the words of another.

This section is about how to get algorithms to repeat steps without your having to specify the steps repetitively.

8.1 For Loops

The fundamental means to instruct the computer to execute the same lines repeatedly is the `for` loop. Here is the square-root function modified to remove redundancy:

	[1]	`function res = squarerootV2(x)  %Version 2`
Set up the interval	[2]	`if x > 1`
	[3]	`    lower = 1;`
	[4]	`    upper = x;`
	[5]	`else`
	[6]	`    lower = x;`
	[7]	`    upper = 1;`
	[8]	`end`
	[9]	
Loop 20 times	[10]	`for k=1:20`
First line in the loop	[11]	`    middle = (lower+upper)./2;`
	[12]	`    if middle.*middle > x`
	[13]	`        upper = middle;`
	[14]	`    else`
	[15]	`        lower = middle;`
Last line in the loop	[16]	`    end`
	[17]	`end`
	[18]	
The output	[19]	`res = (upper + lower)./2;`

---

[2]But round-off error might prevent us from achieving such precision.

Although the function is shorter, the answer is more precise because there are more narrowings:

```
≫  squarerootV2(9)
   ans: 3.000
≫  squarerootV2(0.16)
   ans: 0.400
```

The general syntax of a for loop is

```
for _____ = _____
        variable name            list of values
    _____
    _____
    _____
                  block of commands
end
```

Note that the = is part of the for-loop syntax, not the assignment operator.

The list of values can be either a numerical array or a cell array. In either case, the named variable will be given the value of the next element in the list on each go round of the loop. The for and matching end statements wrapped around a block of commands are the equivalent of saying, "Do this block for each element in the list." The same block of statements is executed repeatedly. This is called *iterating* the statements in the block. Note that for is not an operator: It doesn't return a value. Instead, it directs the interpreter to execute a block of commands multiple times.

Key Term

Let's look at this general syntax as used particularly in squarerootV2. Remember that 1:20 produces the vector [1 2 3 4 5 6 7 8 9 10 11 12 13 14 15 16 17 18 19 20]. The line

```
for k=1:20
```

does not mean "set k to the value of 1:20." Instead, it instructs the interpreter to set k to the "first" component of 1:20, namely 1. This done, the interpreter executes the lines inside the loop's block in the normal way. When the end of the loop is reached, marked by the end statement on line 17, k is set to the next component of 1:20 and the lines in the block are again executed just as if they had been retyped. This process continues, with k set successively to each of the components in 1:20. After the last component has been processed, the interpreter moves to the next statement after the end of the loop. Each time the block of statements is executed, we say that we have completed one iteration of the block.

If we decided to change the number of times we iterate the interval narrowing from 20 to 50, only one line of the function needs to be changed:

```
for k=1:50
```

8.2 Accumulators

Key Term

One of the main uses of loops is to build up an output piece by piece. The variables that hold the output while it is being assembled are called *accumulators.*

Consider the following simple computation: Take as input an integer n and return as output the sum of all the integers 1 to n. Here's a possible implementation:[3]

Compute $\sum_{k=1}^{n} k$	[1]	`function res = sumToN(n)`
Set up the accumulator	[2]	`res = 0;`
	[3]	`for k=1:n`
Update the accumulator on each iteration	[4]	`    res = res + k;`
	[5]	`end`

Trying it out,

```
≫  sumToN(10)
  ans: 55
```

The accumulator is initialized before the loop and updated within the loop. The appropriate form of initialization and updating depends on the computation being done. For a program to multiply the integers 1 to n (i.e., $1 \times 2 \times 3 \times \ldots \times n$), the accumulator should be initialized to 1.

Compute $\prod_{k=1}^{n} k$	[1]	`function res = productToN(n)`
Set up the accumulator	[2]	`res = 1;`
	[3]	`for k=1:n`
Update the accumulator on each iteration	[4]	`    res = res .* k;`
	[5]	`end`

In sumToN and productToN the accumulator is a single number. But an accumulator can be more than one value; the values can be stored as a vector or in multiple variables. For example, in the squarerootV2 function, the accumulator is the two variables, lower and upper. The output of squarerootV2 is not the accumulator itself, but something computed from the accumulator after the iterations have been completed: (upper + lower)./2.

Although the word "accumulator" suggests that the result is being built up gradually, this is not always the case. Chapter 1 gives a simple algorithm for finding the smallest number in a list of numbers. Here is that algorithm

---

[3] A better implementation is sum(1:n), which is much more efficient. But our purpose here is to show how to construct loops.

translated into a MATLAB function that finds the minimum value of a numerical vector:[4]

```
[1]  function S = minimum(L)
[2]  S = L(1);

[3]  for k=2:length(L)
[4]      if L(k) < S
[5]          S = L(k);

[6]      end
[7]  end
```

Initialize accumulator to the first number in the list *(annotation for line [2])*

We found a smaller one, so update the accumulator *(annotation for line [5])*

In this case, the accumulator, S, isn't building up the answer gradually but rather holding a tentative answer, one that becomes definitive only at the end of the loop. This is a common use for accumulators. Note that the loop is starting with the second element of the input; the first element was handled in the initialization of the accumulator. SumToN and productToN could also have been written this way.

8.3 Nested Loops

A loop is said to be nested when its statements reside within another loop. One common use for nested loops is to examine combinations of events. For instance, consider dice rolling. There are six possible ways that a die will come up, each of which is equally likely in a fair die. For two dice, there are $6 \times 6 = 36$ different possibilities. Since the dice are independent of each other, each of these 36 possibilities is equally likely. A random quantity often used in games is the sum of two dice. Most people learn to calculate the probability of any given dice sum by counting the number of ways the sum can arise and dividing this by the total number of possibilities. For instance, the sum 3 can arise in only two ways—$1+2$ and $2+1$—so the probability of the dice sum being 3 is $\frac{2}{36}$. Although this calculation can easily be done by hand, here is a computer program for calculating it for any given outcome:

```
[1]  function res = dicesum(score)
[2]  count = 0;
```

Counts number of times sum matches score *(annotation for line [2])*

---

[4]Note that the built-in operator, min, works faster and better.

Does the sum match?

```
[3]     for die1 = 1:6
[4]         for die2 = 1:6
[5]             if score == die1+die2
[6]                 count = count+1;
[7]             end
[8]         end
[9]     end
[10]    res = count./36;
```

And some results from that program:

```
≫  format rat
≫  dicesum(3)
  ans: 1/18
≫  dicesum(7)
  ans: 1/6
```

(Exercise 8.12 elaborates on the calculation a little more.)

Perhaps more interesting to serious players is the use of dice in the game Risk. This is a game of global domination and, fittingly, players attempt to conquer other player's pretend countries by military means. They do this by pitching armies against one another. To decide which army wins, the aggressor rolls two dice and the defender rolls a single die. If the defender's die matches or exceeds the number shown on the higher scoring of the aggressor's dice, then the defender wins. Here we have three dice, labeled a1, a2 and d in the program; there are $6 \times 6 \times 6 = 216$ possibilities altogether. The condition for the defender winning is d >= max(a1,a2). Since there are three dice, the enumeration of the possibilities involves three nested loops:

```
[1]   function res = risk2
[2]   wincount = 0;
[3]   for a1 = 1:6
[4]       for a2 = 1:6
[5]           for d = 1:6
[6]               if d >= max(a1,a2)
[7]                   wincount = count + 1;
[8]               end
[9]           end
[10]      end
[11]  end
[12]  res = wincount./216;
```

The program indicates that the defender is at a slight disadvantage; the probability of the defender winning is

```
≫  risk2
  ans: 0.42129629629630
```

8.4 Example: Optimal Matching with Nested Loops

You have four tasks to perform and four people to do them. Each person has different talents and skills and can perform some tasks better than others. You have some experience, and so you have quantified each person's capabilities into a table showing how well they will perform each task. Your job is to assign each person to a task in a way that will maximize the total quality; that is, the sum of the qualities on the individual jobs.

Person	Job			
	1	2	3	4
Allison	7	4	2	4
Basil	6	8	5	2
Clyde	4	7	1	3
Daisy	6	5	2	1

For instance, if you assigned Allison to job 1, Basil to job 2, Clyde to job 3, and Daisy to job 4, the total quality would be $7 + 8 + 1 + 1 = 17$.

One way to solve this problem is to consider every possible assignment of a person to a job. In this case, there are $4 \times 3 \times 2 \times 1 = 4! = 24$ possible assignments; the first person can be assigned to any of the four jobs, the second person to any of the three remaining jobs, the third person to either of the two jobs left unassigned to the first two people, and the last person gets the single job left unassigned. It's certainly feasible to compute the total quality for each of these 24 possible assignments.[5]

The first issue we face is how to represent the problem on the computer. One way is to have four variables a, b, c, and d that hold the job assigned to each of the people. The job itself could be represented by a number from 1 to 4. Thus, a=1, b=2, c=3, and d=4 would be the job assignment of Allison to job 1, Basil to job 2, and so on. If the table is represented by a matrix of numbers in a variable called q (short for quality),

```
>> q = [7 4 2 4; 6 8 5 2; 4 7 1 3; 6 5 2 1];
```

The total quality for any assignment is then

```
totalquality = q(1,a) + q(2,b) + q(3,c) + q(4,d);
```

---

[5]The problem does get out of hand, though. If there were 20 jobs and 20 people, the number of possible assignments would be 20! =2,432,902,008,176,640,000, an impossibly large number to handle.

Given this representation, we can tackle the matter of going through each of the 24 possible assignments in turn. It's certainly tempting to try something like

```
for k=1:24
    check possibility k
end
```

but how do we turn a number k into a set of one-to-one assignments of jobs to people?

Another approach is to use nested loops, assigning each person to each of the tasks and checking to see if the assignment is better than the best encountered so far. Here is a first attempt that is flawed:

```
bestassignment = [];
bestqualitysofar = -Inf;
for a=1:4
    for b=1:4
        for c=1:4
            for d=1:4
                totalquality = q(1,a) + q(2,b) + q(3,c) + q(4,d);
                if totalquality > bestqualitysofar
                    bestqualitysofar = totalquality;
                    bestassignmentsofar = [a, b, c, d];
                end
            end
        end
    end
end
```

Inside the innermost loop, each of the variables has been assigned a value, and we can compute the total quality of that set of job assignments and see if its better than the best encounter so far.[6]

The flaw in the preceding scheme is that there is nothing to ensure that two or more people aren't assigned to the same job. For instance, the first time through the innermost loop, everyone is assigned to job 1.

---

[6]We initialize bestqualitysofar to be -Inf to make sure that the first set of job assignments becomes the best one so far. This trick allows us to avoid having to initialize bestqualitysofar to correspond to an actual assignment.

One way to handle this is with an `if` statement; evaluate the total quality only if a, b, c, and d all have different values:

```
if a ~= b & b ~= c & c ~= d & a ~= c & a ~= d & b ~= d
    totalquality = q(1,a) + q(2,b) + q(3,c) + q(4,d);
    if totalquality > bestqualitysofar
        bestqualitysofar = totalquality;
        bestassignmentsofar = [a b c d];
    end
end
```

This is a complicated condition; it's hard to verify that it is correct.[7]

Another more transparent way of handling the situation is to loop over only those possibilities that are allowed. For instance, if a is assigned the value 3, then b should be looped only over the set [1 2 4]. We can exploit a helper function `exclude` written to allow us to easily take out the undesired elements of a set:

```
[1]  function res = exclude( set, undesired )
[2]  %exclude the undesired elements from the set
[3]  res = set;
[4]  for k=undesired
[5]      res = res( k ~= res );
[6]  end
```

Loop over the undesired — line [4]
Exclude from the set — line [5]

Using `exclude`, the program can be written thus:

```
[1]   function bestassignment = matchpeople(q)
[2]   bestassignment = 0;
[3]   bestqualitysofar = -Inf;
[4]   jobs = 1:4;
[5]   for a = jobs
[6]       jobsforb = exclude(jobs,a);
[7]       for b = jobsforb
[8]           jobsforc = exclude(jobsforb,b);
[9]           for c = jobsforc;
[10]              jobsford = exclude(jobsforc,c);
[11]              for d = jobsford
[12]                  totalquality = q(1,a) + q(2,b) + ...
[13]                      q(3,c) + q(4,d);
[14]                  if totalquality > bestqualitysofar
[15]                      bestqualitysofar = totalquality;
[16]                      bestassignment = [a b c d];
```

---

[7]The mathematically inclined will note that to check for any duplicate assignments out of four possibilities requires 4 take 2 comparisons $\frac{4!}{2!(4-2)!} = 6$.

```
[17]                    end
[18]                 end
[19]              end
[20]           end
[21]        end
```

Using this program, the best assignment is

```
≫  matchpeople(q)
  ans: 4 3 2 1
```

Allison gets job 4, Basil job 3, and so on.

Writing the program in this style makes it relatively straightforward to edit the m-file to rearrange the function to handle a different number of cases. In Chapter 12, we will see how to write it in a style so that a single function can handle any number of people and jobs.

8.5   Element-by-Element Operators

A function is often intended to be applied to each element of an array in turn, producing an output for each element. Many MATLAB operators do this automatically. For instance,

```
≫  [3 4 5] + 2
  ans: 5 6 7
≫  [7 8 2 1] == 2
  ans: 0 0 1 0
≫  sqrt([1 4 9])
  ans: 1 2 3
≫  upper('hello')
  ans: HELLO
```

In order to achieve this behavior, the operators contain a loop, and each element of the input is processed one at a time, the individual results being packaged together for the output as a vector, character string, or matrix, depending on the input.

The ease with which the built-in operators act on vectors can mislead you into thinking that the behavior is automatic for all operators. It is not. For instance, our home-grown square-root function, squarerootV2, does not work properly on vector inputs:

```
≫  squarerootV2([1 4 9])
  ans: 1.0000000000 1.0000014305 1.0000038146
```

The problem stems from the conditionals on lines 2 and 11 in squarerootV2: The if-else-end statement must execute either the YES block or the NO block. But in this case, the different elements of x need to have different blocks evaluated; there's no way for a single if-else-end construction to do this. Or, rather, we need to involve a loop so that the elements are handled one at a time.

The most straightforward way to create operators that work on arrays is to use a loop. Here is version 3 of the square root function: It handles vectors properly.

The output has the same number of elements as the input

Version 2 for each element singly

```
[1]  function res = squarerootV3(vec)
[2]  res = zeros(size(vec));

[3]  for k=1:length(vec)
[4]      res(k) = squarerootV2( vec(k) );
[5]  end
```

This function exploits the fact that squarerootV2 handles scalar values properly; version 3 simply uses version 2 on each element of the input, one at a time.

```
≫ squarerootV3([1 4 9])
  ans: 1.000000 2.00000047 3.0000038
```

In squarerootV3, each element of the input vector will produce a single element of the output vector. In such cases, it makes sense to set up the accumulator to have the same shape and size as the input. Line 2 accomplishes this by using zeros(size(vec)) to create a vector of the right size and orientation (that is, row or column). The individual elements are then filled in one at a time in line 4.

For processing arrays, as we shall see, a similar single-loop construction can be used. But in languages other than MATLAB, such as FORTRAN, C++, or Java, it's more common to use nested loops to handle arrays, and this technique can be used in MATLAB as well.

To illustrate, consider a version of squareroot that will operate on matrices. Matrix elements are referred to with two indices, the row number and the column number. To handle the whole matrix, we can loop over the rows, treating each row one at a time. Within a row, we process the elements, one at a time from each column of the matrix. To illustrate, here is version 4 of squareroot, one that works properly with matrices and, since vectors are a narrow form of a matrix, on vectors as well:

Works on matrices	[1]	`function res = squarerootV4( mat )`
	[2]	`[nrows, ncols] = size(mat);`
	[3]	`res = zeros(nrows, ncols);`
Outer loop start	[4]	`for j=1:nrows %  loop over rows`
Inner loop start	[5]	`for k=1:ncols %  loop over columns`
Scalar calculation	[6]	`res(j,k) = squarerootV2( mat(j,k) );`
Inner loop end	[7]	`end`
Outer loop end	[8]	`end`

There is nothing new here besides an application of tools we have already encountered: no new special syntax, no new operators. All of the statements are evaluated in a perfectly ordinary way. The terms "outer" and "inner" are used merely to give the human reader a way to refer to the structure of the loop.

The outer loop causes the block of statements from lines 5 to 7 to be executed nrows times. This block of statements is, of course, another loop: the inner loop. The inner loop causes its block of statements (in this example just a single line, 6) to be executed ncols times for each invocation of the inner loop. Since the outer loop invokes the inner loop nrows times, this means that line 6 is executed ncols × nrows times. This is why the nested-loop construction is able to process all of the nrows × ncols elements of the matrix.

Single-Number Indexing of Arrays

MATLAB provides a much better way of looping over arrays, one element at a time. As will be described in Section 8.10, any array can be indexed as if it were a vector, with a single-number index. This is true even for arrays with more than two dimensions. To illustrate, here is a final version of `squareroot` that will work for any size or shape of matrix:

	[1]	`function res = squarerootV5(mat)`
The output has the same shape as the input	[2]	`res = zeros(size(mat));`
Note the `(:)`	[3]	`for k=1:prod(size(mat))`
Single-number indexing	[4]	`res(k) = squarerootV2( mat(k) );`
	[5]	`end`

A single loop will handle all of the elements in the array. The key, on line 3, is to compute the number of elements in the array to set the number of iterations of the loop. This can be done with `prod(size(mat))`. There must be one iteration for each element in the array.

8.6 Outputs of Unknown Size

In the case of squarerootV3, we know ahead of time that there will be one output for each element of the input. Thus, we can "preallocate" the space for the output, as was done on line 2.

In some situations, however, the size of the output cannot be determined ahead of time. Consider, for example, the problem of translating a text string into Morse code. As we saw in Section 7.6, a single character can be translated into a vector of numbers that represent the Morse code symbols of dots, dashes, and spaces between them. However, the vector is not the same length for all of the characters: for instance, E consists of a single dot and T consists of a single dash, whereas M is dash, space, dash and Q is dash, dash, dot, dash.

The difficulty that the variable-length encoding poses to us is that we cannot tell in advance how long the Morse code output of a text string will be; the length of the output depends not just on the number of characters in the input but on what those particular characters are.

This is a common situation and can be dealt with by using "concatenation" to build up the output; as each element in the input is processed, the output for that element is concatenated to the already produced output. Here is the strToMorse program that loops over the elements in a character string to produce the Morse code output for the entire string.

```
[1] %strToMorse(str) : string to Morse Code
[2] function res = strToMorse(str)
[3] bcspace = 4; %code for space between characters
[4] res = [];                          ← Initialize the accumulator
[5] for k=1:length(str)
[6]    c = str(k); %The character being processed
[7]    m = charToMorse(c); %Convert to Morse
[8]    res = [res, bcspace, m];        ← Append to the accumulator
[9] end
```

Note that the accumulator has been initialized to be the empty vector, and the concatenation operator is used to add the result for each letter in the input to the end of the accumulator. In addition, the numerical representation of the Morse code space between characters is inserted before each character is placed on the accumulator. Note that charToMorse returns a row vector. If it had returned a column vector, the concatenation would have had to be in the vertical sense using the semicolon (;) rather than the comma. Exercises 8.22 and 8.23 are concerned with making graphical and audible representations of the Morse code.

8.7   Loop Termination

Imagine programming a computerized robot to find a missing left sock. You might give the robot a list of places to look and details about what the sock looks like. You could use a loop that might look like this:

```
placesToLook = {'top drawer', 'middle drawer', 'bathroom', ...
    'bedroom', 'hallway', 'hamper', 'washing machine'};
for k = 1:length(placesToLook)
   res = lookForSock( placesToLook{k} );
end
```

With luck, the robot would return with your sock.

This would be a pretty silly robot. Any sensible robot, or person, would stop looking once the sock was found; it's not necessary to consider the possibility that the sock is in the washing machine if we have already found it in the bathroom. This section is concerned with the language constructs break and return, which allow us to terminate a loop when it has done its job.

Since we don't have a sock-finding robot at hand, let's illustrate termination of a loop with another problem: finding the smallest factor of an integer. Remember, a factor of an integer is a whole number (other than 1) that evenly divides the integer. For example, the factors of 36 are 2, 3, 4, 6, 9, 12, and 18 because each of these number divides 36 evenly. The smallest factor of 36 is 2. The prime numbers have only themselves as factors. The smallest factor of a prime number is the number itself.

Here's an algorithm for finding the smallest factor of an integer n: Check all of the integers from 2 to n and take as output the smallest that divides n evenly. A faster algorithm, based on the idea that the smallest factor of a nonprime number must be no larger than $\sqrt{n}$, checks only the integers from 2 to $\sqrt{n}$. If any of these divide n evenly, take the smallest as the output. Otherwise, take n as the output. A program implementing this algorithm is listed below: smallestfactor. A faster program, using only prime numbers, is described in Exercise 9.5.

```
[1]  %smallestfactor(n) - smallest integer factor
[2]  function res = smallestfactor(n)
[3]  res = n;

[4]  for k=2:sqrt(abs(n))
[5]     if rem(n,k) == 0
[6]        res = k;
[7]        break;
[8]     end
[9]  end
```

Assume n is prime until we find a factor $\leq \sqrt{n}$ [3]

Is k a factor? [5]
Yes, so record it and [6]
don't look any further [7]

The boolean expression `rem(n,k) == 0` checks whether k divides n evenly. If it does, then the job is finished; we have the result. There is no point in continuing checking the remaining cases. On line 7, the `break` statement means "don't continue with the loop anymore" or "break out of the loop." The `break` statement causes evaluation to continue after the end of the loop. Of course, before breaking, we need to save the result that we found; this is done with the assignment on line 6. In case no k is found that evenly divides n, then the assignment on line 6 is never executed; line 3, before the loop, provides the appropriate output.

Sometimes, at the point where we would use `break` to terminate a loop, the output is completely determined and we would like our function to return the output. This is the case in `smallestfactor`. In such cases, it is appropriate to use `return` rather than `break`. Return will terminate the entire function, causing the appropriate values to be returned. `Break` allows the function execution to continue at the point where the loop would have ended.

When `break` is encountered in a nested loop, it terminates only that loop; any outer loops continue as normal. `Return` terminates the function evaluation entirely, so all nested loops will be terminated simultaneously.

8.8 Conditional Looping

All of the loops we have seen thusfar involve iterating a computation a known number of times (for instance, refining the interval in `squareroot` or handling all of the elements in a vector or matrix) or considering each of the members in a set (for instance, finding a factor of n by considering the integers less than $\sqrt{n}$). Some tasks involve something a little bit different: iterating a computation an unknown number of times until some condition is satisfied.

To illustrate, consider an algorithm for finding all of the factors of a number n. The `smallestfactor` function finds only one of the factors of n, the smallest. But we can use this in an algorithm to find all of the factors. The input is an integer n, the output is a vector containing the factors of n:

1. Initialize the output vector to be the empty vector.
2. Initialize `remaining` to be n. The variable `remaining` keeps track of how much of n has not yet been accounted for by the factors already found.
3. Find the smallest factor in `remaining`, and call this `sf`. Concatenate `sf` onto the output vector. Update `remaining` to be `remaining/sf`.

4. If `remaining` is 1, then we have found all of the factors of n. Terminate the loop, returning the output vector. Otherwise, continue at step 3.

In principle, this involves nothing new; we can arrange a `for` loop to go a very large number of times and use `break` or `return` to terminate the loop when the condition is satisfied. But there is a more readable construction provided by the `while` form of looping:

```
[1]  % primeFactors(n) - prime factors of n
[2]  function factors = primeFactors(n)
[3]  factors = [];
[4]  remaining = n;
[5]  while remaining > 1
[6]      sf = smallestfactor(remaining);
[7]      factors(end+1) = sf;
[8]      remaining = remaining/sf;
[9]  end
```

Condition for continuing — [5]

Update accumulator — [7]

The while statement does not involve a looping variable. Instead, `while` is followed by a boolean expression and the looping continues so long as the value of the boolean expression is TRUE.

Other than these differences, `for` and `while` are similar; any program that can be written with `for` and `break` can be written with `while` and vice versa. The only reason to choose one over the other is style. The stylistic rule of thumb is this: Use `for` whenever a complete, finite list of possibilities can be determined at the start of the loop. Use `while` whenever the loop will be cycled through an unknown number of times until a condition is satisfied.

One idiom sometimes encountered is this

```
while 1
    ─────────────
        block of statements
end
```

Since the boolean expression 1 always has the value TRUE, the preceding program fragment means "keep looping forever over the block of statements." Taken literally, this style would lead to functions that don't work: Looping forever means never returning a value. However, if the block of statements contains a terminating conditional `break` or `return`, then the style is viable. The style is used when the terminating condition is too complicated to express on the `while` line or if the `break` statement uses quantities computed within the loop.

A very simple modification makes `primeFactors` more efficient. See Exercise 8.14.

8.9 Example: Measuring Information

We describe the size of a computer's memory in bytes. The unit most commonly used nowadays is megabytes, abbreviated MB, which is approximately one million bytes.[8] Since computer memory is hardware, it seems natural to measure it in physical terms: how much hardware there is.

Measuring information itself seems like another issue. It's tempting to think about information in terms of its value to us. The eager college applicant wants to know "Did I get in?". The hopeful lover anxiously awaits the answer to the traditional question "Will you marry me?". But the value of such information is entirely subjective and lies really in the consequences that follow from the information, not the information itself.

The engineering approach to measuring information is simple and objective and wholly divorced from the meaning of the information: How much hardware does it take to store the information?

The amount of information conveyed by the answer to "Did I get in?" or "Will you marry me?" is, in the engineering sense, 1 bit. A single bit is sufficient to store the answer to any boolean question.

The engineering approach to measuring information makes the most sense in the following setting: A single "message" is to be selected from a set of N possible messages. The possible messages are written down in a phrase book, with the messages numbered $0, 1, 2, \ldots, N - 1$. Rather than writing down the actual phrase as the selected message, it's necessary only to write down the phrase number.

As we saw on page 36, to store a number in the range 0 to $N - 1$ requires k bits where

$$k = \log_2 N$$

This number can be rounded up to the next whole number using the `ceil` operator: `ceil(log2(N))`.

This definition of the amount of information in a message doesn't take into consideration the need to store the phrase book itself. This makes sense if a large number of messages are being sent referring to the same phrase book; the storage requirements for the phrase book itself are negligible compared to the requirements for the large number of messages.

The most common phrase book in widespread use is for single-character messages: the ASCII code. There are $N = 128$ ASCII characters so the amount of information in each message is $\log_2 128$ or 7 bits.

A character string can be thought of as a train of single-character messages. The information content is 7 bits per character. For example, the 13 characters of `'Yes, darling.'` convey $13 \times 7 = 91$ bits of information.

---

[8]The word "approximate" is appropriate here. Even though the metric prefix "mega" means exactly one million, when applied to computers the convention is that mega means $2^{20} = 1048576$. The reason for this relates to the manner in which computer hardware is designed.

The Morse code provides another way to measure information: How long does it take to transmit the information by telegraph? This alternative formulation makes important economic sense, particularly in the time of Morse, when a telegraph cable was an expensive bit of capital. The less time it takes to send a message, the more messages can be sent.

Morse took this into account when designing his code. The fastest character to transmit, E, takes a single time unit. E is the most common letter in typical English text. The next fastest to transmit are I and T, each of which takes three time units. These are also common letters. In contrast, letters like Q and Z take a much longer time to transmit but are infrequently used.

Key Term

The Morse code is a *variable-length code*. We might wonder whether there is a variable-length code using bits that would exploit the fact that some characters are used commonly and others rarely.

To show that this might be possible, consider a simplified case: a language whose alphabet has only three letters, say E, Z, and a dash (–). In typical words in this language, we imagine, E is used 50% of the time, and – and Z each 25%. Since there are three characters, we need k bits, where $2^k = 3$. This can be solved for k using logarithms: $k = \log_2 3 = 1.585$ bits. We can round this up to 2 bits to get an ASCII-type encoding, perhaps 00 for E, 01 for –, and 10 for Z. This is a bit wasteful because one of the four patterns possible with 2 bits, 11, is not being used. But there is another source of waste as well, the failure to use a shorter code for the commonly used E.

The only bit patterns shorter than 2 bits are, obviously, one bit long. There are two of these, 0 and 1. We can select 0 to represent E. This leaves only one pattern, 1, to represent both Z and –; clearly we're going to need at least two patterns for the two characters. We therefore choose a 2-bit pattern 10 for Z and a different 2-bit pattern, 11 for –. Altogether, our encoding is

E	Z	–
0	10	11

Using this encoding, the 20-character string

EZEE–ZEZZ–EEE–ZEE–E–

corresponds to the bit sequence

01000111001010110001110001101011

Thirty bits are being used to store 20 characters, or 1.5 bits per character. This is more efficient than the ASCII-like encoding that takes 2 bits per character, and slightly more efficient even than the hypothetical 1.585 bits per character that we calculated were needed to store three different patterns.

It may seem extremely odd to speak of 1.5 bits or 1.585 bits—bits clearly come in integer quantities. But it's really no stranger than saying something like, "The average size of a family is 3.6 people." Each individual family has some integer size, 1, 2, 3, 4, and so on, just as each character's bit pattern has an integer number of bits, 1 or 2 in the E–Z example. It's the averaging process that produces the fractional number of bits.

The theory of information [12] offers a way of measuring the theoretical information content of a sequence of characters. If there are N characters, labeled from 1 to N, and the relative frequency of character i is p_i, then the average information, in bits per character, is

$$-\sum_{i=1}^{N} p_i \log_2 p_i \qquad (8.1)$$

Key Term

This quantity is called the *Shannon information*, after Bell Labs researcher Claude Shannon (1916–2001).

In our example, the three letters have relative frequencies

```
≫ p = [.5, .25, .25];
```

and the Shannon information is

```
≫ -sum(p.*log2(p))
  ans: 1.5
```

exactly the same as we found in our example message (which, it must be admitted, was contrived so that each character occurred exactly the right number of times to produce an in-message relative frequency of p).

Note that the decrease from 2 bits to 1.5 bits is not just because there are three symbols rather than the 4 that could naturally be stored in 2 bits. If the three symbols were equally likely, the information would be somewhat larger than 1.5 bits, namely,

```
≫ p = [1 1 1]./3;
≫ -sum(p.*log2(p))
  ans: 1.585
```

In Chapter 12, we'll consider the general case of how to construct a variable-length code that matches the Shannon information. Here, we examine a simpler question: What is the Shannon information of English text?

In order to compute the Shannon information, we need to find the relative frequencies of the characters. The approach we take is to assemble a corpus of actual English text and count how often each letter appears in it. The relative frequency of a given character is the count for that character divided by the total number of characters.

This is a somewhat complex programming task. It's worthwhile to try to break up the problem into simpler pieces, each of which can be addressed separately.

The first component of the problem is to collect the text to be analyzed. Rather than writing special-purpose software for this, it's worthwhile to exploit existing software and data. In particular, we can exploit Web resources and download text (novels, Web pages, and so on). Conventional software makes it easy to save such text in individual files but not to concatenate it into one large file. So, in order to use conventional software, we store the text as a potentially large number of individual files.

Next, we need a way to count the number of occurrences of each character in the text. In the spirit of keeping the parts of the problem as separate as possible, we avoid issues of how to read in the text files and how to accumulate the result over all of the files. Instead, we consider a program that takes a text string as input and returns the number of times each character appears in the string. We call this program countchars. Since not all text strings will contain every character, we give countchars a second input listing every character to look for in a file—a count of zero will be returned if that character doesn't appear.

Another component of the system is a program that reads text files into character strings, invokes countchars, and totals the results. We call this program countCharsInFiles. Again in the spirit of keeping components of the problem separate, we avoid incorporating in countCharsInFiles logic about how to find which files are in our text database and which files aren't. Instead, we simply give countCharsInFiles a list of files.

A final component of the system is a means to generate a list of names of the text files in our database. Exercise 8.16 gives one way of doing this.

There are two overall outputs from the system: a list of the characters and a vector of the corresponding counts for each character.

In the following implementation of countchars, there are two input arguments: a set of letters to be counted and a list of letters in the character set. The algorithm of countchars is simple but a bit subtle. The idea is to loop over all of the elements in letters, not the string whose letters we want to count. For each of these elements, we count how many times that element appears in the string under analysis. Here's the program:[9]

The accumulator

```
[1]  %countchars(str, charset): count how often
[2]  %each char in <charset> appears in <str>
[3]  %Example:
[4]  %countchars('hello', 'aeiou') => 0 1 0 1 0
[5]  function cnts = countchars(str, charset)
[6]  cnts = zeros(1,length(charset));
[7]  for k=1:length(charset)
[8]      cnts(k) = sum(str == charset(k));
[9]  end
```

---

[9]Note that despite the simplicity of the idea of counting characters, the program starts with help lines that include an example.

Line 6 initializes the accumulator, which is a vector with one number for each of the elements in the set of characters to be calculated. Line 8 uses the relational operator `==` to create a boolean vector that is 1 whenever the corresponding letter in `str` matches the character under consideration and 0 otherwise. The `sum` operator counts how many 1s there are. Note that the `==` operator judges equality of characters based on their ASCII values, so e is different from E.

A key aspect of the program is line 7. Note that the looping is over the set of numbers `1:length(characters)` rather than over the set of characters themselves. This means that the looping variable `k` can be used as an index in a vector and provides the means to reference the vector `cnts` on line 7 in order to fill in the count corresponding to `character(k)`.

To illustrate the use of `countchars`, we count the characters in the string `'hello'`. First we have to define a character set of interest—for this example we take just the lowercase letters

```
≫   letters = 'abcdefghijklmnopqrstuvwxyz';
```

and start the count

```
≫   countchars('hello', letters )
    ans:  0 0 0 0 1 0 0 1 0 0 0 2 0 0 1 0 0 0 0 0 0 0 0 0 0 0
```

Although `countchars` provides only one vector as output, the character string `letters` serves to tell us which letter each count refers to.

The program `countCharsInFiles` uses `countchars` to examine one or more text files. Since we're interested in individual files, the `fread` file reading operator provides the functionality we need. (See page 100 and Appendix C for details on using `fread`.)

CountCharsInFiles loops over a cell array list of file names, reading each in turn and counting the characters in the file.

```
[1]  function [cnts, lets] = countCharsInFiles(fnames)
[2]  lets = char(0:255);
[3]  cnts = zeros(size(lets));
[4]  for k=1:length(fnames)
[5]      fid = fopen(fnames{k},'r');
[6]      while 1
[7]          str = char(fread(fid,10000,'uchar'));
[8]          if isempty(str)
[9]              break;
[10]         else
[11]             cnts = cnts + countchars(str, lets);
[12]         end
[13]     end
[14]     fclose(fid);
[15] end
```

Line annotations (left margin):
- [2] The character set
- [3] Initialize the accumulator
- [5] Open for reading
- [7] Read in a maximum number of characters at a time
- [8] We've read everything
- [10] Accumulate the counts

On line 7, a maximum of 10,000 characters are read at a time, but a loop ensures that files larger than this are read completely. If some limit were not placed on the number of characters read in a single statement, very large files might cause memory problems since a single file easily can be larger than the computer's random access memory (RAM).

For example, here is a count of the characters in a single file:

```
≫ [cnts,letters] = countCharsInFiles({'mobydick.txt'});
```

To display the 10 most commonly used letters and their counts, we find the indices that would arrange the counts into descending order,

```
≫ [trash, inds] = sort(-cnts);
```

and then pull out the first 10 of these:

```
≫ letters(inds(1:10))
  ans: _ etaonshir
≫ cnts(inds(1:10))
  ans: 17657 9484 7100 and so on
```

Reformatting the results gives a more meaningful display: Here are the 20 most commonly used characters in mobydick.txt:

Character	Count	Character	Count
[space]	17657	l	3448
e	9484	d	3305
t	7100	u	2134
a	6361	m	2005
o	6098	g	1828
n	5505	w	1719
s	5145	f	1681
h	5021	c	1662
i	4992	[comma]	1542
r	4256	p	1473

To compute the Shannon information of mobydick.txt based on the observed counts of the letters, we need to compute the relative frequencies. For each letter, this is merely the count of that letter divided by the total number of letters:

```
≫ p = cnts./sum(cnts)
  ans: 0 0 0 and so on
```

The first several values are zero because the corresponding characters (ASCII values 0, 1, 2, and so on) never appear in the text. This poses a problem for computing the logarithm in Eq. (8.1) since

```
≫  log2(0)
   ans: -Inf
```

and

```
≫  0.*log2(0)
   ans: NaN
```

One way to address this issue is to exclude from the computation those characters that never appear:

```
≫  p1 = p(p>0)
   ans: 0.0139 0.1741 0.0008 0.0005 and so on
```

For these characters, the Shannon information is

```
≫  -sum(p1.*log2(p1))
   ans: 4.3967
```

This number may not be representative of English text generally since it is based only on *Moby Dick*.

8.10 Dimensions and Arrays

A familiar way to specify location is this: "Our house is five blocks past the firestation." Assuming that you already know what street the firestation and house are on, a single distance—in this case five blocks—suffices to give the location relative to the firestation landmark. This works because a street is a kind of one-dimensional structure; one coordinate suffices to specify location on a curve.

In a trackless land, describing location is more difficult. "1.5 miles north north-east of the hilltop." Two quantities—in this case a distance and a direction—are needed, as well as the landmark. On a map, or with global positioning system (GPS) receivers, the two quantities are latitude and longitude. The landmark in this system is always the same: a watery point in the Gulf of Guinea south of Ghana, selected because it is the intersection of the earth's equator and a prime meridian standardized as the line of longitude passing through the Royal Observatory in Greenwich, England. On the two-dimensional surface of the earth, as on any two-dimensional surface, two coordinates suffice to specify location.

In three dimensions, three coordinates are needed. For an airplane, we use latitude, longitude, and altitude.

To describe the location of an event (a happening localized in time) we need to specify the time in addition to latitude, longitude and altitude: four quantities altogether.

There are simple and familiar geometric situations with a dimension greater than 4. For instance, the location of a line segment requires six dimensions: the three coordinates of one end plus the three coordinates of the other end.

A library offers a way to visualize high-dimensional structures. A line of text is a one-dimensional array: a vector of characters. A page of text is a two-dimensional array: a matrix. A book is a three-dimensional array; to specify the location of a single character, we need to give the page number, line number, and location in the line.

The same logic applies to the books themselves. A shelf of books is one dimensional. A bookcase (an array of shelves) is two dimensional, a row of bookcases is three dimensional, a library floor with its many rows of bookcases is four dimensional, a library building five dimensional, a consortium of libraries is six dimensional, and so on. To find the book from the library consortium, we need to specify six quantities: which library, which floor, which row, which bookcase, which shelf, and which book in the shelf. Change any one of these quantities and we end up with a completely different book.

Thinking of the library system as a collection of characters, we need nine quantities to specify the location of a single character: the six quantities to locate the book and the three quantities to locate the character within the book.

Perhaps you protest. Each book in the library system can be given a unique ID number. This is one quantity, so the collection of books is only one dimensional. We can even go further and string out all of the characters in all of the books as a single long vector. The library collection of characters is really one dimensional, not nine!

It's instructive to imagine how such an ID could be constructed for each letter. In real-world libraries, books are given IDs based on their topic and date. The books are stored with the IDs in alphanumeric order; the ID itself doesn't tell where the book is, only the location relative to other books. But we could construct an ID that would tell us exactly where to find any book, simply by assigning a number to each library, floor, row on the floor, bookcase in the row, shelf in the bookcase, and book on the shelf. Using L to stand for library, F for floor, R for row, C for bookcase, S for shelf, and B for book, we could write the ID as a number in the following format:

LLFRRCCSBB

This format leaves space for 100 libraries numbered 0 to 99, 10 floors numbered 0 to 9 per library, 100 rows of shelves, 100 bookcases in each row, 10 shelves in each bookcase, and 100 books on each shelf. For ease of reading, we might choose to punctuate the number in the way that telephone numbers or government ID numbers are punctuated; for example, 09-323-05-342 for the 42nd book on the 3rd shelf of the 5th case of the 23rd row of the 3rd floor of the 9th library. The scheme could be extended with numbers to locate characters within a book. We might prefer to use a pair of numbers—the book ID and a PPPLLLCCC for Page, Line, Character—or combine all of the information into one grand number, perhaps LLFRRCCSBB.PPPLLLCCC.

Note that this collapsing of dimensions works for the library because books and characters are discrete units that can be represented by integers. For specifying the location of an airplane, in contrast, each of the coordinates is a continuous quantity without a fixed number of digits of precision.

Computer memory can be arranged in the same way as library books. If we like, we can think of each byte as identified by a single index number. Or we might prefer to organize the bytes in another way. For instance, in a two-dimensional matrix of floating point numbers, the bytes are collected into groups of eight, and the groups are organized into rows and columns.

This flexibility comes from the fact that, for rectangular arrays, any location can be converted into a single number and vice versa. To illustrate, consider the vector `vec` set up so that the value of each element is the element's index.

```
≫ vec = 1:15
  ans: 1 2 3 4 5 6 7 8 9 10 11 12 13 14 15
```

For instance, the value 6 is stored at location 6.

The same values can be stored as a matrix with three rows and five columns:

```
≫ mat = reshape(v,3,5)
  ans: 1 4 7 10 13
       2 5 8 11 14
       3 6 9 12 15
```

In `mat`, the value 6 is stored at location $(3,2)$: row 3, column 2.

There is a simple connection between a location i in the vector `vec` and a location (m,n) in the matrix `mat`:

$$i = (n-1)*r+m \tag{8.2}$$

where r is the number of rows in the matrix.

This works for any size matrix. For instance, consider

```
≫  matb = reshape(v,5,3)
  ans:  1  6  11
        2  7  12
        3  8  13
        4  9  14
        5  10 15
```

In matb, the value 6 is stored at location $m = 1$, $n = 2$. There are $r = 5$ rows in matb, so that Eq. (8.2) gives $i = 6$.

Note that Eq. (8.2) gives a relationship between (m,n) and i that is one to one; there are no values of k for which there is more than one possible (m,n) pair (so long as $1 \leq m \leq r$, which is a natural enough requirement for a row index). This equivalence between the single number i and the pair (m,n) means that the data in a rectangular matrix can be stored in vector form and accessed with a single index i. If it is desired to use matrix coordinates (m,n) to access the elements, Eq. (8.2) is used to convert (m,n) into i.

The same approach applies as well to higher-dimensional arrays. For example, element (m,n,k) in a three-dimensional array with r rows, c columns, and p planes is

$$i = rc(k-1) + r(n-1) + m$$

For an array with dimensions D and size $(d_1, d_2, \ldots, d_D)$, the element at coordinate $(c_1, c_2, \ldots, c_D)$, the relationship is

$$k = 1 + \sum_{j=1}^{D} \left((c_j - 1) \prod_{k=0}^{j-1} d_k \right)$$

where $\prod_{k=0}^{0} d_k$ is defined to be 1. This is a generalization of the way the single-number library book ID was constructed. In that case, the d_i were selected to be 10 or 100, and the terms $\prod_{k=0}^{j-1} d_k$ serve to position the digits $(c_j - 1)$ in the appropriate positions in the number as a whole.

One important consequence of the storage of matrices in vector form is that one can loop over all of the elements in an array array, regardless of shape and size, by indexing it with a single index. The index will range from 1 to prod(size(array)).

This single-index approach is useful when the array elements will be handled one at a time. But it's often the case that a computation on one element will refer to one or more of the element's neighbors. (See, for example, the Cellular Automata Project, Section 8.12.) The coordinates of an element's neighbors are easily computed from the array coordinates of the element itself. For example, in a matrix, the neighbors of element (m,n) are the four elements $(m+1,n)$, $(m-,n)$, $(m,n+1)$, and $(m,n-1)$. (On the edges of the array, for instance when $m = 1$, there are fewer neighbors.) When neighbors are used in the computation, it makes sense to use nested loops.

8.11   Exercises

Discussion 8.1:

Consider the following three functions:

```
function res = m1(n)
res = 0;
for j=1:n
   for k=1:n
       if k==2
          break;
       res = res+1;      ,
       end
   end
end

function res = m2(n)
res = 0;
for j=1:n
   for k=1:n
       if j==2
           return;
       else
           res = res+1;
       end
   end
end

function res = m3(n)
res = 0;
for j=1:n
   for k=1:n
       res = res+1;
   end
   if j==2
       break;
   end
end
```

Without using the computer, give the return value for each of the following:

```
>>  m1(2)
>>  m1(3)
>>  m2(2)
```

```
≫  m2(3)
≫  m3(2)
≫  m3(3)
```

Discussion 8.2:

The smallestfactor function computes the smallest factor of its integer argument. For example,

```
≫  smallestfactor(11111111111111111)
   ans: 11
```

Explain why the program gives the following result for a number that is clearly odd:

```
≫  smallestfactor(111111111111111111)
   ans: 2
```

Exercise 8.1:

Write repeat(m) that takes a two-column matrix m as an argument and returns a vector where each element in the first column is repeated a number of times specified in the second column. For example,

```
≫  repeat([3,2; 1,3; 2,1])
   ans: 3 3 1 1 1 2
```

Exercise 8.2:

An algorithm sometimes works only for limited values of the input. For example, the program squarerootV2 on page 151 won't work if it's argument is negative and is unnecessarily slow if the argument is zero or one. Write a function squarerootV3 that includes conditions that check the value of the argument and

- Immediately give an answer if the argument is 0 or 1.
- Return an imaginary number if the argument is negative. Remember that $\sqrt{-x}$ is $i\sqrt{x}$.

Exercise 8.3:

Write a function coinAges(datafile) that takes a data file in the format of Exercise 8.1 as an argument and returns two vectors: one giving the mean age of all of the coins of each of the demoninations and the other the denominations. The two vectors should each have a length that is the number of different denominations. You may find the unique operator helpful for this.

Exercise 8.4:

Write a function, myfind, that mimics the behavior of the built-in function find. That is, it takes as input a boolean vector of 1s and 0s (e.g., 0 0 1 1 0 1 1 1) and returns the indices of the 1s.

Exercise 8.5:

Write a function to compute the nth Fibonacci numbers. The zeroth Fibonacci number is 0, the first Fibonacci number is 1, the second is 1, the third is 2. Writing the kth Fibonacci number as f_k, then $f_k = f_{k-1} + f_{k-2}$.

Exercise 8.6:

Write a boolean function `alphabeticallyBefore(first,second)` that takes two character strings as arguments. If the `first` comes before `second` when alphabetized, return 1; otherwise return 0. Remember that uppercase letters are equivalent to lowercase letters for alphabetization purposes.

Exercise 8.7:

RNA plays an intermediate role in translating the genetic information stored in DNA into the form of a protein. An RNA molecule is a sequence of nucleotides. A protein molecule is a sequence of amino acides.

There are many amino acids, with abbreviations like Phe, Ser, Gly, and so on. But there are only four nucleotides, represented by A, C, G, or U. The RNA code is based on sets of three adjacent nucleotides; such a triple is called a codon.

Each of the $4^3 = 64$ possible codons corresponds to one amino acid or to the instruction "stop." Here's a translation table:

First Position	Second Position				Third Position
	U	**C**	**A**	**G**	
U	Phe	Ser	Tyr	Cys	U
U	Phe	Ser	Tyr	Cys	C
U	Leu	Ser	Stop	Stop	A
U	Leu	Ser	Stop	Trp	G
C	Leu	Pro	His	Arg	U
C	Leu	Pro	His	Arg	C
C	Leu	Pro	Gln	Arg	A
C	Leu	Pro	Gln	Arg	G
A	Ile	Thr	Asn	Ser	U
A	Ile	Thr	Asn	Ser	C
A	Ile	Thr	Lys	Arg	A
A	Met	Thr	Lys	Arg	G
G	Val	Ala	Asp	Gly	U
G	Val	Ala	Asp	Gly	C
G	Val	Ala	Glu	Gly	A
G	Val	Ala	Glu	Gly	G

The system that translates RNA to a protein begins at a starting point on the RNA molecule and reads off codons one at a time, translating each codon into the corresponding amino acid. This means that given an RNA nucleotide sequence, one can calculate what will be the amino acid sequence of the resulting protein. For example, 'GUCACCUAA' would translate to ValThrStop.

Write a function `rna2amino` that takes two arguments: a character string giving a sequence of RNA nucleotides (e.g., 'UAUCUAUCUAU-CUAUCUAUCUAUC') and a starting point. The starting point can be a numerical index into the character string. As output, the function should return the amino-acid sequence that would be translated.

Exercise 8.8:
Write a function `intToBits` to take an integer number as input and return as output a collection of bits representing that number in base 2. Since there is no bit type in MATLAB, use a vector of 1s and 0s to stand for a collection of bits. (*Hint:* For performing the conversion to base 2, the least significant bit is 1 if the number is odd and 0 if it is even.) The next bit can be found using these steps:

1. Subtract a 1 if the number is odd.
2. Divide by 2.
3. Check whether the resulting number is odd or even.

Also, write a function `bitsToInt` that takes a vector of 1s and 0s and converts it to an integer number. The function should work so that `n==bitsToInt(intToBits(n))` is always true for any integer n.

Exercise 8.9:
Write a function `makechange` that takes an amount of money as an input and returns instructions for making up that amount of money using currency. Let us suppose that the denominations of the different coins and bills are stored in a vector. In the United States, this vector would look like

```
>> usdenoms = [100 50 20 ...
10 5 1 .25 .10 .05 .01];
```

in the United Kingdom like

```
>> ukdenoms = [100 20 10 ...
5 2 1 .50 .20 .10 .05 .02 .01]
```

or in the European Union like

```
>> eudenoms = [500 200 100 ...
50 20 10 5 2 1 .50 .20 .10 .05 .02 .01]
```

Your function should take as arguments the amount to give in change and the denominations vector. It should return a vector of the same length as the denominations vector containing the number of items of that denomination to return in change. For example,

```
>> makechange(19.37, usdenoms )
ans: 0 0 0 1 1 4 1 1 0 2
```

```
>> makechange(19.37, ukdenoms )
    ans: 0 0 1 1 2 0 0 1 1 1 1 0
```

Of course, there is usually more than one way to make change. Write your function so that the larger denominations are used as much as possible in order to minimize the total number of items given as change.

Exercise 8.10:

Write a function `biggestTwo` to return two values, the first and second largest elements in a vector. Your function should loop over the elements in the vector, updating accumulators appropriately.

Exercise 8.11:

Write a function `minimum` that works properly on matrices. This raises the question of what "properly" means. It could mean to find the smallest number of all the numbers in a matrix. Alternatively, it could mean to find the smallest number in each column or even the smallest number in each row. Write all three, calling them `minAll`, `minCol`, and `minRow`.

Your `minimum` function should use loops, simply because this is an exercise in using loops. However, you can also use the the built-in function `min` to accomplish the same task. For each of your three functions, write a one-line expression that accomplishes the same task using the built-in `min`.

Exercise 8.12:

The simple program `dicesum` calculates the probability of a specified outcome when summing two dice. Modify the program to return two vectors: (1) a list of the possible outcomes and (2) a list of the corresponding probabilities. Then modify the program to give the vector of all outcomes for the sum of five dice. Draw a graph of this probability versus the outcome; it should look like the bell-shaped curve called a "normal" or "gaussian" distribution.

Exercise 8.13:

The game Risk also allows an aggressor to roll three dice and the defender to roll two dice in return. The higher of the aggressor's dice is compared to the higher of the defender's in order to determine the fate of one army. A second army's fate is set by the comparison of the second highest dice to each other. In both cases, a tie means that the defender wins. There are three possible outcomes: The defender loses no armies and the aggressor two; the defender and aggressor each lose an army; and the defender loses two and the aggressor none. Write a function that returns two values: the probability of the defender losing two armies and the probability of losing just one army.

Exercise 8.14:

The `primeFactors` program on page 165 can be improved by taking into account that once a smallest factor `sf` has been found, the program `smallestfactor` does not need to search for factors smaller than `sf`. Modify `smallestfactor` so that it takes two arguments: the number to factor and a starting number to look for factors. Then modify primeFactors, using sf as the starting number for factoring. It will be necessary to initialize sf before starting the loop.

Exercise 8.15:
Write a function primesieve that finds the prime numbers from 1 to n using the algorithm called the "Sieve of Eratosthenes." (Eratosthenes, a Greek, lived from 275 to 194 B.C.) To illustrate the algorithm, consider this description that instructs a human how to find the primes in the range from 1 to 16.

1. Write down the numbers 1 to 16.
2. Start at 2. Circle 2 as prime and then count by twos through the list, crossing out each number in turn; that is, 4, 6, 8, 10, 12, 14, and 16.
3. Circle the next number that is neither crossed out nor circled. Then count forward in steps of this number, crossing out each number in turn. For example, when the circled number is 3, cross out 6, 9, 12, 15.
4. Continue step 3 until all of the numbers are crossed out.

Try this by hand, using the list 1 to 25, to make sure that you understand the algorithm before thinking how to translate it into an algorithm on the computer. (*Hint:* Rather than "crossing out" numbers on the computer, you might want to have a vector that takes on the value 0 for numbers that are "circled", 1 for numbers that are crossed out, and 2 for numbers that haven't yet been handled.)

Exercise 8.16:
One way to organize a body of text for the character-counting problem is to set up a designated directory to hold the text files. The directory might have subdirectories holding files of different types (e.g., novels, technical documentation, and so on).

Write a program, listOfFiles, that takes a directory name as an argument and returns a cell array containing the names of all files in that directory or in subdirectories. (*Hint:* See the dir operator.)

Exercise 8.17:
Find the perfect numbers less than 10,000. A perfect number is a number whose prime factors (including 1) add up to the number itself.

Exercise 8.18:
Find all of the integer Pythagorean triples up to 10,000. A Pythagorean triple is a set of three numbers a, b, and c such that

$$a^2 + b^2 = c^2$$

(*Hint:* Loop over a^2 and b^2, looking for those cases where the square root of their sum is an integer.)

Exercise 8.19:
Write a function that takes two matrices as inputs and returns the matrix product. Rather than using the built-in matrix multiplication function, use looping. Make sure to put in adequate error statements to catch situations like wrongly shaped matrices.

Exercise 8.20:

The home-made matrix product function that you wrote for Exercise 8.19 is much slower than the built-in matrix multiplication operator (*). Make a graph of how the time it takes to multiply two square matrices of size $n \times n$ depends on n. To do this, loop over different values of n, generating two square random matrices whose product to take. Use `tic` and `toc` around the multiplication statement to have MATLAB measure the elapsed time for the multiplication. For example,

```
tic;
y = mymatrixmultiply(a,b);
elapsedtime = toc;
```

Do the same thing for the built-in MATLAB operator.

Exercise 8.21:

Write a function, `num2word`, that takes an integer number up to 1000 as input and returns a character string giving that number in a natural language such as English.

```
>> num2word(5)
  ans: five
>> num2word(76)
  ans: seventy-six
>> num2word(892)
  ans: eight-hundred ninety-two
```

Start simple and build up, perhaps writing first a function that works for numbers less than 10, then one that works for numbers less than 100, then finally up to 1000. Store the basic words in cell arrays (e.g., `{'one', 'two', 'three', ...}`), and index these arrays to obtain the parts of the number word. (*Hint:* Some mathematical functions you might find useful: `rem`, `floor`, `log10`.)

Once you have done this, try to extend your program to handle much larger numbers.

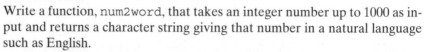

```
>> num2word(2343234230)
  ans: two billion, three-hundred forty-three million, two-hundred
       thirty-four thousand, two-hundred thirty
```

This is reasonably hard and requires good testing of the program to be sure that you have it right.

Exercise 8.22:

Convert the numerical representation produced by `strToMorse` into a vector that, when played on the speaker, sounds like real Morse code. We'll consider the representation of sounds on the computer in Chapter 13, but for now it will suffice to give you a recipe: Here is a program `morseCodeSound` that will take a single number and return an appropriate sound vector.

```
 [1]   function res = MorseCodeSound(num)
 [2]
 [3]   timeunit = 0.05;
 [4]   soundfreq = 400;
 [5]   sampfreq = 8192;
 [6]
 [7]   switch num
 [8]   case 1
 [9]      duration = timeunit; amp = 1;
[10]   case 2
[11]      duration = 3*timeunit; amp = 1;
[12]   case 3
[13]      duration = timeunit; amp = 0;
[14]   case 4
[15]      duration = 3*timeunit; amp = 0;
[16]   case 5
[17]      duration = timeunit; amp = 0;
[18]   otherwise
[19]      duration = 0; amp = 0;
[20]   end
[21]
[22]   if amp == 0
[23]      res = zeros(floor(duration*sampfreq),1);
[24]   else
[25]      res = amp*sin(2*pi*soundfreq*...
[26]           [(1/sampfreq):(1/sampfreq):duration]');
[27]   end
```

Not 7 — line [17]

A row vector — line [23]

This program produces a sound vector output for each of the numerical symbols that we are using to represent the dots, dashes and spaces of Morse code. Your job is to write a function that loops over all of the symbols in a vector produced by strToMorse and concatenates them appropriately into one long sound vector. Once this is done, the sound can be played on the computer's loudspeaker with a command like

```
>> sound( soundvector )
```

(*Hint:* Due to the way that the sound function works, it is best to concatenate the various dots, dashes, and spaces together into one long signal and then to play the long signal with a single call to the sound function rather than calling sound for each dot or dash separately.)

Exercise 8.23:
The Morse code was originally transmitted by wire: An electrical voltage was applied to the wire during the dots and dashes. Convert the numerical representation produced by strToMorse into a graph of voltage versus time. Let the resting voltage of the wire be 0 V and the active voltage be 1 V.

Exercise 8.24:

Generate a three-dimensional array using reshape [e.g., a = reshape (1:24,4,3, 2)]. Find the index (m,n,k) of the element with a value of 23. Then, defining v=a(:);, show how to convert (m,n,k) to the appropriate index into v.

8.12 Project: Cellular Automata

A cellular automaton is a device that consists of a collection of "cells" that can take on any of two (or more) states. Each cell is connected to some of the other cells and changes its state depending on the state of itself and the connected cells. Cellular automata are capable of complicated behavior; a computer is one type of cellular automata. (See [24] for a comprehensive description of cellular automata.)

In this example, we consider some very simple cellular automata that are nonetheless capable of interesting behavior. The cells in our automata are boolean: The state is either zero or one. The cells are arranged into a circle with each cell connected to the next-door neighbor on each sides. At the tick of a clock, each cell computes a new boolean state that depends on its own state and on the state of its two neighbors. Each cell then takes on that new state and waits until the next tick of a clock, at which time the process is repeated. In other words, the updated state is a function of the boolean triplet (LEFT NEIGHBOR, SELF, RIGHT NEIGHBOR) of the current state.

Figure 8.1 shows the evolution of such a cellular automaton. The cells are arranged horizontally (the circle has been cut in order to do this, but cell 100 and cell 1 are indeed neighbors in the computation). The state of the whole array of cells is one row of the figure. The top row is the "initial condition" of the array. Each successive row is the state computed from the previous row. In this case, the rule is simple: The new state of a cell will be on only if the cell was off in the previous row and the neighboring cell to its left was on previously. In all other cases, the next state of the cell will be off. That is, the triplet $(1,0,0) \rightarrow 1$, but all other triplets $\rightarrow 0$. Apply this rule by hand to Figure 8.1, starting at the top, and make sure you understand how things are working.

The result of this simple rule is to produce a diagonal stripe. Note that at about time 50 the stripe goes off the right edge of the figure and comes back in on the left edge. This is because the cells are connected in a loop.

Exercise 8.25:

Write a function newstate = caiter(oldstate) that takes a boolean vector of any length > 2 and computes the new state after one iteration, applying the rule for Figure 8.1 to each cell. Keep in mind that the first cell and the last cell in the vector are neighbors.

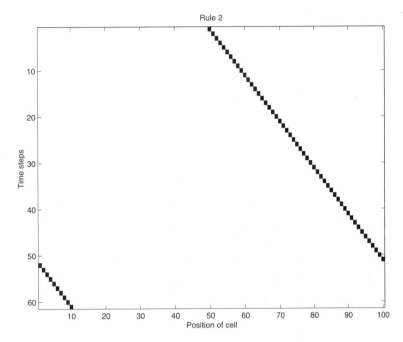

Figure 8.1.   An automaton of 100 cells iterated 60 times. The rule is that the triplet $(1,0,0) \rightarrow 1$, but all other triplets go to zero.

The following is a function, showca(initialstate), that uses caiter to make a plot such as that shown in Figure 8.1. In order to make this work, you will need to make sure that caiter returns a state that is in the proper format (Should it be a column vector or a row vector?) and gives an initial state vector that also has the right format.

Create a matrix to hold the evolving state

Fill in the first row

Iterate the CA

Update the state

Fill in the appropriate row

Plot out the state over time

Draw 1 as black and 0 as white

```
[1]  function res = showca(initial, niters)
[2]  % showca(initial, niters)
[3]  % Iterate a 1-dim cellular automaton, plotting the result
[4]  res = zeros(niters+1, length(initial));
[5]  res(1,:) = initial;
[6]  for k=1:niters
[7]      initial = caiter(initial);
[8]      res(k+1,:) = initial;
[9]  end
[10] pcolor(1:length(initial), 0:niters, res);
[11] colormap(1-gray);
[12] shading('flat');
[13] xlabel('Position');
[14] ylabel('Time steps');
[15] axis ij % Arrange for 0 to be at the top.
```

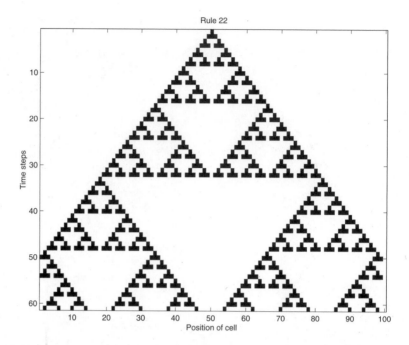

Rule 22

Figure 8.2. An automaton of 100 cells iterated 60 times. The rule is number 22: the triplets $(1,0,0)$, $(0,1,0)$ and $(0,0,1)$ all $\rightarrow 1$, but all other triplets go to zero.

The figure will be made by the `pcolor` function, which takes a matrix and displays it as an image, drawing a small, colored rectangle for each element of the matrix. The `colormap` function sets the range of colors to be used for the display.

Once you have your cellular automaton working, try it out with different initial conditions. For example, you can create a random initial condition of 100 cells with the statement

```
≫  initial = rand(100,1) > 0.8;
```

which will turn on about 20 cells.

Now it is time to explore other rules. Start with rule 22—a cell is on in the next generation only in three cases: The cell itself was on and both its neighbors were off; the cell was off and one, and only one, of its neighbors was on. The result should resemble Figure 8.2. This rule produces a rather complicated, fractal global pattern from a simple rule acting locally.

Try making up your own rule that produces interesting behavior. One way to implement the different rules is using the `if/elseif/else` or `switch/case` constructs. This is certainly legitimate, but it means that each time you want to try a new rule you have to write a new program or modify your old program.

A better way to explore different rules is to exploit a property that is specific to cellular automata. Note that if we write the states of the cell and its two neighbors in a group as a triplet (LEFT, SELF, RIGHT), then there are only eight possible triplets, as indicated in the following table. A rule associates each triplet with the subsequent state of the middle element of the triplet. Thus, a rule can be specified by a vector of eight booleans.

Triplet	Index into Rule	Rule 2	Number 22
$(0,0,0)$	1	0	0
$(1,0,0)$	2	1	1
$(0,1,0)$	3	0	1
$(1,1,0)$	4	0	0
$(0,0,1)$	5	0	1
$(1,0,1)$	6	0	0
$(0,1,1)$	7	0	0
$(1,1,1)$	8	0	0

Of course, a collection of eight binary elements (e.g., 01101000) can also be used to specify an integer between 0 and 255. This gives us a way to name rules with numbers, using the convention that the states are ordered as in the table with the least significant bit corresponding to the top triplet and the most significant corresponding to the bottom triplet.

If the rule is represented as a vector of eight boolean elements (e.g., [0,1,1,0,1,0,0,0]), the subsequent state can be computed by converting each (LEFT, SELF, RIGHT) triplet into an index 1 through 8 and using this index to pull the appropriate outcome from the boolean rule vector.

Exercise 8.26:
Modify caiter so that it takes two arguments, the state and a rule expressed as an eight-element boolean such as [0 1 1 0 1 0 0 0] in the case of rule 22. Make a few plots of interesting-looking rules and confirm (by hand) that your program is working as desired.

Exercise 8.27:
Your caiter function might possibly have if statements in it. This is fine, but if statements cannot be vectorized, so you have to have a loop to cover all of the elements of the state vector. Write caiter in a way that doesn't involve any if statements. [*Hint:* Use arithmetic operators to translate (LEFT, SELF, LEFT) into the appropriate index 1 through 8 into the boolean rule vector.]

8.13 Project: The Mandelbrot Set

[This project involves the use of complex numbers, but you don't need any previous exposure to them to do the project.]

Consider a very simple economic system: a worker who deposits money in a retirement account. At month m, the account has balance z_m. The worker is on a fixed budget and deposits a fixed amount c in the account for 25 years. The account pays 0.2% interest per month when the balance is positive; the worker pays 1.0% interest per month when the balance is negative. The equation describing this situation is

$$z_{m+1} = \begin{cases} 1.002z_m + c & \text{when } z_m \geq 0 \\ 1.010z_m + c & \text{when } z_m < 0 \end{cases} \tag{8.3}$$

Although we usually assume a deposit adds to the account, if the value c is negative the worker is really making a withdrawal each month: not a very sensible retirement strategy but certainly a mathematical possibility.

If the account starts with a balance of zero—that is, $z_0 = 0$—how much money will be in the account after 25 years? The answer, of course, depends on the amount c. We can compute the account balance with a simple program:

```
[1]  function z = bankbalance(c,nmonths)
[2]  z = 0;
[3]  for m = 1:nmonths
[4]      if z >= 0
[5]          z = 1.002.*z + c;
[6]      else
[7]          z = 1.010.*z + c;
[8]      end
[9]  end
```

Initial balance — [2]
Positive balance — [5]
Negative balance — [7]

A savings of $1 per month, after 25 years—300 months—accumulates a bit:

```
≫ bankbalance(1, 300)
 ans: 410.5137
```

and, since the interest rate paid on debt is higher, a withdrawal of $1 per month incurs a larger debt:

```
≫ bankbalance(-1, 300)
 ans: -1878.8
```

We can make a graph of the account balance for a range of values of c:

```
≫ fplot(inline('bankbalance(c,300)', 'c'), [-3,3])
```

Figure 8.3 shows two different qualitative behaviors. If $c < 0$, the worker is in debt at the end of the 25 years. If $c > 0$, the worker has a positive savings. This is probably completely obvious as a matter of intuition.

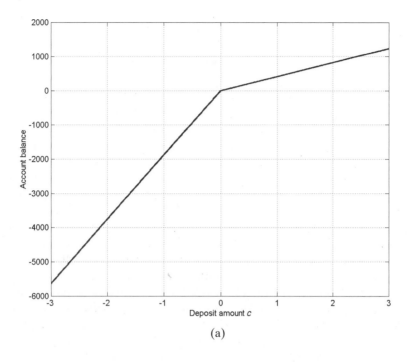

(a)

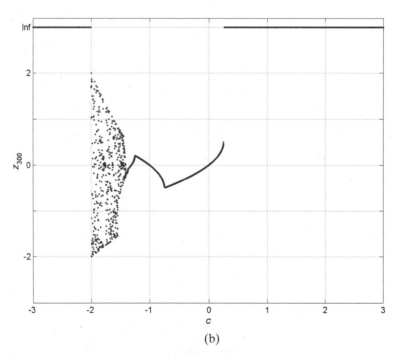

(b)

Figure 8.3.   (a) The value of z_{300} for the bank-account system [Eq. (8.3)] for different values of c. (b) The value of z_{300} for the quadratic system [Eq. (8.4)].

But things aren't always so obvious. Consider a slightly different system, a mathematical abstraction that is no longer a bank account but almost as simple mathematically:

$$z_{m+1} = z_m^2 + c \tag{8.4}$$

This is called the *quadratic system*.

Plotting this out as a function of c shows that the situation is complicated. Notice that for $c > .25$ or $c < -2.0$, the value of z_{300} is huge: Numerically, it is `Inf`. But for $-2 < c < .25$ the value is small, between -2 and 2. There is some very rich mathematical behavior here; for example, for c near -1.5 the graph is very irregular.

We explore one aspect of the behavior of the system: whether z_m goes to ∞ or stays small as m increases. We will indulge in mathematical elaboration: Rather than c being a real number, we will let it be complex. The goal of this project is to make a graph indicating whether z_m goes to ∞ or not for different values of c. The set of c for which z_m stays finite is called the *Mandelbrot set* and is considered to be one of the most complicated structures in mathematics.

A complex number consists of two ordinary numbers paired together: One of the numbers is called the "real part" and the other is called the "imaginary part."

To denote a numeral as representing an imaginary number, we write a normal real number followed by the letter i without any intervening space:

```
≫  c = 5.2i
  ans:  0 + 5.2i
```

`5.2i` is a single token in MATLAB and is different from the following expression, which has two tokens separated by a space:

```
≫  5.2 i

??? 5.2 i
        |
Error: Missing operator, comma, or semicolon.
```

To create a complex number, we add together a real and an imaginary number:

```
≫  c = -3 + 5.2i
  ans:  -3 + 5.2i
```

Note that the printed representation of a complex number shows both of the paired numbers with the imaginary component marked with an i.

Complex numbers follow rules for addition, multiplication, and other operations that are closely related to the familiar operations for real numbers. MATLAB automatically handles complex numbers in arithmetic operations.

```
≫  c+2
  ans: -1 + 5.2i
≫  c + 2i
  ans: -3 + 7.2i
≫  c.*2
  ans: -6 + 10.4i
```

For multiplication by a complex number, the fundamental rule that makes complex numbers distinctive is that

```
≫  1i.*1i
  ans: -1
```

This results in behavior that is a bit nonintuitive:

```
≫  c.*1i
  ans: -5.2 - 3i
≫  c.*c
  ans: -18.04 - 31.2i
```

In order to graph the Mandelbrot set for complex c, we need to plot the behavior of the quadratic system as a function of both the real and the imaginary part of c. We can do this by setting up a coordinate system with a horizontal axis giving the real part of c and a vertical axis giving the imaginary part of c. Any point in the coordinate system is a single, complex value of c. We'll indicate the behavior of the quadratic system using shades of gray. Figure 8.4 shows the Mandelbrot set in this way. Values of c in the set are shown in black; values near the boundary of the set are shown in gray.

Exercise 8.28:

Write a function `drawMandelbrot` that takes arguments specifying the range of c and draws a figure like Figure 8.4. For example, to make that figure, the command would be

```
≫  drawMandelbrot([-2.5, 2.5], [-2, 2])
```

The first argument gives the range of the real part of c; the second argument gives the range of the imaginary part of c.

You can follow these steps:

- Write a function, `mandelbrotIterate`, that takes two arguments: `c` and `niters`. It should iterate Eq. (8.4) `niter` times, starting at $z_0 = 0$. Your function will look somewhat like `bankbalance` but with some differences. First, it will use Eq. (8.4) rather than Eq. (8.3). Second, rather than returning the final value of z (which will be `Inf` for

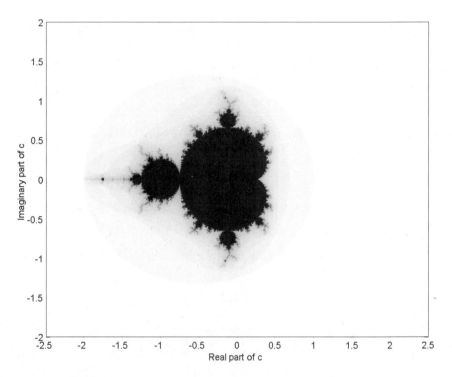

Figure 8.4. The Mandelbrot set.

any value of c not in the Mandelbrot set), at each iteration check whether $abs(z) > 2$. If it is, break out of the iteration loop, and return the number of iterations executed up until the break divided by `niters`. This returned number will be near zero for points that are heading rapidly to `Inf`, it will be equal to 1 for points in the set, and it will be close to 1 for points near the set.

- The function `drawMandelbrot` will construct a matrix of values c, apply `mandelbrotIterate` to each of the elements of c, and plot out the results. The result matrix—which we'll call s—will have the same shape as c.

In order to create the matrix c, it is helpful to create two auxiliary matrices, x and y, which contain, respectively, the real and imaginary components of c. The two matrices will also be helpful in making the plot.

To illustrate, suppose we were making a very coarse figure, considering only a 3×3 array of values between 2 and -2. The matrices would look like this:

```
x =
        -2      0       2
        -2      0       2
        -2      0       2


y =
        -2     -2      -2
         0      0       0
         2      2       2
```

x will be the real part of c, and y will be the imaginary part. Putting them together produces c:

```
≫   c = x+1i*y


c =
  -2.0000 - 2.0000i     0 - 2.0000i      2.0000 - 2.0000i
  -2.0000               0                2.0000
  -2.0000 + 2.0000i     0 + 2.0000i      2.0000 + 2.0000i
```

Computing mandelIterate(c(k),300) for each of the elements in c will produce the following matrix:

```
s =
        0.0033      0.0067      0.0033
        1.0000      1.0000      0.0067
        0.0033      0.0067      0.0033
```

- Draw the plot with two commands:

  ```
  ≫   pcolor(x,y,s); shading('flat');
  ```

  A 3×3 matrix gives too little detail; you will want to use a considerably larger matrix. But, in developing your software, it's sensible to use small matrices for testing purposes. Remember that the time it takes to compute s is proportional to the number of elements in the c matrix: A 300×300 matrix will take about 10,000 times longer to process than a 3×3 matrix.

One of the fascinating aspects of the Mandelbrot set is that it contains many miniature copies of itself (Figure 8.5): It is a *fractal*. In Chapter 10, we will build a graphical user interface that allows us to zoom in on parts of the Mandelbrot set.

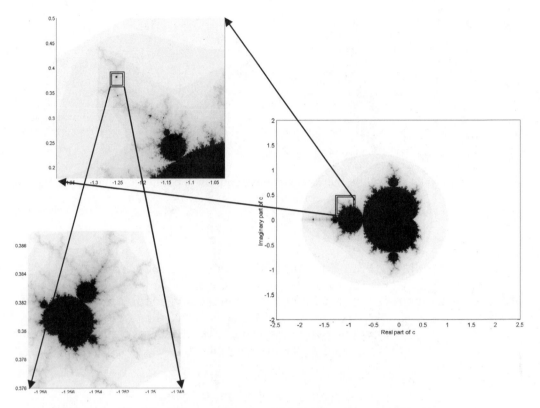

Figure 8.5. The Mandelbrot set contains within itself small copies of itself.

9

Scope

The main programming tool for organizing computations is the function. The function provides a simple and powerful implementation of the idea of a computation as a transformation of inputs to outputs.

But there is an alternative view of a computation, as a change of state in response to events. Functions, at least so far as we've seen them, don't work very well in this regard. Although we easily can change the state of a variable inside a function by using assignment, this change is forgotten when the function returns. In this chapter, we explore ways to have functions remember their state between invocations and ways to have functions communicate with one another.

Suppose, for instance, that we would like to keep track of the number of times a particular event has occurred. There are small, hand-held counters that help with this task; the press of a button increments the count by 1. We would like to make a computerized version of such a counter, which increments each time a function keepcount is invoked.

```
≫  keepcount
   ans:  1
≫  keepcount
   ans:  2
≫  keepcount
   ans:  3
```

Writing such a function is a challenge at this point. Note that there appears to be no assignment and so no change of state. But clearly, the state is changing.

9.1 Environments and Scope

In Chapter 6, the process of invoking a function was described by analogy, imagining the inputs to the function being sent off in an envelope for processing in a remote site with the outputs coming back by return post. It is time now to be a bit more precise.

Statements inside a function make references and assignments to variables. Often, these variables have simple and commonplace names like x, y, temp, and foo. There may be many hundreds of different functions that use variables with such names. The names are the same, but we expect that the variables in different functions are distinct from one another.

Key Term

The link between variable names and values is set by *scoping rules*. The most important scoping rule concerns the creation of *environments*. An environment is a sort of storage closet for variables. Variables themselves can be thought of as bins in the storage closet. In this analogy, the value of the variable is the contents of the bin, and the name of the variable is the label on the bin.

Statements at the prompt line refer to the "base environment," sometimes called the "base workspace." But when a function is invoked, a new environment is created to hold the variables assigned values in the course of executing that function.

> **Each time a function is invoked, a new environment is created to store the variables referred to in the function for the duration of that invocation. All ordinary variables refer to that environment.**
> **The environment is destroyed when the function returns and all of the variables in the environment are erased from memory.**

The different environments are completely separate from one another. An assignment done in one environment can have no impact on the variables in any other environment. But it is possible to send a value from one environment to another for assignment in the receiving environment. There are two mechanisms for doing this:

Invocation and initialization: When a function is invoked, copies of the arguments are made. These copies become the values of variables in the newly created environment of the invoked function. The names of these variables are the names in the function declaration line that correspond to the arguments.

Return and assignment: When an invoked function returns, copies are made of the values listed as outputs in the function declaration. Then the environment of the invoked function is destroyed. The copies of the returned values are used if the statement with the invocation contains an assignment.

Although each invocation of a function creates a new environment, this is not the case for scripts:

Execution of a script does not create a new environment. Statements in script files are in the scope of the environment where the script was invoked.

This means that scripts executed from the command line refer to the base environment. Scripts executed from a function do their work in the environment created for the invocation of that function.

9.2 The Debugger

As functions are invoked and returned, environments are created and assignments create and modify variables in those environments. As the function returns, the variables and their environment are destroyed. The returned variables (if any) have their values sent back to the calling environment where they can be assigned to variables there. Ordinarily, all this is invisible to us; we see only the output: the end result of the computation.

Key Term

To help in visualizing the process, we can use a special-purpose facility called the *debugger*. The debugger allows a programmer to stop the execution of a program at any point and to examine the environments that exist at that point. Although the name "debugger" suggests that it can eliminate bugs in programs, it is merely a tool for interrupting and resuming execution and examining environments and the variables in them. But it is a powerful tool, well worth learning.

In MATLAB, there are two interfaces to the debugger. One is a graphical interface that uses the text editor to show the progress of execution. Another is a command-based version. We'll use the command version here, only because it lends itself to the static, printed format of a book.

Using the debugger involves three steps:

Key Term

1. Set one or more *breakpoints* where the debugger will stop execution. Such breakpoints can be a line in an m-file, or a condition such as stopping whenever a value of `Inf` or `NaN` is encountered.
2. Execute the program in the normal way.
3. When the debugger stops the execution at a breakpoint in (1), give commands to examine the environment, set new breakpoints or delete existing ones, continue execution, and so on.

Before showing an example of using the debugger, we introduce several types of commands relevant to the debugger.

- Commands to instruct the interpreter where to stop the execution. These are as follows:

dbstop sets a "breakpoint." When the interpreter reaches a breakpoint, it stops the execution. To set a breakpoint to stop in refine at line 6, we could use the command

> » dbstop in refine at 6

Once a breakpoint is set, the interpreter will always stop when the execution reaches the breakpoint. You can set more than one breakpoint; execution stops whenever you reach any of the breakpoints.

dbstatus lists all of the breakpoints currently in effect.

dbclear allows you to remove one or more breakpoints so they are no longer in effect.

- Commands to continue the execution. These are as follows:

dbstep instructs the interpreter to execute the current line and move to the next line, stopping there. There are a number of variants of dbstep that allow us to tell the interpreter to move to the end of the function, follow the execution into an invoked function, and so on.

dbcont instructs the interpreter to move to the next breakpoint. If there are no more breakpoints encountered, the execution continues to the end of the computation.

dbquit aborts the current computation and returns us to the command prompt.

- Ordinary commands. These are the regular commands you can give to look at the values of variables, make assignments, or whatever. But there is one very important difference: These commands work in the environment you are currently visiting.
- Commands to list the existing environments or to set which environment you are visiting; that is, which environment is currently in scope for ordinary commands. Whenever a function is invoked, a new environment is created. All of the statements inside the function work in that environment, but the environments that already existed when the function was invoked still exist.

dbup moves to the environment that was in scope when the environment currently being visited was created.

dbdown undoes the move of the previous dbup.

dbstack lists the environments that currently exist.

In addition to dbup and dbdown, any of the stepping such as dbstep or dbcont commands will reset the environment being visited to the environment in scope when execution next stops.

To illustrate, we trace the creation of environments in the execution of the primefactors from Chapter 8 (on pages 163 and 165).

```
≫  primefactors(30)
   ans:  2 3 5
```

From the following m-file listings, the course of the computation can be traced by hand. Primefactors will invoke smallestfactor three times in succession, with arguments of 30, then 15, then 5.

To activate the debugger, we set breakpoints so that execution stops each time primefactors or smallestfactor is invoked:

```
≫  dbstop primefactors
≫  dbstop smallestfactor
```

The ≫ prompt indicates that execution is in the base environment. For latter reference, we make a variable in this environment:

```
≫  one = 1';
```

The who command lists the names of all the variables in the environment:

```
≫  who
   Your variables are:  one
```

Invoking primefactors will trigger the debugger

```
≫  primefactors(30)
```

On many computers, two things will happen at this point. First, the editor window will pop up with a yellow arrow and a red dot at line 3 of primefactors.m, as shown in Figure 9.1. The red dot means that a breakpoint has been set at that line; the yellow arrow indicates that execution is currently at the beginning of that line, that is, that the line has not yet been executed.

Second, the prompt will change form in the command window from ≫ to something like K≫. (You may have to use the mouse to uncover the command window if the editor window has obscured it.) This new prompt reminds us that we are in debugging mode. *Debugging mode* means that the commands we type at the prompt occur within the scope of the environment

Key Term

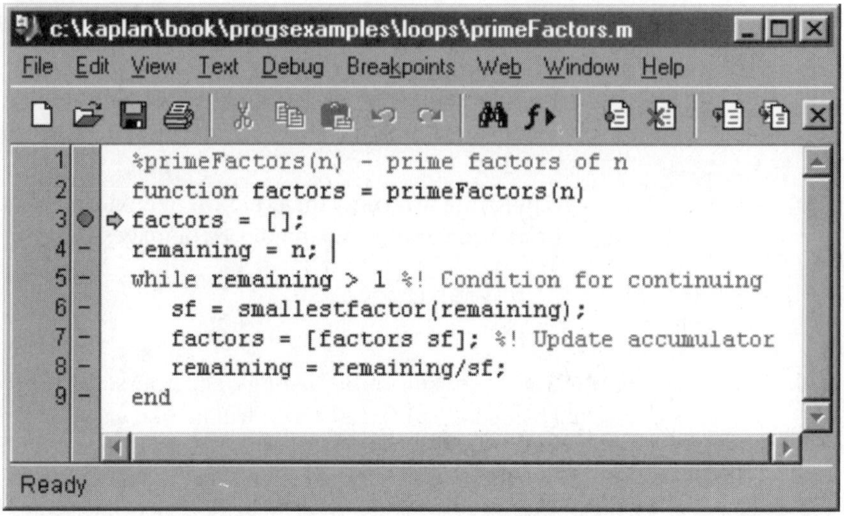

Figure 9.1. The graphical debugger as it appears on a Windows computer. The dot at line 3 indicates that a breakpoint has been set there. The arrow shows which line will next be evaluated. The arrow moves in the course of execution.

currently being visited. We can see which environment this is by using the `dbstack` command:

```
K≫ dbstack
  > In c:/book/primefactors.m at line 3
```

At this point we can use any commands we like, remembering that they occur within the scope of the environment being visited. For instance, we can list the existing variables:

```
K≫ who
  Your variables are: n
```

The value of this variable is

```
K≫ n
  ans: 30
```

For future reference, we create a new variable in this environment.

```
K≫ two = 2';
K≫ who
  Your variables are: two n
```

Dbstep allows us to execute one line at a time.

```
K≫ dbstep
  ans: 4 remaining = n;
```

Now we are at line 4 of `primefactors`; in line 3 the variable, `factors` was created and it now shows up in the variable list for the environment:

K≫ who
    Your variables are: *factors n two*

We continue the execution, jumping to the next breakpoint:

K≫ dbcont
    3 res = n; %! Assume ...

The debugger seems to have stopped the execution almost instantaneously, but in fact several lines have been executed:

K≫ dbstack
    > In c:/book/smallestfactor.m at line 3
    In c:/book/primefactors.m at line 6

The stack lists all of the function environments that exist. There are two: the function invocation environment for `primefactors` and that for `smallestfactor`. The environment we are currently visiting is marked with a >; we are visiting the `smallestfactor` environment. Just to make a point, we create a variable in this environment:

K≫ three = 'We are in smallestfactor env.';
K≫ who
    Your variables are: *n three*

Notice that the one and two variables that we created earlier aren't in this environment. But we can visit them. One level is the `primeFactors` environment:

K≫ dbup
    ans: *In workspace belonging to primeFactors.m.*
K≫ who
    Your variables are: *factors n remaining two*

Another level up and we are visiting the base environment:

K≫ dbup
    ans: *In base workspace.*
K≫ who
    Your variables are: *ans one*

We continue with the execution:

K≫ dbcont
    3 res = n; %! Assume ...

We are back in the `smallestfactor` environment.

K≫ who
    Your variables are: *n*

Notice that the variable three, which we previously created, is gone. That's because that variable was created in the environment created for the first invocation of smallestfactor, the one with $n = 30$. But that environment is gone, destroyed when smallestfactor(30) returned its value to primeFactors. Now the environment we are in is the invocation environment for smallestfactor(15):

```
K≫  n
  ans: 15
```

For reference, we'll create a new variable in this environment:

```
K≫  four = 4;
K≫  who
  Your variables are: four n
```

Two other environments exist at this point in the execution, the prime-Factors invocation environment and the base environment.

```
K≫  dbstack
  > In c:/book/smallestfactor.m at line 3
  In c:/book/primefactors.m at line 6
```

We can visit them with dbup. Instead, let's skip to the next breakpoint:

```
K≫  dbcont
  3 res = n; %! Assume ...
```

Now we are in yet another environment, that created for the third invocation of smallestfactor.

```
K≫  who
  Your variables are: n
```

The variable four is gone, destroyed along with the environment for the second invocation of smallestfactor.

```
K≫  n
  ans: 5
```

This is the last of the three invocations of smallestfactor prompted by the command primeFactors(30). Continuing the execution will finish the computation and return us to the prompt line and the base environment:

```
K≫  dbcont
  3 res = n; %! Assume ...
≫  who
  Your variables are: ans one
≫  one
  ans: 1
```

Base		Global		primefactors invocation 1	primefactors persistent	smallestfactor invocation 1	smallestfactor invocation 2	smallestfactor invocation 3	smallestfactor persistent
				factors 30		n 30	n 15	n 5	
				remaining		res	res	res	
				sf		k	k	k	

Figure 9.2.   The environments that exist due to the invocation of `primefactors(30)`. The value of the variables defined as arguments is shown, along with the other variables created in each environment. The base environment is completely out of scope during the evaluation. The global environment is always in scope (but contains no variables in this instance); persistent environments are within the scope of their own functions.

9.3   Shared Environments

Communication between environments is not the only means of communication between functions. Functions can share a common environment. The shared environments are not destroyed when the function returns; the values of variables in the shared environments are potentially available to other functions or to other invocations of the same function.

There are two shared environments that are within the scope of a statement inside a function:

Key Term

1. The function's *persistent* environment. This is an environment that is created the first time that the function is invoked in each MATLAB session. The environment persists until the MATLAB session ends.[1] The function's persistent environment allows data to be communicated between invocations of a function. Each function has its own persistent environment that is completely inaccessible from any other function. Variables to be stored in the function's persistent environment are declared in the function with the keyword `persistent`.

Key Term

2. The *global* environment. This is a single environment that is within the scope of any statement at any time. Essentially, it is a shared environment. The global environment is created when the MATLAB session is started and destroyed when the session ends. By appropriate `save` and `load` commands, the global environment can even be preserved between sessions. Variables to be stored in the global environment are declared with the keyword `global`.

The complete set of environments involved in the execution of `primefactors(30)` is shown in Figure 9.2.

---

[1]More precisely, the function's persistent environment is recreated on the first invocation after the function's m-file has been saved and stays in existence until either the MATLAB session ends or the m-file is saved once again.

Persistent Variables: Maintaining State between Invocations

We can't use the ordinary environment created for each function invocation in order to maintain the state between invocations. That environment is destroyed when the function returns. Instead, we need to use the persistent environment.

In order to inform the interpreter that a variable is intended to be stored in the persistent environment, we use the persistent keyword. Here, for example, is a working version of keepcount:

Define variable count in
persistent environment

Initialize count

```
[1]  function res = keepcount(arg)
[2]  persistent count;

[3]  if isempty(count)
[4]      count = 0;
[5]  end
[6]  count = count+1;  %Update the counter
[7]  res = count;      %the returned value
```

The basic idea is to define the variable count in the persistent environment for keepcount. This means that count will be retained between invocations of keepcount.

One small complication is that the persistent variable count still needs to be initialized. When the persistent environment is created, all of the variables in it are initialized to the empty vector. We can test whether the persistent environment has been freshly created or whether count is the empty vector (line 3). If so, we set count to be 0, which properly sets things up to be used for the counter. The use of the empty vector to indicate that the persistent environment has been freshly created is clever and makes it fairly easy to reinitialize persistent variables to some other value. However, it seems somewhat wasteful that an if-end statement must be interpreted each time the function is invoked, even though it only will be used immediately after the persistent environment is created.

Global Variables: The Tragedy of the Commons

Each m-file function has its own persistent environment; no other function can access the variables in another function's persistent environment. The global environment, in contrast, is a single, shared environment that any function can access. To identify a variable as being in the global environment, use the global keyword. To illustrate, here are some simple counting functions that provide a reset facility:

```
[1]   function res = counter
[2]   global countdata;

[3]   if isempty(countdata)
[4]       countdata = 0;

[5]   end
[6]   countdata = countdata+1;
[7]   res = countdata;
```

Declare a variable to hold the state of the counter

Initialize the variable when first created

Update it

And the resetting function:

```
[1]   function resetCounter
[2]   %resetCounter--reset counter to zero
[3]   global countdata;

[4]   countdata = 0;
```

Declare that countdata is in the global environment

Update it

The resetCounter function is able to access the same countdata variable as the counter function, because countdata has been declared in each function to be in the global environment. If the countdata variable had not been declared as global, then it would be in the function's invocation environment and therefore inaccessible to any other function.

```
≫   counter
  ans: 1
≫   counter
  ans: 2
≫   counter
  ans: 3
≫   resetCounter
≫   counter
  ans: 1
```

The same global countdata variable can be accessed from the command prompt. The variable first needs to be declared:

```
≫   global countdata
```

Then it can be accessed:

```
≫   countdata
  ans: 1
```

This is an extremely simple and powerful way of communicating between functions. However, it has a severe flaw. Another programmer, unknown to you, might have written another function that uses a global variable called countdata for some other purpose. Your own count will be disrupted by this other function. There are no debugging tools within

MATLAB that would allow you to trace such a situation. This problem can arise even after you test your own function successfully; at any point in the future a new function can be added that uses a global variable for a conflicting purpose.

With persistent variables, this problem cannot arise. The only statements that can change a persistent variable are those within the function itself.

If you use global variables, you expose yourself to the whims and personal styles of the writers of all the other functions that might be run in your workspace or that might be added in the future. By keeping your use of global variables to a minimum, you limit the ability of other software to cause you harm and at the same time limit the harm you might inadvertently cause other software.

Another limitation of the use of global variables in a system like `counter` and `resetcounter` is that it provides only a single counter. To implement additional counters for other purposes, new global variables would have to be created. In Section 9.5, we'll see a way to communicate between functions allowed by some languages, but not MATLAB.

9.4   Scoping of Functions

We have seen two ways to create functions:

1. Write an m-file containing MATLAB commands whose first line is a declaration starting with the `function` keyword.
2. Use `inline` to generate a function from a text string.

Functions created with `inline` are values that can be assigned to variables; they are scoped in the same way as any other variable.

But functions created as m-files are scoped in a different way. This is because variables are created either within a function or on the command line; in either event, it is clear which environment a variable is a part of. M-file functions are created by editing a file; this can be done completely outside of any MATLAB session; no environment is involved. This limits the flexibility for scoping m-file functions. The rule used is simple:

> **Scoping of m-file functions: All m-file functions that are available are in scope everywhere. There are two exceptions that we will deal with later: subfunctions and private. There is also a third exception, operator overloading for objects, that we don't cover in this book.**

The fact that all m-file functions are in scope everywhere means that there is no way to have two different functions with the same name: Each m-file in use must have a unique name.

The need to have a unique name for every function causes some problems. First, it means that a computation that was working perfectly well might stop working because someone else, or even yourself, adds a new function with the same name as some other function used in your computation. This is called a *name conflict*. Second, because it's possible to create functions with conflicting names, there needs to be a way to choose which one of conflicting functions will be used.

Key Term

Resolving Name Conflicts

MATLAB has a simple mechanism for resolving name conflicts. Function m-files are stored in ordinary directories in the computer's file system. MATLAB maintains a list of directories to look in whenever a function name is encountered in an executable statement. This list of directories is called the *search path*. When trying to locate the m-file that corresponds to a function name, MATLAB looks in the first directory in the search path list. If the function is found, then searching stops: MATLAB won't even see another directory containing a conflicting m-file. If the function isn't found in the first directory, MATLAB looks in the second directory in the search path. The search continues, looking in each successive directory until the earliest directory is reached that contains an m-file with the appropriate name. When two m-files have the same name, the conflict is resolved in favor of the m-file that is in a directory appearing earlier in the search path list.[2]

Key Term

The command `path` is used to examine which directories are on the search path and to add or delete directories from the search path. There's also a graphical way of doing this using a tool called the "path browser."

Another command, `which`, identifies which directory in the search path a given function name is identified with.

Key Term

The first directory in the search path is always a directory referred to as the *working directory*. (See Chapter 5.) One technique that is extremely useful when developing new functions is to create a new directory to hold the new m-files. Use the `cd` command to set the working directory to be the new directory. Since the working directory is always first in the search path, you are guaranteed that your new functions will be the ones accessed.

The path-searching mechanism is arranged so that each function name refers to a unique m-file, even if there is more than one m-file with that name. If a new m-file is created in a directory earlier in the path than an existing m-file with the same name, then the existing m-file will no longer be invoked. For this reason, it is important to construct unique function names. To help in checking whether a proposed function name is unique, the `exist`

---

[2]You don't need to worry about having identically named files in a single directory; the computer's operating system won't allow this.

command can be used. This tells whether a function (or variable) is already defined with a given name.

One potential source of name conflict is between variables and functions. When an assignment is made to a variable that shares the same name as a function, the function is effectively hidden as long as that variable is in scope. For example,

```
≫  cos(pi) % the usual cosine function
   ans: 1
≫  cos = 1:10; % re-using the name
≫  cos(pi)
   ans: 3
      Warning: Subscript indices must be integer values.
```

Since the scope of variables is limited to the function in which they are assigned, we can look locally within the function to see if there are any name conflicts.

Advanced: Avoiding Conflicts with New Software

Although the function name resolution system guarantees that any function name conflict will be resolved, it is not always resolved in the way you intended. Depending on where your functions are located in the search path, they may not be seen in the event of a name conflict.

Avoiding name conflicts entirely by using unique names is a good way to avoid conflicts, but it is not always a workable strategy. Consider what happens when you load someone else's software onto your computer: M-files are copied that might conflict with ones already there. A similar risk is posed when you copy your own software onto another computer or even just change the path. You can only guarantee that a function name is unique at the time you create an m-file and on the computer where it was created and given the existing search path.

MATLAB provides two facilities for helping to avoid name conflicts even when software is copied or the path is changed.

Key Term

Subfunctions: Many times a function is used purely as a helper to a single function; it is not needed on its own. In such cases, it is best to make the helper function a *subfunction* of the one it is helping. This is done by including it in the same m-file as the main function. The main function is always the one declared on the first statement of the m-file.

A single m-file can contain more than one subfunction. All of the subfunctions are out of scope; that is, invisible, except from the primary function and the other subfunctions in that m-file. Before looking in the current directory or on the search path, MATLAB checks to see whether a function being invoked is declared as a subfunction. This allows subfunctions to avoid any conflicts introduced by the installation of new software or changes in the search path. (But a built-in function's name cannot be used for a subfunction.)

For instance, the matchpeople function on page 158 used a helper function, exclude, that is used in matchpeople and nowhere else. It would be appropriate to define exclude as a subfunction to matchpeople rather than as an independent m-file.

Another example: In Chapter 8, we used a function squarerootV2 that computes the square root of a single number, but since squarerootV2 gives erroneous results when given a vector or matrix as an argument, we "wrapped" it in another function that handled the looping over arrays appropriately. Since there is no use for squarerootV2 outside of this wrapper function, it is unnecessary—and even dangerous—to make it available for general use. SquarerootV2 should properly be made a subfunction. (See Exercise 9.7.)

Private functions: Sometimes two or more functions share the same helper functions, but the helpers are not needed outside of the primary functions. In this case, the helper functions need to be placed in their own m-files. To keep the m-files hidden and to allow them to be accessed properly even if a name conflict is introduced later, the helper m-files can be collected into a subdirectory named private of the directory where the primary function m-files are stored. The private directory should not be placed on the search path.

A Matter of Style:

When is it worthwhile to create a new function to encapsulate a series of statements? Any of the following might suffice as a justification:

Repetition: The statements to be included in the function are found typographically in more than one place. By "typographically," I mean that the same statements are written in more than one place in the program. In the squareroot example, the statements appear typographically in only one place, although there is a loop that causes these same statements to be executed more than once.

The difficulty with typographically repeated statements is that it can be difficult to make sure that all of the copies are identical. The cut-and-paste facilities of editors make it easy to create identical statements, but over the course of debugging and refinement, it is easy for initially identical statements to drift apart. By including the statements in a function, we guarantee that each invocation refers to an identical set of statements.

Segregation: If there are variables in the statements that are not intended to be inputs or outputs of the computation, then by placing the statements in a function, we ensure that there won't be conflicts caused by later or earlier use of those variable names. In the `squareroot` example, there is one such variable, `middle`.

Abstraction: There might be more than one algorithm that could accomplish the same task as the statements in question. By packaging the statements as a function, we make it easy to "unplug" one algorithm and replace it with another.

9.5 Pass by Reference

The usual mechanism for function invocation and return provides limited means for transferring information from one scope to another. Whenever information is being transferred between scopes, a copy of the information is made. Function invocation copies the values of variables in the calling environment given as arguments into new variables in the new invocation

Key Term

environment. Such a situation is called *pass by value* because only the value of a variable, not the variable itself, is passed between environments.

The functionwise persistent environment and the global environment offer other ways to communicate between environments. The variables in these environments can be referred to directly by any statement that has them within its scope. All statements in any invocation of a function have that function's persistent environment within scope. All statements anywhere have the global environment within scope.

Key Term

Many computer languages provide an additional means to communicate between environments called *pass by reference*. This provides a mechanism whereby the variable itself is passed by the function invocation so that statements within the invocation environment can change the passed variable's value in a way that reaches outside the invocation environment. (The pass-by-reference mechanism often involves passing an actual address of bits within the computer memory. Given the address, any statement can access the memory directly to examine or change the bits at that address—the entire scoping mechanism is essentially bypassed. In languages like c or

Key Term

c++, such an address in memory is called a *pointer*.)

MATLAB's absence of a pass-by-reference capability often tricks programmers. Consider a simple counter constructed out of a structure:

```
≫ x.count = 0;
```

The `setfield` program allows a field of a structure to be changed:

```
≫ setfield(x,'count',1)
ans: count: 1
```

Judging from the name `setfield` and the output, it looks like the value of `x.count` has been changed. But it has not:

```
≫ x
  x: count: 0
```

Since MATLAB arguments are passed by value, it's impossible for a function like `setfield` to change the value of a nonglobal variable defined outside of its invocation or persistent environment.

To change the value of x, it's necessary to assign x a value in the same environment in which the variable is defined.

```
≫ x = setfield(x,'count',1)
  x: count: 1
```

Although MATLAB doesn't have a built-in general-purpose type for providing pass by reference, it does use a pass-by-reference approach to dealing with graphics. A graphic created by one function can be updated by others. The special-purpose graphics type provided by MATLAB is called a **Key Term** *graphics handle*. The use of graphics handles is illustrated in the Section 9.10 Project.

9.6 Warnings and Errors

Some events are generated not by mouse clicks and keystrokes but by the computer hardware and software in response to problems. MATLAB has been set up so that these events can influence how a computation can proceed. In this section, we'll examine how to generate and control these events.

The mathematical operation of division takes two arguments, conventionally called a numerator and a denominator. The traditional notation is

$$\frac{\text{numerator}}{\text{denominator}}$$

The operation isn't defined when the denominator is zero. "Isn't defined" is a nice way of saying that we don't know what to do in this case. The idea of ∞ helps us to think about cases like $\frac{1}{0}$, but unless we know about the theory of limits (generally covered as part of calculus), ∞ is just a name that stands for something really big. Indeed, without knowing about limits, it's hard to make much sense out of $\frac{0}{0}$: Perhaps this should be 1, perhaps it should be ∞, 0, −∞, or even something in between.

The people who wrote the MATLAB division operator (./) had to decide what to do in those cases where the denominator is zero. They might simply have told people, "Don't use ./ when the denominator is zero," which is sensible enough advice but doesn't deal with the reality that such things will happen. Alternatively, they could have arranged to punish anyone who has the temerity to use ./ with a denominator of zero: Perhaps they could have made appear the infamous blue-screened "General System Error" familiar to Windows users. This has the substantial disadvantage of making it hard for the programmer to find where the problem came from, which is the second step in fixing the problem. The first step is realizing that there is a problem!

Instead, the people who wrote ./ decided to take a two-pronged approach:

Key Term

- When it's possible to have ./ return something sensible when the numerator is zero, do so. But, to indicate that something out of the ordinary has happened, provide a message to the programmer to this effect. Such messages are called *warning messages*.
- When no sense can be made of the situation, abort the calculation entirely. The message sent in this case is called an error message. Perhaps "malfunction" would be a better term here, but "error" seems to convey the idea that the problem is not due to ./ but to the user.

For instance,

```
≫  x = 1/0;
ans: Warning: Divide by zero.
```

gives the warning message, but x has a sensible value:

```
≫  x
ans: Inf
```

Recall from Chapter 3 that Inf is the printed form of a bit pattern that behaves numerically much like ∞ does:

```
≫  Inf + 0
ans: Inf
≫  Inf./7
ans: Inf
≫  1./Inf
ans: 0
```

Another such special numerical-bit pattern is printed as NaN, standing for "Not a Number."

NaN shows up in cases like these:

```
≫  0./0
ans: NaN
≫  Inf - Inf
ans: NaN
```

Any operation that involves NaN returns NaN:

```
≫  0.*NaN
ans: NaN
```

Inf and NaN show up as appropriate in vector computations:

```
≫ x = [ 1 2 2 3 0 4]./[5 1 0 3 0 1];
  ans: Warning: Divide by zero.
≫ x
  ans: 0.20 2.00 Inf 1.00 NaN 4.00
```

In those cases where nothing sensible can be done with the arguments to ./, an error message results. For instance, suppose s is a structure:

```
≫ s.name = 'George'; s.age = 23;
```

Here, it's unclear what might be meant by 1./s, and so an error message is appropriate:

```
≫ 1./s
  ans: ??? Error using ==> ./
       Function './' not defined for variables of
       class 'struct'.
```

The most important thing to keep in mind about error and warning messages is that you, the author of a function, have complete control of the circumstances in which errors and warnings are generated. As an example, let's consider a simple function that computes the arithmetic mean of a vector.

```
[1]  function res = average(vec)
[2]  res = sum(vec)./length(vec);
```

This program appears to work well enough when the inputs are sensible:

```
≫ average([1 2 3])
  ans: 2
≫ average(1:100)
  ans: 50.5
```

It might occur to the aware reader that there is a problem lurking. What will happen when the empty vector ([]) is given as an argument to average? The length of [] is zero, so we will have a case of division by zero.

If we wanted, we could check for this possibility and handle it explicitly. Perhaps we would like to return NaN as the average of the empty vector but give a warning. While we're at it, we can check for nonnumerical arguments to average and generate an error message. Here's the modified program:

```
                      [1]  function res = averageV2(vec)
Now with error checking [2]  %averageV2(vec)---the arithmetic mean
there's a sensible answer [3]  if isempty(vec)
```

```
[4]        res = NaN;
[5]        warning('Empty argument. Returning NaN.');
[6]    elseif isnumeric(vec)
[7]        res = sum(vec)./length(vec);
[8]    else
[9]        error('Invalid argument type.');
[10]   end
```

No sensible answer

In this program, we're somewhat mimicking the error-handling behavior of the division operator. If no data are provided as input, the average is not defined, and returning NaN sensibly signals this. But to amplify the signal, a warning to the user is also provided using `warning` operator, which takes a character string as an argument. Other than sending the warning message, `warning` has no effect on the execution of the program.

In contrast, the `error` operator sends a message but then immediately stops execution. One says that `error` "throws an error." We will see how to "catch" it. Unless the thrown error is caught, `error` will terminate all of the functions currently being executed and return to the interpreter's command prompt. This is often a very reasonable thing to do, and many thrown errors are never caught.

The messages sent by both the `warning` and `error` include stack information that enables you to trace the functions being executed at the time the error was generated.

It turns out that the non-error-checking version, `average`, worked reasonably well in the error conditions. For instance, the empty vector as an input generates an warning and returns NaN:

```
≫ x = average([])
  ans: NaN
       Warning: Divide by zero.
       > In average.m at line 2
```

Similarly, if `average` is given a nonnumerical argument, an error message results:

```
≫ average(s)
  : ??? Error using ==> sum
    Function 'sum' not defined for variables of
    class 'struct'.
    Error in ==> average.m on line 2
```

All that the warnings and errors in version 2 have done is to provide somewhat more informative error messages. This might be helpful, but it is not essential. In both the cases trapped in version 2 of `average`, the built-in error and warning behavior of the MATLAB operators accomplished what needed to be done.

This is not always the case. Consider the `squareroot` program of Chapter 8. Version 2 of that program calculated the squareroot of a scalar number in a reasonable way but gave completely wrong results when the input is a vector or matrix.

```
≫  squarerootV2(4)
  ans: 2.0000
≫  squarerootV2([1 4 9])
  ans: 1.0000 1.0000 1.0000
```

The situation here is that the logic of `squarerootV2` isn't set up to handle a vector input, but each of the individual steps makes sense to MATLAB so that no errors or warnings are generated. In such cases, it's imperative to check for the error condition and respond appropriately.

Catching Errors

The `error` operator is properly used in those situations when there is no reasonable action to be taken. But sometimes it's possible to respond to an error at a higher level. For instance, in processing the census data from Chapter 4, it might happen that a data-entry error for one family results in a malfunction. What we want to do in such a case is compute the answers for those families with correct data and skip over the incorrect data. Even better, we should make a note that there was a problem so that it can be dealt with.

One way—a bad way—to do this is to pass the data analysis functions the information that they need to respond to the error: perhaps giving them the family's ID number so that this can be printed in an error report.

A much better way is to catch the error. This is done using a special syntax, `try-catch-end`, that is analogous to the `if-else-end`, syntax. For the census data, the error catching might look like this:

```
for k=1:nfamilies
  try
      res(k) = myfunction(families(k));
  catch
      res(k) = NaN;
      problems = [problems k];
      warning('Found a problem family!');
  end
end
```

The `try` block contains the statements we want to execute in the normal course of events. The `catch` block contains the statements that we want to execute if there is some error thrown—it doesn't matter what function or operator throws it. In this case, when `myfunction` works without an error, we simply store the result in the accumulator `res`. If there was some error, then it seems sensible to put some indicator in the accumulator that the value is not valid. In this case, NaN. We might also save some information about where the problem came from or give a warning message.

It's possible for an error to be caught at a lower level. For example, if myfunction uses a try-catch-end construct, errors caught by myfunction will not be passed along to a higher level (unless the catch block itself generates an error).

The try-catch-end context is useful particularly when dealing with abnormal conditions. It allows the statements in the try block to be written under the assumption that nothing goes wrong.

It's often a judgment call about when to give a warning and when to give an error. Here's a rule of thumb: Give a warning when the special case can be handled locally with considerable certainty. Give an error when the special case needs to dealt with at a higher level. Then, at the higher level, a try-catch-end construct can be used to catch any errors.

9.7 Testing Functions

A critical issue whenever writing a function is making sure that the output of the function is correct. An error might arise from either a programming mistake or an incorrect algorithm. In the case of primefactors or matchpeople, how do we know that the program is giving the right answer? It's terribly easy to make either a logical mistake or a programming mistake.

An obvious but important idea is to check the output of the programs against answers that we know. In the case of primefactors, for instance, we can try out the program on numbers known to be prime (e.g., 7, 13) and verify that we get the right answer. We also know that the product of the prime factors must equal the number being factored (this is the definition of factors), so we can try statements like

```
≫ num = 83723423424323;
≫ f = primefactors(num)
  ans: 31 109211 24729703
```

Check whether f really does have the primefactors of num:

```
≫ prod(f) == num
  ans: 1
```

It does.

A very effective strategy is to use random inputs. Surprisingly, often there are hidden or implicit assumptions in an algorithm that the author of the program knows, perhaps not even being aware of this knowledge. In such cases, the author may tend to generate inputs that satisfy the hidden assumption, but another user (or the programmer himself or herself at some later date) can easily generate inputs that break the program.

Key Term

Also important are *edge cases*. These are values of inputs that resemble typical inputs but that the algorithm might not have been designed to handle. Often, edge cases are the very simplest cases (it's surprising how often algorithms are designed to work for complicated cases but not simple ones). Here are some edge case examples from `primefactors`:

```
≫  primeFactors(2)
   ans:  2
≫  primeFactors(1)
   ans:  []
≫  primeFactors(0)
   ans:  []
≫  primeFactors(-10)
   ans:  []
≫  primeFactors(4.1)
   ans:  4.1
```

The `primeFactors` function returned an answer for each of these cases; the question is whether the answer is what we want. There are choices to be made here. For example, it's not unreasonable to return the empty vector as the prime factors of 1; typically 1 is not included in a list of factors and there is considerable consistency since `prod([])` gives 1. But, by the same token, possibly the answer to `primefactors(0)` should be 0, not `[]`. It certainly doesn't seem right that the prime factors of −10 should be the same as those of 1 and, since 4.1 isn't prime, 4.1 should never be returned as a prime factor of any number. Perhaps we want to throw an error for noninteger inputs, or perhaps round to an integer and provide a warning. But whatever the choices are, we should make them ourselves and not rely on the unanticipated actions of an algorithm working with inputs for which it was not designed.

The following function tests the `primefactors` function against a few edge cases and some random cases. Since the prime factors of k multiplied together should equal k, we test for correctness of the answer using `prod(factors) == k`:

```
[1]  %test primefactors on n random inputs
[2]  %return inputs for which the answer was wrong
[3]  %or an error was generated
[4]  function res = testprimefactors(n)
[5]  res = [];
[6]  edgecases = [-1 0 1 2 -10 4.1];
[7]  randomcases = floor(1000000000000*rand(1,n));
[8]
[9]  for k = [edgecases randomcases]
[10]     try
[11]        factors = primefactors(k);
[12]        if prod(factors) ~= k
[13]           res = [res k];
[14]        end
[15]     catch
[16]        res = [res k];
[17]     end
[18] end
```

Round to integers — [7]

Test for wrong answer — [12]

An error was generated — [16]

Note that testprimefactors is set up to return those inputs for which the output of primefactors was wrong or for which an error was generated. This enables the programmer to start debugging those cases.

```
≫ testprimefactors(100)
  ans: -1 0 -10 4.1
```

Four cases were found for which the program returns an incorrect answer. Of course, a random trial like this does not prove the primefactors works correctly, but it certainly helps in detecting any gross errors in the algorithm or program.

In many cases, it's difficult to verify the output of a program because the program itself may be the only practical way of computing the output. Such situations require great care. There usually are some aspects of the output that feasibly can be checked. For instance, a program like matchpeople is difficult to check. Certainly, the programmer can, and should, generate some simple cases that can be checked by hand. It's also easy to make random input matrices but tedious to check these by hand. Sometimes, when it's very difficult to check whether the answer is correct, it's relatively easy to check whether the answer meets certain known conditions. In the case of matchpeople, an important thing to check is whether different job assignments are being made to the different people. This can be automated; for instance,

```
≫ a = matchpeople( rand(4,4) )
  ans: 2 4 1 3
```

finds the output for a random input. By examining the output, we can see that each person is assigned a job. To be more systematic, we can write a simple boolean statement that evaluates this for any set of assignments a:

```
≫  all( sort(a)  ==  [1 2 3 4] )
    ans:  1
```

While this isn't a check that the answer a is absolutely correct, it does check one aspect of a.

A complementary approach is to insert verification statements inside the program itself. For instance, in matchpeople we might incorporate a counter in the innermost loop to make sure that we are indeed handling the right number of cases; we know that there should be 4! = 24 cases. Similarly, we might use the boolean expression all(sort(a) == [1 2 3 4]) to make sure that each case is of the proper form: a unique assignment to each person. Such statements can be commented out once testing of the function is complete.

There is often a reluctance by programmers to test and validate their work. Although this might reflect a natural desire to avoid seeing one's labors discredited, there is a huge penalty to be paid when incorrect functions are used to build higher-level functions. I have seen many cases where weeks or months of research effort were wasted because of reliance on a program that produces incorrect answers and where the only test of the program that had been made was to check that no error messages were being generated. If there is a way to check the output of your programs, you should do so before accepting the program for productive work. If there is no way to check the output of your program, in using that program, you have left the realm of scientific computation and entered that of mysticism, numerology, and the occult.

9.8   Optional and Default Arguments

Consider the modest cosine function, packaged up in the cos operator:

```
≫  cos(0)
    ans:  1
≫  cos(pi)
    ans:  -1
≫  cos(180)
    ans:  -0.5985
```

Evidently, the argument to cos is an angle in units of radians, not degrees. In a sense, there is no problem here; anyone who wants to use degrees can easily convert from degrees to radians using a factor of $\pi/180$. The help documentation for cos doesn't say that the argument is in radians, but the designers of the cos function assumed, reasonably, that most experienced scientific programmers would simply assume that the argument is in radians.

This was a design choice that reflected how the designers expected the cos argument to be used. But the designers, if they had students in mind, might have taken a different approach: Let the user specify which units to use. We can imagine a whole family of friendly trigonometric functions that allow this flexibility; for instance,

```
>> nicecos(180, 'degrees')
   ans: -1
>> nicecos(pi, 'radians')
   ans: -1
```

Such flexibility can be nice, but it does become tiresome after a while. When using radians, we might prefer to be able to say simply

```
>> nicecos(pi)
   ans: -1
```

This is an example of a situation where a function has more than one input, but we would like some of the inputs to take on default values. In this nicecos case, the default value for the units is to be radians.

In order to write a function like nicecos, we need to have a mechanism to determine how many arguments were given when the function was invoked. MATLAB provides this with the nargin operator, which returns the number of arguments that were given to the calling function.

Here is nicecos:

```
[1] function res = nicecos(angle, units)
[2] %nicecos(angle, units)
[3] %cosine with explicit angle units
[4] %Ex: nicecos(pi,'radians') or nicecos(180,'degrees')
[5] if nargin == 1
[6]     units = 'radians';
[7] end
[8]
[9] switch lower(units)
[10] case { 'radians', 'rad', 'r'}
[11]     res = cos(angle);
[12] case { 'degrees', 'deg', 'd'}
[13]     res = cos(pi.*angle./180);
[14] otherwise
[15]     error('Unrecognized units.');
[16] end
```

The default for units [6]

Note that when nicecos is called with a single argument, the value of the units argument is undefined. Before units can be referenced, it must be assigned a value. This is done on line 6 of the function.

Named Arguments

There are many functions that have several arguments, each with its own default value. For example, the MATLAB differential equation solvers (such as `ode23`) are interfaces to a complex technology that includes many adjustable parameters and options. It's possible to use `nargin` directly to check for such optional arguments, but the interface becomes very complicated since one needs to keep track of which optional argument comes first, which second, and so on.

A better scheme involves *named arguments*. Some languages (unfortunately not MATLAB) have a special syntax for identifying optional arguments by name and assigning them a value. A very successful syntax is that used in the S and R statistical languages. For example, the `rnorm` function in S and R is analogous to the `randn` gaussian random number generator in MATLAB. Rnorm allows you to override the default values of the mean and standard deviation by using named arguments:

```
R> rnorm(10, sd=7.5, mean=2)
```

will generate 10 random numbers from a population with a mean of 2 and a standard deviation of 7.5. Since MATLAB has no built-in support for named arguments, a variety of systems are in use, which creates some confusion among users. Possibly the best of these schemes is used in the "handle graphics" routines (see Section 9.10), where the optional arguments come in pairs: A quoted string is used to identify the argument and is followed by the value to be assigned. For example, to create a set of axes for a plot that is 2 inches horizontally by 3 inches vertically,

```
>> a = axes('Position', [.5 .5 2 3], 'Units', 'inches')
```

Typically, the optional arguments are not the only arguments to a function. When the number of optional arguments can vary widely, it becomes inconvenient to assign a name to each of them as the standard function syntax requires. MATLAB therefore provides an alternative approach where the argument list consists of name and value pairs. This approach, which uses the `varargin` argument syntax, will be illustrated in Section 9.10.

Flexible Arguments

Aside from default arguments, `nargin` can be used as part of a flexible function invocation user interface. Consider the `plot` function, which can be called with various numbers of arguments to produce different effects:

```
>> plot(y)
>> plot(x,y)
>> plot(x,y,'.',x,y2,':')
```

Functions such as `plot` use `nargin` to provide an effective and easy user interface. Such interfaces require careful planning if they are to be successful. They have to follow a simple and uniform scheme if the user is to be able to remember how to use them and avoid errors. For example, here is a horrid use of `nargin` that does little but confuse any potential user.

```
[1]   function res = horridfunction(arg1,arg2,arg3)
[2]   switch nargin
[3]   case 0
[4]      res = rand(1,1);
[5]   case 1
[6]      res = cos(arg1);
[7]   case 2
[8]      res = atan2(arg1,arg2);
[9]   case 3
[10]     res = (arg1.^2).^arg3 + arg2;
[11]  end
```

This function merely packs four unrelated operations into one function. It would be far better to define the four functions separately.

9.9 Exercises

Discussion 9.1:
Here is a program for computing the difference between successive elements in a vector (like `diff`).

```
>> mydiff([1 2 4 1]
   ans: 1 2 -3

function res=mydiff(vec)
if length(vec)==1
   res='ERROR!';
else
   res=vec(2:end)-vec(1:(end-1));
end
```

Keeping in mind that it's best when a function returns always the same type of value, in what ways could this program be improved to handle the exceptional cases where there are no successive elements to be differenced?

Exercise 9.1:
Write a function, `parallelresistors`, that computes the total electrical resistance of two parallel resistors, R_1 and R_2. The formula is

$$R_{\text{total}} = \frac{1}{\frac{1}{R_1} + \frac{1}{R_2}}$$

When one of the resistances is zero, the physical situation is clear: There is a short circuit and the total resistance is zero. But division by zero in the formula might cause a problem.

Show that the formula behaves properly as one of the resistances becomes very small, even zero. This illustrates an advantage of the Inf system for dealing with division by zero.

Exercise 9.2:
Write an asktime command that takes no arguments and returns the correct time. But unlike the now and datestr functions, asktime gets bothered if you ask the time too many times in close succession and gives the sort of sarcastic response you might expect from an impatient or annoyed person.

Exercise 9.3:
In Chapter 2, we looked at a sequence of statements to reverse two variables, a and b:

```
≫  tmp = a;
≫  a = b;
≫  b = tmp;
```

An auxiliary variable was needed in order to retain the value of a for later assignment to b after a had been changed.

Write a function swap that works when invoked like this:

```
≫  [a,b] = swap(a,b);
```

It is possible to write such a function with only two statements, without using an auxiliary variable like tmp. More advanced: Write swap as a function with no statements at all.

Exercise 9.4:
You can't write a function swap that works in the way illustrated in the following statements:

```
≫  a=1; b=2;
≫  swap(a,b);
≫  a
   ans: 2
≫  b
   ans: 1
```

Explain why the preceding can't possibly work with regular variables, but show how it can be made to work with references.

Exercise 9.5:

Rewrite the prime factors programs in Chapter 8 so that rather than looping over all of the numbers from 2 to $\sqrt{n}$, only the *prime* numbers from 2 to $\sqrt{n}$ are used. Keep in a persistent variable a moderate sized list of prime numbers so that these do not need to be generated each time the program is called. First, loop over these and if a factor is found, then stop. If additional primes beyond those stored with the program are needed, generate them. If you were doing many calculations with large numbers, you might want to add the newly generated primes to the persistent list for later use; this would speed things up but would take up more computer memory.

Exercise 9.6:

Write a `keepbest` function that keeps track of the "best" items encountered so far. Of course, "best" depends on what you are looking for. Let's assume that you have a score for each item that indicates how good it is. You also have a value (that might or might not be a number). `Keepbest` will save the best n items. It should work like this:

```
>> keepbest('setup', 2)
```

Set it up to keep the best two items.

```
>> keepbest(3, 'George')
```

George has a score of 3.

```
>> keepbest(7, 'Katie')
```

Katie's score is higher.

```
>> keepbest(1, 'Frank')
```

To report the best so far, return the scores as a vector and the values as a cell array.

```
>> [scores, values] = keepbest('report');
>> scores
  ans: 7 3
>> values
  ans: Katie George
```

Break ties in favor of the ones already encountered.

Exercise 9.7:

Rewrite the `squarerootV4` function from Chapter 8 using a subfunction for the scalar calculation of the square root handled by `squarerootV2`.

Exercise 9.8:

Write a special counter to keep track of a baseball game. At each event, the counter would be called with a character string describing the event. You will have to be thoughtful in making up a list of events and in deciding what should be the state of the system. Some examples of events are ball, strike, foul, batter out, runner on first out, runner on second out, runner on third out, single, double, triple, advance from first, advance from second, advance from third, walk. At the end of an inning, the program should automatically switch over to scoring for the second team. At the end of a game, the program should indicate the score of each team.

Exercise 9.9:
Write a counter program divided into several functions: `cincr`, `creset`, `cset`, `creport`. Write the program in two different ways:

1. With global variables to communicate among functions.
2. With a `memory()` function in a private subdirectory to handle to communication.

9.10 Project: Precision Graphics and the Electrocardiogram

The electrocardiograph (ECG) is a widely used medical instrument. It records the small differences in electrical potential between pairs of points on the body surface. These potential differences are generated by the electrical activity of muscle fibers in the contraction and relaxation processes. In a clinical ECG, multiple simultaneous recordings are made from different spots on the body; each such recording is called a "lead."

The file `ecg.dat` holds a two-lead recording of an ECG signal from a human patient [26]. The data are stored as a matrix with one lead in each column. The ECG signal, which is continuous in time, has been sampled 128 times per second. The entire recording is 6 seconds long and consists of 768 samples (768 samples ÷ 128 samples/s = 6 seconds). Each row in the matrix is the two simultaneous samples at each point in time. The amplitude of the recording is in millivolts (mV).

Figure 9.3 shows a plot of the two ECG leads. This plot is easily made with the following commands:

```
ecg = load('ecg.dat');
subplot(2,1,1)
plot(ecg(:,1))
subplot(2,1,2)
plot(ecg(:,2))
```

Such a simple plot is perfectly acceptable for many purposes. The plot would be considerably improved if the *x*-axis showed time in seconds rather than in samples. This could be accomplished by creating a proper *time base* for the signals; that is, a vector that gives the time in seconds for each of the samples in the signal:

```
≫  timebase = (1:length(ecg))/128;
```

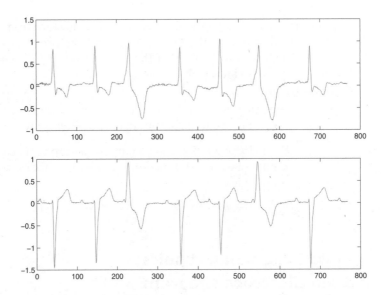

Figure 9.3. A rough plot of a two-lead ECG. Each of the sharp peaks corresponds to a heart beat. The third and sixth beats shown are abnormal, "ectopic" beats.

Then a properly scaled plot can be made:

```
≫ plot(timebase, ecg(:,1)
```

In this project, we want to go rather further in making a proper plot; we seek to plot out the ECG signals in a format that follows the standards for clinical electrocardiography. These standards are important when presenting information to physicians. Standardization helps to reduce the possibility of error and misinterpretation. Figure 9.4 shows a plot of the same ECG signals in the standard format.

There are several things to note about the standard format:

- There is a grid consisting of a square array of small red dots and thin red lines. The dots are spaced by 1 mm and the lines by 5 mm.
- The axes are not labeled with tick marks. Instead, the distance between adjacent solid vertical lines represents a time span of 0.2 seconds.
- Calibration pulses (the top-hat-shaped signals on the left of the plot) are used to indicate the vertical scale of the plot. The vertical scale on an ECG is usually 1 mV/cm, but it may be altered to be 2 mV/cm or 0.5 mV/cm. The calibration pulse shows which scale is being used; the scale may be different for different leads.
- The signals are plotted one above the other with the vertical location of each signal selected to space the signals nicely. That is, an arbitrary constant value is added to each signal. (Note for electrical engineers: The ECG is an ac-coupled signal; meaning that on average, it has a value of zero.)

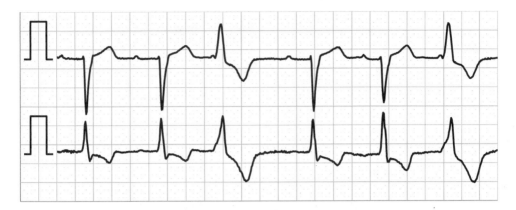

Figure 9.4.   A plot of the ECG signals in the standard clinical format used by physicians. (On a color display, the grid lines and dots are red.)

The origin of these conventions for displaying the ECG predates computers. Early ECGs were drawn on chart recorders that scrolled a roll of graph paper past the wiggling pens driven by the amplifiers connected to the electrodes attached to the patient. Today ECGs are recorded using a different technology, but the graphical format remains a standard to be followed. It's well to remember that any graph is intended as part of an interface to humans and, as such, following presentation conventions is extremely important even if it does impose an extra workload on the programmer.

In order to produce a graph that follows the standard for clinical ECGs, we have to instruct the computer to generate plots that meet our precise specifications. In particular, the physical layout of the plot on the screen is important: The plot must have an appropriate physical size on the screen so that the grid lines are separated by 5 mm and the dots by 1 mm. We also need to arrange that the time scale is 0.4 s/cm and the vertical scale is 1 mV/cm.

In order to accomplish this, we need to use three of the many types of graphical objects provided by MATLAB:

Figure: A *figure* is a window on the screen in which graphics can appear. It is created with the `figure` command.

Axes: An *axes* is a region of a figure in which graphics can be drawn.[3] The axes determines the plotting scale. An axes is created with the `axes` command.

Line: A *line* is a collection of points to be drawn in an axes. The points may be connected or may be plotted as symbols.

---

[3]The plural word axes is used in MATLAB to refer to a pair: the *x*-axis and *y*-axis which together define the plotting region.

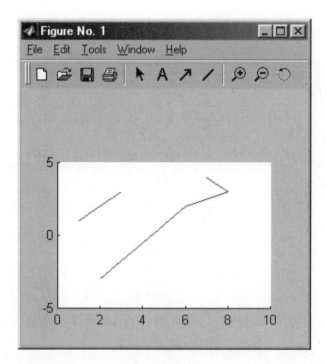

Figure 9.5. The graphic created with the following sequence of commands:

```
[1]   f = figure(1);
[2]   set(f, 'units', 'centimeters', ...
[3]       'position', [5 6 8 7])
[4]   figure(f)
[5]   a = axes('units', 'centimeters', ...
[6]       'position', [1 1 6 4], ...
[7]       'Xlim', [0 10], 'Ylim', [-5 5])
[8]   axes(a)
[9]   lineA = line([1 2 3], [1 2 3]);
[10]  lineB = line([7 8 6 2], [4 3 2 -3]);
```

The details of figures, axes, and lines (as well as other graphical objects in MATLAB) can be established at the time of their creation or modified thereafter using the `set` command. Each of the creation operators returns **Key Term** a value, called a *graphics handle*, that allows the object to be referred to and altered. For example, to create a figure numbered as figure 1 (Figure 9.5)

```
≫  f = figure(1);
```

To change the properties of the figure (for example, the size, location on the screen, or background color), the `set` operator is used. The first argument to `set` is the graphics handle of the graphical object that is to be

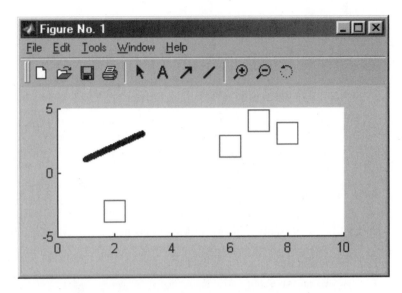

Figure 9.6.   The result of modifying the graphic in Figure 9.5 as follows:

```
[1]   set(f, 'units', 'inches', ...
[2]       'position', [1 1 4 2]);
[3]   set(a, 'position', [1 1 8 3.5])
[4]   set(lineA, 'color', 'black', ...
[5]       'linewidth', 5);
[6]   set(lineB, 'linestyle', 'none', ...
[7]       'marker', 'square', ...
[8]       'markersize', 20);
```

modified. Graphics handles are special entities built into MATLAB that are analogous to references. But, unlike references, only the set operator can modify them.

The remaining arguments to set are pairs of values. The first member of each pair is a character string that specifies which property is to be set; the second member of each pair is the value that the property is to have. For example, the following sets the position of figure f to be 5 cm from the left edge of the computer screen and 6 cm from the bottom. The height of the figure is 7 cm and the width 8 cm.

```
set(f, 'units', 'centimeters', ...
    'position', [5 6 8 7]);
```

(The size refers to the graphics area of the window not including the title bar.) There are many other aspects of figures that can be modified using set (Figure 9.6); to see a list, give set a single handle as an argument [e.g., set(f)].

A similar procedure is used to create an axes. An axes is created within a figure. The specification of the figure is done implicitly, by designating a figure as "active" using the `figure` command using as an argument the handle to the desired figure; for instance,

```
>> figure(f);
```

In creating an axes one can specify the location and size relative to the figure and also information such as scale limits (i.e., the range of the *x*-axis and the *y*-axis), where tick marks are located, and so on. An example is given in the caption of Figure 9.5.

At this point, lines can be plotted by making the desired axes active and using the `line` command to specify the details of the lines, as shown in Figure 9.5.

Even after the graphics objects have been created, they can be modified using the `set` command. Figure 9.6 shows how the plot in Figure 9.5 can be modified.

Exercise 9.10:

Write a function that takes as input a matrix of ECG signals measured in mV and sampled at 128 samples/s, and produces a standard ECG plot as shown in Figure 9.4.

Here are some hints:

- Start with just a single lead.
- To make the grids, use the `line` command separately for the small 1-mm dots and the 5-mm lines. You may find it advantageous to write a function that returns a handle to an `axes` to which the grid lines and dots have already been added. The coordinate units should be in mm. The function would take as arguments the desired location of the `axis` in the figure window and the size of the `axis`.
- Convert the ECG lead to have units of mm in both the vertical and horizontal scales. This will entail separate conversions for the amplitude and the time base. Then the converted signal and time base can be plotted directly on the `axes` with the lines and dots.
- Write a function that takes a single ECG lead, an `axes`, and a vertical location as arguments and plots the ECG on the `axes` centered around the specified location.
- Use separate `axes` for the calibration pulses and synthesize a square pulse with the correct vertical and horizontal scaling. The synthesized pulse will be represented by *x*- and *y*-coordinate vectors that specify the location of the endpoints and turning points of the pulse.

CHAPTER 10

Events

In the preceeding chapters, we have used the computer in several ways:

- As a machine for transforming inputs to outputs
- As a storage medium for information in the form of files and variables
- As a display tool, generating graphics, text, and sounds

In this chapter, we exploit another capability:

- As a device for collecting data, interacting with its environment, and making measurements

Of course, we have always used the machine in this way. Consider the keyboard, for example. From the user's perspective, it is a way to tell the computer what to do. But from the computer's point of view, the keyboard is a sensor: a means of collecting data from its environment as an array of switches whose position is monitored. That the computer takes our keystrokes literally, collecting them into MATLAB expressions or assembling them into text for documents, is entirely a matter of design. A different design might process the keyboard data in another way, perhaps recording the timing of key presses rather than the letter.

The computer mouse is another sensor: one that measures movement and, like the keyboard, the position of a set of switches. That this measured movement is translated to the position of a cursor displayed on the screen is also a matter of design; only software connects movements of the mouse to the cursor position.

This chapter introduces some programming techniques for dealing with sensors: activating them, responding to the events they generate, and collecting data from them. Beyond the keyboard and mouse, computers can have additional sensors as well: microphones, joy sticks, and laboratory instrumentation. There is even a clock from which the system can read the current time.

The wide variety of such devices makes it impractical to discuss in any detail how they can be controlled and accessed from software. But the broad principles involved are well demonstrated by the keyboard, mouse, and clock; we use these as examples since they are available on almost every computer.

10.1 Activating Input Devices

The connection between a physical input device such as the keyboard and the MATLAB interpreter is long and complicated; there are multiple hardware and software systems involved, and, in principle, new software can interact at a variety of different levels, some close to the hardware, some distant. We will interact at a fairly high level, using commands that are built into MATLAB.

A basic MATLAB operator for soliciting keyboard or mouse input is waitforbuttonpress. The discussion of this makes the most sense if you can follow it with a computer handy, having started up a fresh session of MATLAB. Give the command

```
>> a = waitforbuttonpress
```

In response to this command, MATLAB displays a figure window. If this was a fresh MATLAB session, the figure window has a large blank area. Otherwise, an existing figure, with its existing contents, is displayed. On Windows computers, the figure window is highlighted to show that it is the active window.

Use the mouse to move the cursor into the main area of the figure where graphs are usually displayed. Press the mouse button. Two things will happen. One is that a value of 0 is returned; in the command window, the line

```
a =
     0
```

appears. The other is that another command prompt appears in the command window; the interpreter is now waiting for another command.

Repeat the command

```
>> a=waitforbuttonpress
```

and move the cursor again into the main area of the figure. But this time, press a key. This time, `waitforbuttonpress` returns the value 1:

```
a =
     1
```

One more time, but different. Give the command

```
≫  a = waitforbuttonpress
```

but move the cursor away from the figure window. You can place it in any innocuous area of the screen, or even in the title bar of the figure window. For example, you might put the cursor over the MATLAB command window or a word processor. Click a mouse button. What happens now is different and depends on where the cursor was placed; perhaps the command window is activated. The mouse click, in contrast to the preceding cases, does not cause a value to be returned or the command prompt to reappear. Neither will a key press.

Now, move the screen cursor over the figure window in the same way as before. (Depending on how your computer is set up, you may have to move other windows out of the way.) At this point, either a mouse click or a key press will cause `waitforbuttonpress` to return.

As the name suggests, `waitforbuttonpress` activates the two button-oriented sensors—the keyboard and the mouse—and waits. When a button is pressed, `waitforbuttonpress` returns a 0 or a 1, depending on whether a key or a mouse button was pressed.

But `waitforbuttonpress` is competing with other software on the computer for information about the keyboard or mouse activity. `Waitforbuttonpress`, or any other software that we write to use the computer sensors, does not operate in isolation. It exists in a social environment, where it inevitably interacts with other software running concurrently on the computer. The mouse clicks and keyboard strokes that we want to sense are used by the operating system and other software to control the display: bringing windows to the front, closing them, invoking programs, and providing input. We need to operate within the limits imposed by this society of software as governed by the computer's operating system. In particular, we need to ensure that the measurements made by the computer's sensors are directed by the operating system to our software rather than to some other software. In order to do this, `waitforbuttonpress` displays a figure window and instructs the operating system that any mouse or keyboard events that occur while the cursor is over this window be directed to `waitforbuttonpress`. So long as this window is the one in view and activated by the operating system, and the screen cursor is over the window, sensor information will be directed to `waitforbuttonpress`. But if another window is activated, even the MATLAB command window, the keyboard and mouse data may be directed elsewhere. For instance, keystrokes given while the MATLAB command window is active will be directed to the command window, although characters may not appear immediately since the interpreter is waiting for `waitforbuttonpress` to return.

`Waitforbuttonpress` collects more information than the 0/1 indicating whether a key or a mouse button was pressed. This information can be accessed via the handle for the figure window in which the event occurred. (This window handle is returned by the `gcf` operator.)

For example,

```
≫ waitforbuttonpress
```

and press a key when inside the figure window. This key is stored in the `CurrentChar` attribute of the figure handle. To retrieve it,

```
≫ get(gcf,'CurrentChar')
```

Similarly, if a mouse button is pressed, the position is stored in the `CurrentPoint` attribute.

Menus and Other Choices

From software like `waitforbuttonpress`, more specialized programs have been written. For example, consider the following two lines:

```
≫ items = {'Car', 'Truck', 'Motorcycle', 'Delete', 'QUIT'};
≫ menu('Next vehicle:', items)
```

The `menu` operator displays a small figure window with boxes in it: one for each of the character strings in the cell array second argument. It works somewhat like `waitforbuttonpress` but is more choosey: It will only return when a mouse click is made over one of the small boxes. It returns an index indicating which box the click occurred in. For example, clicking over "Truck" will produce a return value of 2.

Perhaps it would be friendlier and more familiar to describe `menu` as making "buttons" and the mouse as selecting a choice. But "button" and "select" are only metaphors for what is happening. `Menu` has been written to draw graphics that look like buttons, with patterns of light and dark to suggest the shadows of a real, three-dimensional button. It even arranges to change the graphics, inverting the shading as the mouse button is clicked to give the visual impression that one of the buttons on the screen is being depressed. But using a less metaphorical description, `menu` is displaying information in a format that easily communicates to humans and processes the mouse sensor input in an appropriate way to mimic the selection of an item. In reality, nothing is being selected; the mouse input is simply being translated to a single number in a sensible way.

This same sort of functionality can be packaged in other ways. For instance, try

```
>> questdlg('Does this make sense?', 'A question for you')
```

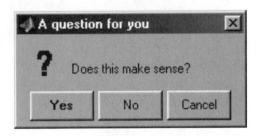

Soliciting Character Input

Buttons are not the only way to make choices. The listdlg operator makes another familiar graphical format for selection. To illustrate, we'll take the first 100 words from *Moby Dick*:

```
>> foo = textread('mobydick.txt', '%s',...
   'delimiter', ' ');
>> listdlg('liststring', foo(1:100))
   ans: 3 10 11 17
```

In the resulting display shown on the left, we have selected several items.[1]

Such devices are not restricted to collect input from the mouse. Here is one that provides spaces for the user to type entries:

```
>> persondata = {'Your Name', 'Date Of Birth', 'Your Height'};
>> a = inputdlg(persondata, 'Personal Information')
```

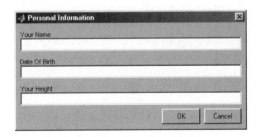

---

[1]On Windows computers, holding down CNTL while clicking the mouse adds an item to the selected list.

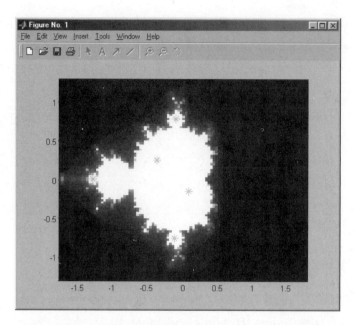

Figure 10.1. Using [x,y] = getpts to collect the location of mouse-click events within a figure window. Each point is marked with a *. Pressing the return key or a double-click of the mouse button causes getpts to return vectors containing the *x*- and *y*-coordinates of the points.

Mouse Position and Movement

Mouse input is not restricted to which button was pressed. Detailed coordinate information can be solicitied. In order to do this, the figure window must contain an axis system as made with plot, pcolor, axes, and so on. For example, Figure 10.1 shows a window in which the Mandelbrot set has been displayed. (See Section 8.13.) The getpts operator detects mouse clicks and returns the location of the mouse where the click was made, scaled to the same coordinates as the plot.

```
≫  [x,y] = getpts
```

Closely related operators are ginput, which allows points to be marked with the keyboard as well and returns a third value describing which button made each selected point, and getrect, which allows a rectangular region to be marked by click-dragging the mouse.

Other Input

All of the preceding inputs have been from the mouse and keyboard, and, to coordinate with the computer operating system, all of them have involved creating a figure window and placing the screen cursor in that window. However, not all input devices need to be coordinated in this way. For example, we can access the system clock to find out what time it is:

```
≫  a = now
   a:  7.318833201119213e+005
```

This information is encoded as a date number, which is the number of days since 1 Jan 0000 (according to the Gregorian calendar). This includes fractional days, giving the hour, minute and second. It can be converted to a familiar format in several different ways; for example,

```
≫  datestr(a)
   ans:  29-Oct-2003 07:40:58
```

Many computers are equipped with a microphone and appropriate recording hardware and software. On some operating systems, MATLAB can activate this. For example, the following makes a recording of three seconds duration:

```
≫  fs = 11025;
≫  y = wavrecord(3*fs, fs);
```

This returns a vector containing digitized sound; whatever sound impinged on the microphone during the three seconds after the command was executed; in this case, the author saying "Mary had a little lamb."

```
≫  plot(y)
```

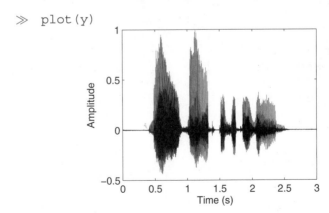

The `fs` variable sets the sampling rate of the recording. We'll explore this in Chapter 13.

10.2 Example: Recording the Times of Events

Perhaps your window overlooks a roadway; you often sit pensively and watch the cars go by. What better opportunity could there be to collect data? Let's find out how much traffic the road carries and whether the cars are spaced randomly or according to some pattern.

A professional approach to this would involve specialized hardware: rubber hoses stretched across the road, attached to a transducer that senses when a vehicle crosses and records the time. As part of a computerized data-collection system, the transducer would be arranged to signal the computer whenever a vehicle is detected; the computer would record the time of the detection.

Without such hardware, we will use a readily available data-collection system: You will be the transducer and will send a message to the computer by typing a button. We'll take advantage of this interface to elaborate: Type C for car, T for truck, and M for motorcycle. When you're done, type ESC to quit.

Here is a program `trafficTimer` implementing this system:

```
[1]    function [times,letters] = trafficTimer
[2]    % [times, letters] = trafficTimer
[3]    % --- record the times of events
[4]    % type a character key for each event
[5]    % ESC to quit
[6]    times = [];
[7]    letters = [];
[8]
[9]    f = figure(1);
[10]
[11]   while 1
[12]       a = waitforbuttonpress;
[13]       % Process the event.
[14]       time = now;
[15]       letter = get(f,'CurrentChar');
[16]       if letter == 27 % ESC character
[17]           break;
[18]       else
[19]           times(end+1) = time;
[20]           letters(end+1) = letter;
[21]
[22]       end
[23]   end
[24]
```

Initialize the state [6]
Additional initialization goes here [8]

Loop: catch events and process them [11]
Wait for the next event [12]

Collect the time [14]
Collect the key [15]

Exit [17]
Update the state [18]

Additional updating goes here [21]

Post-processing goes here [24]

The basic structure of this program, like many other such programs, consists of three parts:

1. Initialing the state of the system. This involves setting up arrays to store the data.
2. Looping, catching each event, and processing it. In our case, *processing* means interrogating the computer's clock to find out what time the event occurred and finding which key triggered the event.
3. Postprocessing. In this simple program, there's nothing to do, just pass the collected data as return values.

Using the program is straightforward:

```
≫  [t,c] = trafficTimer;
```

A figure window pops up. Position the screen cursor in that window and start typing. When you are done, press the ESC key.

The event times, collected with now, are stored as date numbers. This format contains a lot of information: the year, month, and day, as well as the time of day with a resolution of a fraction of a second. Here is a typical recording from trafficTimer:

```
≫  t
  ans:  731446.49477 731446.49480 731446.49481
≫  char(c)
  ans:  cctc
```

If desired, the recorded times can be converted to a more convenient format for human readers:

```
≫  datestr(t)
  ans:  18-Aug-2002 11:52:29
        18-Aug-2002 11:52:31
        18-Aug-2002 11:52:32
        18-Aug-2002 11:52:35
```

But this nice printed format is not always the best. For instance, to look at the time intervals between events, in seconds, it suffices to compute

```
≫  diff(t).*(24*3600)
```

where 24*3600 is the conversion factor from days to seconds.

The system clock has surprisingly low resolution: worse than 1/20 of a second on many systems. This is probably good enough for the purposes of timing traffic; it's hard to type that fast. But for other purposes (for instance, detecting the times of radioactive decay events), the system clock might be too poor. In such cases, special hardware is needed that records a high resolution time in addition to the event.

TrafficTimer is a simple program. It does the job, but it could be improved in a number of ways. For example, the program doesn't have a way to deal with data-entry mistakes. Perhaps we could have the user press the DELETE key to indicate that the previous entry was an error and update the state accordingly, dropping the last event from the times and letters arrays.

One particular problem is that the program doesn't provide any feedback to the user so that the user knows the keystrokes are being recorded. This can be a serious problem if the user forgets to put the screen cursor in the window and so the keystrokes are never sent to trafficTimer.

A simple way to deal with this problem is to use disp or similar operators to display a message in the command window. This might not be effective, though, since the command window isn't activated when input is being given. Alternatively, a pinging sound might be played at each input.

Another way is to replace the waitforbuttonpress event collector with a menu system; for instance,

```
menu('What kind of vehicle?', ...
{'car', 'truck', 'motorcycle', 'Mistake', 'FINISHED'})
```

This is easy, and might work well, but it does slow down the entry of data.

The most familiar approach is to have the data-entry figure window alter its display to give feedback. For example, a word processor responds to key strokes by updating the displayed text. There are two steps to arranging for the figure window to display information:

Initialization: Set up the window and place graphics objects in it to display the desired information.

Update: As the state is updated, similarly update the information displayed.

Here is a simple example. The function trafficFig creates a figure window with a single text area. These entities are created with the handle graphics commands that were introduced in Chapter 9. The uicontrol command is the general-purpose method of constructing items in a graphics window. These items are called *user interface controls*. TrafficFig invokes uicontrol to make a user interface control containing text, specifically the numeral 0.

```
[1]    function [fig,counter] = trafficFig
[2]    %  user interface window for traffic counter
[3]    fig = figure(1);
[4]    set(fig,'units', 'centimeters', ...
[5]        'position', [5,5,5,5], ...
[6]        'menubar', 'none');
[7]    counter = uicontrol('style', 'text', ...
[8]        'units', 'centimeters', ...
[9]        'position', [.5,1,4,3], ...
[10]       'string', '0', ...
[11]       'fontsize', 40);
```

Create figure — [3]

Set size and position — [5]

Create text display — [7]

Set the text — [10]

To use `trafficFig`, change line 8 of `trafficTimer` to read `[f,c]
= trafficFig`. The return value `c` holds a handle to the text display
object in the figure. To update the display in the course of data entry,
we use `set(c,'String',` _____ `)`. For example, to display the
number of vehicles, we could place the following statement on line 20 of
`trafficTimer`:

```
set(c, 'String', num2str(length(times)))
```

The resulting display, after 16 events have been entered, is shown in the
margin.

Another useful elaboration on `trafficFig` would be to save the collected data to a file. It would sensible to use a standard file format such as a
spreadsheet so that the data can be read by other software.

10.3   Example: Exploring the Mandelbrot Set

In Chapter 8, we wrote a program to draw the Mandelbrot set within a delimited region.

```
≫  xlims = [-2.5, 2.5];
≫  ylims = [-2, 2];
≫  drawMandelbrot([xlims, ylims)
```

Here we will write a program, `mandelbrotGUI`, that is based on
`drawMandelbrot` and allows a user to zoom in on interesting parts of the
set, exploring it in more detail.

The basic idea will be to display the Mandelbrot set and allow the user
to mark a small region of interest with the mouse. Then, `drawMandelbrot`
will be used to draw the set within this small region. The sequence will be
repeated so that the user can zoom in on more and more detailed areas of
the set. It's also helpful if the user can pan out, displaying the set over a
larger region. And, of course, the user needs to be able to quit the program.

There are two types of user input required here:

- Which action—zoom, pan, quit—the user wants to take. This information can be acquired with a menu display.
- If zooming or panning is requested, which region to zoom in on or
  pan out to. This information can be acquired with the `getrect` operator.

We can simplify things for the user by arranging for the pan action to undo
the previous zoom action. So, instead of having to specify a pan region, we
need only keep track of what regions were previously displayed.

Key Term

A simple way to keep track of the regions is a *stack*. A stack is a data structure with two operations: *push* and *pop*. Pushing an item onto the stack simply adds it to the stack. Popping an item from the stack removes the last pushed item from the stack and gives that item as the return value. For mandelbrotGUI, on each zoom, we will push the new *x*- and *y*-limits of the display, and on each pan, we will pop the previous limits.

Here is mandelbrotGUI, which implements the interface described previously:

```
[1]   function mandelbrotgui
[2]   % mandelbrotGUI: graphical user interface
[3]   stack = {[-1.8,1.8,-1.3,1.3]};
[4]   f = figure;
[5]   while 1
[6]       coords = stack{end};
[7]       drawMandelbrot(coords(1:2), coords(3:4));
[8]       choice = menu('Which one?', ...
[9]           'Zoom', 'Pan Out', 'Quit');
[10]      switch choice
[11]      case 1
[12]          m = getrect(gca);
[13]          stack{end+1} = newcoords(m);
[14]      case 2
[15]          if length(stack) > 1
[16]              stack = stack(1:(end-1));
[17]          else
[18]              stack{1}=expandview(stack{1});
[19]          end
[20]      case 3
[21]          break;
[22]      end
[23]  end
[24]
[25]  %%%%%%%%%%%%%%%%
[26]  function res = newcoords(m)
[27]  res = [m(1), m(1)+m(3), m(2), m(2)+m(4)];
[28]
[29]  %%%%%%%%%%%%%%%%
[30]  function res = expandview(coords)
[31]  midx = mean(coords(1:2));
[32]  midy = mean(coords(3:4));
[33]  lenx = abs(diff(coords(1:2)));
[34]  leny = abs(diff(coords(3:4)));
[35]  res=[midx+lenx*[-1,1], midy+leny*[-1,1]];
```

Margin annotations:
- Bring up a figure [4]
- Menu input [8]
- Mouse input [12]
- Push [13]
- Is there something to pop? [15]
- Then pop [16]
- Subfunction [26]
- Subfunction [30]

10.4   Inputs without Waiting

The previous programs for activating input devices and handling the events generated by them all used the same form: There is a single function that solicits one event at a time; the function waits for that event (e.g., with `waitforbuttonpress` or `menu`), and when the event occurs the event is processed. While waiting for the event, no computations are performed, and no other actions taken.

This form has the advantage of keeping events in a strict sequence. In `mandelbrotGUI`, for instance, the mouse could not be used for marking a region until the zoom/pan/quit menu selection was pending.

The single-function form of user interface is relatively simple to program. For simple kinds of input (selecting from a set of alternatives or selecting a region in a graphics display) it can provide an effective interface. But there are often situations where inputs take on more varied forms. For instance, in a word processor, you might use the mouse to select a region of text (click and drag), to place the insert cursor at a particular point in the text (simple click), or to activate any of numerous menu items. At the same time, you can be typing (providing input via the keyboard). Other actions might be going on in the background, such as saving text to a backup file or printing.

We can imagine a word processor that uses the one-event-at-a-time style: The user would first indicate whether she wants to insert text, mark a region, or so on. Then the mouse would be activated for that purpose. The first word processors actually worked this way (although without a mouse). It is not necessarily old-fashioned—the professional-level text editor EMACS does not need the mouse at all and simply solicits and handles text input. For instance, to move to the previous line in a body of text, one types CNTL-p. But most users consider programs like EMACS unfriendly and too difficult to use.

Or, consider a computer game. Even when no user inputs are being given, the state of the system (the location of the enemy entities in a shoot-'em-up game, for instance) is changing. If computation needs to wait for the next event, how does the state change between events?

In a no-waiting interface such as used by a word processor or a game, many different types of events might occur at any time. To deal with complex situation, a different style of programming is called for. Rather than a single function handling all the possible events, a set of functions is created with one function for each type of event. When an event occurs, the corresponding function—termed a *callback function*—is invoked. The callback function can change the state, update the display, generate new events, and so on.

Key Term

This form of system, where control is distributed across many functions, can greatly simplify the process of handling events. New types of events can be added by creating a new callback function and without necessarily altering any existing function. Nonetheless, there are significant challenges:

- Maintaining the state of the system in a way that is accessible by each of the many callback functions that are needed to read and update the state. In MATLAB, whose scoping rules don't allow functions to access data in another function's environment, this often means using global variables. In other languages, references would be a more appropriate way to allow access.
- Ensuring that each callback function updates the state in a valid way and that the display is consistent with the state.

A Matter of Style:

Meeting the challenges of implementing a callback system can lead to a system that is easy to use. But the programming difficulties are substantial and the programs can become complex. The single-function approach does create a clumsy and limited user interface, but it usually is much easier to write. The user interface examples in the rest of this book employ the single-function approach.

But the flexibility and power of the callback style have tremendous advantages when implementing a user interface. Commercial software almost always uses this style for user interfaces, and most computer programming systems support the callback style.

In deciding between the two for your own work, consider whether a simple but clumsy interface will do the job for you, whether the interface will be used by others, and whether it will be used frequently or just once in a while. In particular, for prototype software, it may be best to use the single-function style until you have determined exactly what functionality you want to implement. But there is one situation when the callback style is required: when computation must go on all of the time, even when there are no events. This is the case in computer games and other, perhaps more scientifically oriented, simulations.

As a simple example, consider a temperature converter as shown in Figure 10.2. The figure window is constructed by putting together several user interface control objects in the function tempConvFig:

```
[1]  function [f,c,m] = tempConvFig
[2]  % GUI for temperature converter
[3]  fig = figure(1);
[4]  set(fig,'units', 'centimeters',...
[5]      'position', [5,5,5,3],...
[6]      'menubar', 'none');
[7]  f = uicontrol('style', 'edit',...
[8]      'units', 'centimeters', 'position',...
[9]      [.5, 2, 2.5, .5], 'string', '') ;
```

with margin labels:
- Create figure — line [3]
- Set size and position — line [5]
- Create editable display — line [7]

Label for F box	[10]	`flabel = uicontrol('style', 'text',...`
	[11]	`    'units', 'centimeters', 'position',...`
	[12]	`    [3.25, 2, 1, .5], 'string', 'Fahr');`
Create editable display	[13]	`c = uicontrol('style', 'edit',...`
	[14]	`    'units', 'centimeters', 'position',...`
	[15]	`    [.5, 1.25, 2.5, .5], 'string', '');`
Label for C box	[16]	`clabel = uicontrol('style', 'text',...`
	[17]	`    'units', 'centimeters', 'position',...`
	[18]	`    [3.25, 1.25, 1, .5], 'string', 'Celsius');`
Display for messages	[19]	`m = uicontrol('style', 'text',...`
	[20]	`    'units', 'centimeters',...`
	[21]	`    'position', [.5, .5, 4, .5],...`
	[22]	`    'string', '', 'ForegroundColor', [1, 0, 0]);`

This function has been written to return three function handles: the parts of the display that will need to be updated in the temperature conversion process.

```
>> [f,c,m] = tempConvFig;
```

The returned arguments are handles to the fahrenheit and celsius text entry boxes and to a third display for messages.

These text entry boxes already have functionality; you can type and edit inside them in the familiar way. But doing so does not affect the other elements of the display. To implement this, we have to provide callback functions. We need a callback function for each type of event that might be generated. In our case, there are two such events: one for text entry in the fahrenheit box and one for text entry in the celsius box. These events will be triggered when the ENTER key is pressed in the text box, or when the screen cursor moves from inside the text box to outside it.

The first step in writing a callback function is to define the state of the system that the callback function will operate on. The callbacks will need to read the information in their own text entry box, convert this to the format

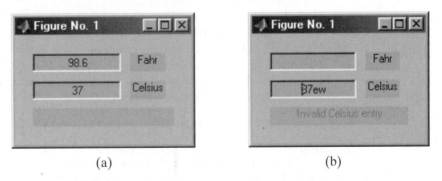

(a)   (b)

Figure 10.2. The temperature conversion graphical-user interface described in the text. (a) A successful conversion. (b) An error.

for the other box, and perhaps set the message text. To do this, the callback needs to be able to access the f, c, and m handles—no other information is needed. So the state will be the f, c, and m handles.

In order for the callback functions to be able to access the state, we will use a global variable, accessible from any function. This is a simple approach but incurs all of the problems described in Section 9.3.

```
≫  global tempConvState;
≫  tempConvState.f = f;
≫  tempConvState.c = c;
≫  tempConvState.m = m;
```

Here is the callback function to process text entry in the fahrenheit box. It reads the text in the fahrenheit box, converts the text to a number and the number from fahrenheit to celsius, converts the celsius number to a text string, and writes this in the celsius box. If the fahrenheit string cannot be successfully converted, a suitable message is displayed.

```
[1]   function fahrenheitEntry
[2]   global tempConvState;
[3]
[4]   s = get(tempConvState.f, 'String');
[5]   ftemp = str2num(s);
[6]   if isempty(ftemp)
[7]       cstr = '';
[8]       mstr = 'Invalid Fahr entry';
[9]   else
[10]      cstr = num2str( 5*(ftemp-32)/9 );
[11]      mstr = '';
[12]  end
[13]  set(tempConvState.c, 'String', cstr);
[14]  set(tempConvState.m, 'String', mstr);
```

Convert to a number — line [5]
Conversion unsuccessful — line [6]
Update the display — line [13]

A similar callback, celsiusEntry, is written to process events arising from the celsius text box. No callback is needed for the message text since no events are generated there.

At this point, the system can be operated manually by invoking the callbacks on the command line. The callbacks are like any other function. In order to invoke them automatically when text is entered, we need to set the Callback property of the text boxes.

```
≫  set(f,'Callback', 'fahrenheitEntry');
≫  set(c,'Callback', 'celsiusEntry');
```

Ordinarily, all of these steps for creating the figure window, setting up the state of the system, and setting the callbacks would be packaged up into a function. Here is one:

```
[1]  function tempConvGUI
[2]  %  a simple temperature conversion window
[3]  global tempConvState;
[4]  [f,c,m] = tempConvFig;
[5]  tempConvState.f = f;
[6]  tempConvState.c = c;
[7]  tempConvState.m = m;
[8]  set(f,'Callback', 'fahrenheitEntry');
[9]  set(c,'Callback', 'celsiusEntry');
```

Set up the state — lines [3]–[7]

Set callbacks — lines [8]–[9]

Invoking this function displays the temperature-conversion window. Note that the command prompt returns immediately after the function is invoked. The temperature-conversion window is still active, and you can also use the command line for any other purpose. Editing text in one of the temperature windows invokes the appropriate callback just as if you had typed the callback at the command prompt. The callback style of interface allows multiple events to be processed, even in different windows.

There are many software tools available to simplify the process of writing a graphical-user interface. MATLAB includes a tool called guide that makes it easy to lay out the user interface controls, set their properties, and set up callbacks. Guide has a particularly clever system for organizing callback functions and passing state information to them. This is well described in the MATLAB documentation and other sources [22].

10.5 Exercises

Exercise 10.1:
The resolution of an ordinary wall clock is one second: The clock ticks once a second, and it's not possible to read off fractional seconds. The system clock on a computer ticks somewhat more often. Write a program to determine the resolution of the system clock. You can do this by calling now repeatedly in a loop and finding the smallest nonzero change in time between successive calls.

Exercise 10.2:
Modify trafficTimer to add

1. A command for saving the collected data to a file.
2. A means to add annotations to the data by selecting from a menu list of standard annotations (e.g., 'sunset', 'rainstorm', 'accident').

Exercise 10.3:
Make a table of the values given by get(gcf,'CurrentChar') for the noncharacter keys (e.g., delete, insert, f5, and so on).

Exercise 10.4:

Modify `mandelbrotGUI` to change the cursor to a watch when the computation is being performed. (*Hint:* Each figure has a `Pointer` property. Use `set` to turn this into a watch before the computationally intensive part of the Mandelbrot calculation. Set it back to its original form when the computation is complete.)

Exercise 10.5:

Add to the Mandelbrot GUI features to change the resolution of the display (the number of points in the *x*- and *y*-directions) and the number of iterations. Save the previous display values so that pan can go faster and allow users to mark a region as interesting and to save the interesting *x*- and *y*-regions, which will then be returned (along with the calculated values?) when the function exits.

Exercise 10.6:

Modify the ECG display program you wrote in Section 9.10 so that a long ECG can be scrolled forward and backward in time by pressing the right-arrow and left-arrow keys. The file `ecglong.mat` contains an ECG sampled at 360 Hz. (This is record 102 from the MIT/BIH database in `www.physionet.org`.)

Exercise 10.7:

Construct a GUI window with a single button labelled "Yes." When the button is pressed, the label should toggle between "Yes" and "No."

Exercise 10.8:

We often need to convert between numbers and character strings when reading or writing formatted information. For example, suppose we have a variable `tempF` that holds a numerical value, and we wish to display a message like "The temperature is 38.3 deg F." This involves both converting the numerical value to an ASCII string and combining it with the other words in the message. The `sprintf` operator provides a powerful way of doing this. `Sprintf` takes two or more arguments: The first is always a character string describing the desired format; the remaining arguments are the values to be included in the formatted string. For example,

```
>> sprintf('The temperature is %6.3f deg F', tempF)
   ans: The temperature is 38.3 deg F.'
```

In the `sprintf` format string, the `%f` indicates where to insert the formatted number and provides certain details about how it should be displayed (in this case, six spaces with three decimal digits). There is a number of special codes that can be used in the `sprintf` format string. These are described in the `help` documentation. Particularly important ones are `%d` for integers, `%s` for strings, and `\n`, which inserts a newline character into a string. For example,

```
>> sprintf('line 1\nline 2')
  ans: line 1
       line 2
```

Even though there are no % conversion characters in the format string, sprintf still parses the format string and replaces the \n with an ASCII newline character.

Use sprintf to write formatted strings in the following cases:

- A variable t containing a numerical temperature and a variable u containing a character string of units ('F' or 'C' formatted as "36 deg C."
- A vector of counts and a corresponding cell array saying what the counts refer to, for example, counts = [4 5 2] and names = {'cars', 'trucks', 'buses'} formatted as

```
    4 cars
    5 trucks
    2 buses
   -------------
   11 TOTAL
```

CHAPTER 11

Arranging Data: Searching and Databases

Scientists and engineers work extensively with data. They are not unique in this respect: Businesses, governments, and public and private organizations all need to keep track of data for their various purposes—billing, taxation, medical records, even something as mundane as locating library books. None of these organizations could work effectively without data and the ability to make use of it. Computers have genuinely revolutionized the ability to collect and store data and to make use of it.

This chapter is an introduction to some of the ways of organizing data and of extracting information from it. The subject is of great importance in everyday modern life, and there is, correspondingly, an elaborate technology of handling data and sophisticated software that implements much of the technology. Much of this software goes under the term database management, and those who work extensively with large sets of data are well advised to learn one of the advanced software packages that are widely available.[1]

The purpose of this chapter is not to provide instruction in a specific database package. It will introduce the concepts of data organization to a degree sufficient to allow you to manage the small- or moderate-sized datasets that individuals commonly encounter in scientific or technical work and emphasize the importance of advanced database systems in managing large datasets.

---

[1]A freeware database package, MYSQL, is available. There are numerous texts and manuals about databases, for example [13, 14, 16].

Consider a telephone book. This common artifact of the early information age, found in just about every household, seems to have been rendered obsolete by computers, which make looking up a telephone number faster and more flexible. But before you contribute your telephone book to a technological recycling heap, let's exploit it for what it tells us about how to organize data. The lessons to be learned are important: It is proper organization of data and efficient algorithms that exploit this organization that allow for the effective searching of data. Without proper organization, a computer search would be every bit as slow as using a telephone book by hand.

A phone book is, first and foremost, a collection of data: a list of names, addresses and phone numbers. A book for a large city might have half a million entries. Each entry looks something like this:

Bassingthwaighte, Curtis E, 11 Northdale 483-555-6815

This entry has about 50 characters, not including formatting. At this rate, there would be 25 million characters in a large phone book: an appreciable amount of data even by the standards of modern computing.

Beyond merely containing all of this information, the entries in a phone book are arranged to facilitate a particular sort of computation: taking a name as input and returning a telephone number as output. Performing this involves first locating the desired name in the book and then reading off the juxtaposed phone number.

The primary skill used in searching a phone book—beyond reading—is being able to alphabetize. The basic computation involved in ordering data in alphabetical order is this: Given two names as input, provide as output one of three conclusions: (1) the two names match; (2) the first name comes before the second; or (3) the first name comes after the second. For instance, given Boyle and Stokes as the two input names, the output is 2 since Boyle comes before Stokes. Human alphabetizers commonly add another output component to the alphabetization computation: how far the first name is from the second—Boyle is far from Stokes. But, as we shall see, this additional information does not improve performance very much.

A second important skill is page flipping. Typically, a person will flip rapidly through the pages, stopping occasionally to compare the searched-for name to the indexed names written in the top corner of each page. Alphabetization is used to decide whether to reverse the direction of flipping. At each direction change, the flipping speed is reduced a bit so that page indexed names are examined more frequently.

Once an appropriate page has been found, a similar procedure is followed with page flipping replaced by scanning a finger along the columns of names on the page, jumping from place to place in a direction set by the output of the alphabetization computation and again with the jumping distances getting smaller with each reversal of direction.

The phone book is designed specifically for this type of searching. Performing the operation of searching for a name takes approximately 30 seconds. This is a remarkable fact for two reasons. One is that 30 seconds is a much shorter time than required to read even a single page of the book; evidently searching a phone book doesn't involve much reading. Even more remarkably, the estimate of 30 seconds is unqualified: It doesn't refer to any particular size of phone book. It takes only a few seconds more to search a very large phone book and only a few seconds less to search a very small one.

Yet to use the information in a phone book for a different purpose, say searching for a name given a telephone number, takes so long as to be completely impractical. All of the required information for such a search is there in the book, but the data have not been arranged in a way that they can be used efficiently.

This chapter is concerned with the principles of organizing and searching data in order to allow it to be used for the purposes you have in mind.

11.1 Datasets

Key Term

A *dataset* is a collection of items that share a common property. Examples include the names of students enrolled at a university; the medical records of patients involved in a clinical trial of a new drug; the set of soil samples collected on a field expedition; the values of wind velocity measured every minute at a weather station; newspaper articles published in the past decade; and answers to questions provided by a randomly selected group of people for a political poll.

There are two basic types of operations that we perform on datasets:

- Selection of subsets. This refers to picking those items out of a collection that satisfy some further criterion. For instance, those newspaper articles containing the phrase "vote fraud" or the political polling answers given by women.
- Summarizing. Condensing the information in the dataset, or a selected subset of the dataset, in some way. Perhaps the simplest form of summarization is counting the number of items, but there is a huge variety of ways of summarizing, and the best choice depends on the particular question you want to answer with your data.

Key Term

A *table* is a dataset for which there is one or more kind of information provided about each element in the set. The vocabulary to refer to a "kind of information" is diverse and often discipline specific; some of the words used in different disciplines are "measurement," "variable," "value," "property," and "column." In this book we use the word "field." Similarly, various words are used to refer to the elements in the dataset: "case," "row," "tuple," and so on. We use the word *item*, understanding that each item is the collection of values for each of the fields in the set.

Key Term

The conventional printed format for a table is a rectangular array: one line for each item and one column for each field. For instance, in a telephone book there are typically three fields: (1) the person's name, (2) the address, and (3) the telephone number.

Key Term

A dataset with a single field is called a *list*. Lists are, of course, common in everyday life. They help us to keep track of what items are in a set (for instance, what we need to buy at the grocery store, which students are enrolled in a course, or who we want to invite to a wedding). As simple as lists are, they provide the foundation for the use of more complicated datasets with more than one field.

Representing datasets on the computer first involves figuring out how to represent the values in each field. Names and phone numbers are simple enough; they can be stored as character strings. But other types of data (for instance, photographs, sound recordings or soil samples) need to be transformed into computer-storable information. Photographs and sound recordings can be digitized, but something like a soil sample requires some thought. The soil sample itself cannot be stored on the computer, only information about it such as where and when it was collected, a description of the sample in geological terms, and perhaps an indication of where the sample itself is stored.

A simple, but important, data-organizing principle is that every element in one field of a dataset should be stored in the same format. For instance, if the field contains telephone numbers and it has been decided to represent telephone numbers as character strings, then every element should be stored as a telephone string. It becomes difficult if different elements have different formats (for instance, if some of the telephone numbers are stored as strings and others as floating point numbers). The precise meaning of "the same format" is actually rather subtle. As we'll see, it ultimately is defined in terms of the operators that act on the elements of the field: So long as the operators appropriately can handle the information stored about every element, the format is effectively the same for all of the elements, even if details differ.

Figure 11.1 shows a small table containing telephone-book data. We've stored these data using a spreadsheet format, which facilitates entering and editing the data. Although we use MATLAB commands to process the data in order to demonstrate basic principles, it should be kept in mind that these same principles of data processing apply in other data-oriented software, and that in many cases it is much more appropriate to use such software rather than MATLAB.

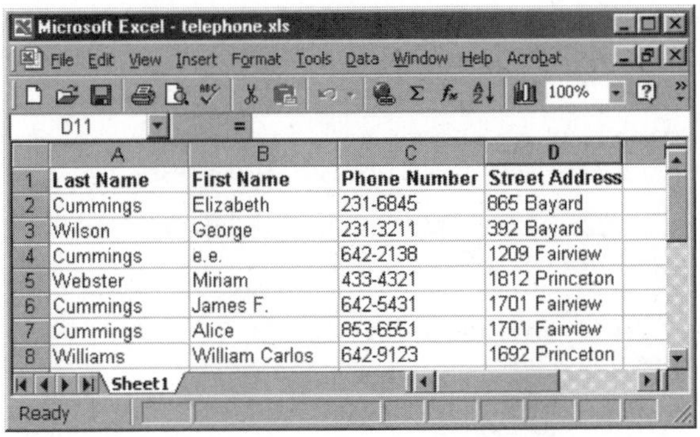

	A	B	C	D
1	Last Name	First Name	Phone Number	Street Address
2	Cummings	Elizabeth	231-6845	865 Bayard
3	Wilson	George	231-3211	392 Bayard
4	Cummings	e.e.	642-2138	1209 Fairview
5	Webster	Miriam	433-4321	1812 Princeton
6	Cummings	James F.	642-5431	1701 Fairview
7	Cummings	Alice	853-6551	1701 Fairview
8	Williams	William Carlos	642-9123	1692 Princeton

Figure 11.1. A table of telephone data stored in spreadsheet format in the file `telephone.xls`.

The telephone table can be imported to MATLAB using the `importdata` operator. In this case, `importdata` will arrange the data as a cell array, with one element for each of the spreadsheet cells:

```
≫ telephone = importdata('telephone.xls')
  ans:
   'Last Name'   'First Name'       'Phone Number'   'Street Address'
   'Cummings'    'Elizabeth'        '231-6845'       '865 Bayard'
   'Wilson'      'George'           '231-3211'       '392 Bayard'
   'Cummings'    'e.e.'             '642-2138'       '1209 Fairview'
   'Webster'     'Miriam'           '433-4321'       '1812 Princeton'
   'Cummings'    'James F.'         '642-5431'       '1701 Fairview'
   'Cummings'    'Alice'            '853-6551'       '1701 Fairview'
   'Williams'    'William Carlos'   '642-9123'       '1692 Princeton'
```

Note that the first line of the spreadsheet has been used for column labels: We separate these labels from the body of the dataset.

```
≫ labels = telephone(1,:);
≫ data = telephone(2:end,:);
```

11.2 Selecting Subsets from Tables

Key Term

Selecting a subset from a table involves applying a *selection criterion*. The criterion describes whether or not to include any given item in the table: It can be implemented as a function that takes a single item as input and returns as output a boolean that indicates whether to select that item.

For instance, suppose it is desired to select all of the people whose last name is Cummings. The following, special-purpose function can be written:

```
[1]  function res = isCummings(item)
[2]  res = strcmp(item{1},'Cummings');
```

The isCummings function takes a single item—one row of the data cell array—and returns a boolean:

```
≫  isCummings(data(1,:))
   ans:  1
≫  isCummings(data(2,:))
   ans:  0
```

Using a criterion function like isCummings to select out a subset from a table involves applying the criterion function to every item in the table, one at a time, and then keeping those items for which the criterion indicates a positive result. This is a very generic operation, so we can write a generic function that will do this for all criteria that we might like to apply.

Here is the function selectFromTable that takes two arguments: a list and a criterion function:

```
[1]   function subset=selectFromTable(table,criterion)
[2]   % selectFromTable(list,criterion)
[3]   % select a subset from a table
[4]   % using the specified criterion
[5]   % Returns the indices of the selected items
[6]   [nitems,nfields]=size(table);
[7]   selected=zeros(nitems, 1);
[8]   for k=1:nitems
[9]       selected(k)=feval(criterion,table(k,:));
[10]  end
[11]  subset = find(selected);
```

Boolean: which items were selected

Fill in the boolean for each item in the list

Find the indices of the 1s in the boolean list

Let's try it:

```
≫  selectFromTable(data,'isCummings')
   ans:  1
         3
         5
         6
```

Note that selectFromTable does not return the subset itself, but rather, a set of indices that can be used to extract the subset. The reasons for this will become clear soon, but for now it's sufficient to note that the indexing brackets can be used to turn the list of indices into the subset of the table:

```
≫ data(selectFromTable(data,'isCummings'), :)
  ans:
    'Cummings'   'Elizabeth'   '231-6845'   '865 Bayard'
    'Cummings'   'e.e.'        '642-2138'   '1209 Fairview'
    'Cummings'   'James F.'    '642-5431'   '1701 Fairview'
    'Cummings'   'Alice'       '853-6551'   '1701 Fairview'
```

The simple function isCummings illustrates one of the most important types of criterion functions: the exact match. Often, though, our interest matches with a more subtle criterion. For example, suppose we wish to select the people living on Princeton Street. The street address field of the table contains the name of the street, along with other information. Using strcmp to compare 'Princeton' to the contents of the fourth element of each item will not do the job. Instead, operators such as findstr need to be used that can look for 'Princeton' within a string that contains other data. (See Exercise 11.2.)

When there are no items in the table that satisfy the given criteria, the output from selectFromTable is the empty vector, which conventionally **Key Term** is called the *empty set*.

11.3   Combining Tables: Union and Intersection

Often we have two or more lists that need to be combined in some fashion. The lists might have come from separate sources, or they may have resulted from separate queries using different selection criteria on a single list.

For example, suppose we wish to find a person whose last name is Cummings, living on Bayard street. One way to do this would be to write a criterion function that checks for both sorts of matches. Another way is to find first the last-name matches to Cummings, and then the street-address matches to Bayard and then combine the two. The indices of the matches to Cummings are [1;3;5;6]; the street-address matches to Bayard are **Key Term** [1;2]. The indices that match both criteria are the *intersection* of the two sets of indices

```
≫ intersect([1;3;5;6], [1;2])
  ans: 1
```

The intersection of two sets is the set of items that appear in both sets.
Key Term The *union* of two sets is the set of items that appear in either of the sets. For instance, if we wanted the list of people whose last name is Cummings

or who live on Bayard Street, we would take the union of the two sets of indices:

```
≫ union([1;3;5;6], [1;2])
  ans: 1
       2
       3
       5
       6
```

Note that although index 1 is in both sets, it does not appear twice in the union.

When more than two sets are involved, the result can be computed by combining the sets in successive pairs. For example, the union of three lists, one, two, and three, can be computed by union(one, union(two, three)). The intersection works similarly.

There are other operations on pairs of sets that are also important. See the built-in operators ismember, setdiff, and setxor, as well as the single-list operator unique.

11.4 Databases

Tables are a wonderfully useful organizing scheme for data, as evidenced by the tremendous success of spreadsheet programs that store data as a rectangular array of cells. Common sense tells us that we need separate tables for separate purposes: It seems silly to integrate a phone book and a grocery list into a single table.

Key Term

A *relational database* is a set of tables, all relating to a common purpose. One of the important skills in designing a database is to recognize when it's best to use multiple tables rather than a single table to represent data. The principal reason for doing this is to avoid whenever possible the redundancy of storing the same information in different places. A small problem with redundancy is that it wastes space. Much more significant is that redundancy introduces the possibility of error and inconsistency. The items intended to be the same may fail to be so, perhaps because one of the items has been updated but the others have not been.

Consider, for example, the census information described in Chapter 4, where we sought to keep track of several items for each household: the address, the income, and the names, ages, sex, and profession of the household members. In Chapter 4, we organized this information as a single table with one row for each household; the fields were address, income, first names, and so on. While all of the information could be stored in this manner, accessing the information was sometimes difficult.

In this section, we show how the census data can be stored as a relational database. This provides a much more flexible way of arranging the data, but to use the database, we must first master special table operators, such as "join," that will be introduced in the next section.

The usual pattern in this book has been to use built-in MATLAB operators or to construct our own operators in MATLAB. We will deviate from that pattern here. Database operators are not included in MATLAB. There is not even a standard, general-purpose data structure for tables that fits in well with relational operators.

In principle, it would be possible to construct such data structures and operators using MATLAB primitives. There is little point in doing this, however, since commercial and freeware database systems are readily available, such as MYSQL. These systems will give superior performance, reliability, portability, and flexibility compared to a home-brewed MATLAB-based system. Database systems are of such great economic importance that literally tens of thousands of man years of development have gone into them. There is even a standardized way of communicating with databases, called

Key Term *Structured Query Language (SQL)*, that works with many commercial and freeware systems.

It would take much less time to learn to use a commercial or freeware relational database system than to program even a simple one in MATLAB. The skills gained will be portable. If one needs to access complicated data from MATLAB, the proper approach is to set up data communication between a database system and MATLAB. This can be done in many ways; there is even a MATLAB toolbox for database communications.

In this section, we do not attempt to teach the syntax of SQL or other skills of database use and management. (There are many texts that cover these matters, e.g., [16].) Instead, we want to introduce a few selected concepts of relational databases. Our main objective is to show how a relational database can provide flexible access to data. We hope to convince you that when it comes to complex forms of data, you should use appropriate database tools.

Operators on Tables

The basic unit of a relational database is a table. The nomenclature that we have been using for tables has been informal: We've used "item" to denote

Key Term one row of a table, and *field* to denote one column. While we will stay with this nomenclature, be aware that database management, like many specialized areas of technology, has its own terminology. In relational databases, a table is called a relation, an item is called a tuple, and a field is called an attribute.

When data in a relational database are accessed, the result is always returned in the form of a table. The transformation of database tables into new tables involves operators: Each operator takes one or more tables as input and returns a table as output.

Key Term

We have already encountered one such operator. When we pull items from a table that meet a specified criterion, we are performing a *selection*. For example, if we select from the telephone number table on page 255 using the criterion that the telephone number matches 433 4321, the returned table is

Last Name	First Name	Phone Number	Street Address
Webster	Miriam	433 4321	1812 Princeton

It may seem grandiose to call such a thing a table, since there is only one item, but it is indeed in the format of a table with four fields and one item. We could even have an empty table; that is, a table with no items.

Key Term

The *projection operator* creates a new table with only the fields specified from the input table. For example, projecting the telephone table onto the Last Name field would return

Last Name
Cummings
Wilson
Cummings
Webster
Cummings
Cummings

Such operators can be combined. For example, looking up Miriam Webster's phone number involves a selection based on first and last name. The output of the selection is then projected onto telephone number, producing a simple table.

Phone Number
433 4321

Key Term

Another operator that takes a single table as input is *grouping*. When grouping, we divide all of the items in a table into sets; within each set, all of the items are identical in some defined way. Then a computation is performed on each of the sets. For example, grouping the telephone table over last name and then computing the count of items in each set, we get the following table:

Last Name	Count
Wilson	1
Webster	1
Cummings	4
Williams	1

In principle, any quantity can be computed on the sets; perhaps the most common quantities are simple ones such as the count, the mean, the minimum, and the maximum. In order to indicate which quantity is to be computed, we'll use the notation group/count, group/mean, group/min, and so on.

Some operators involve more than one table as input. Two such operators that we have already seen are "intersection" and "union." The intersection operator takes two tables as input and produces an output with all of the items that are in both input tables. The union operator is similar, but the output has all of the items that are in either table. Union and intersection cannot operate on any two tables: Both input tables must have the same fields. Such same-field tables are called *union compatible*. For example, the two inputs might be the result of two selections on the same table.

Key Term

A related operator is *minus*. If A and B are two union-compatible tables, then A minus B is a table with all of the items in A except for those that are found in B.

Key Term

Another important operator with two tables as input is *natural join*. When joining two tables, the items in one table are associated in some way with items in the other table. The output table has all of the fields in either table (that is the union of the fields, but this is different from the union of the items). To illustrate, consider the grocery "list," really a table that describes what we need to buy. We'll use the name "Grocery Item" to emphasize the fact that each row in the table is a single item; the set of rows is the grocery list.

Key Term

Grocery Item

Name	Needed	Gotten	Unit
cheese	1	1	lb.
milk	1	1	gal.
oranges	5	0	
kiwi	3	0	
baguette	1	1	
bananas	1	0	bunch
apples	3	3	lb.

An uncommonly organized shopper might also have a Store table that describes where to buy each item. Here is a short version:

Store

Commodity	Store	Price	Aisle
apples	Fruiteria	1.25	1
apples	Shop & Drop	1.49	18
baguette	Boul's Bakery	1.99	
baguette	Shop & Drop	2.49	9
milk	Shop & Drop	3.57	23
oranges	Fruiteria	0.40	1

We can join the two tables, setting as the association between items a match on the name field in the Grocery table and the commodity field in the Store table:

Natural Join: Grocery Item and Its Store Data

Name	Needed	Gotten	Unit	Commodity	Store	Price	Aisle
apples	3	0	lb.	apples	Fruiteria	1.25	1
apples	3	0	lb.	apples	Shop & Drop	1.49	18
baguette	1	0		baguette	Boul's Bakery	1.99	counter
baguette	1	0		baguette	Shop & Drop	2.49	9
milk	1	0	gal.	milk	Shop & Drop	3.57	23
oranges	5	0		oranges	Fruiteria	0.40	1

The joined table contains much redundant information. For example, the fact that we need three apples but have not yet gotten any appears in two different rows in the table. But by selecting and projecting, we can create another table with information in a useful format. For instance, selecting on Store equals Fruiteria and projecting onto Name, Needed, Gotten, Unit, Price and Aisle, we get a table that tells us what we should pick up while in the Fruiteria:

Name	Needed	Gotten	Unit	Price	Aisle
apples	3	0	lb.	1.25	1
oranges	5	0		0.40	1

Key Term

In a natural join of two tables, only those items that have a matching item in the other table are present in the output. In an *outer join* each item in either table appears in the output regardless of whether there is a match in the other table; this involves leaving blanks in the undefined fields of the unmatched items. For example, the outer join of the Grocery Item and Store tables will have rows for kiwi and cheese but blanks in the columns relating to the Store data.

▶

Example: A Census Database

One of the aspects of the census data in Chapter 4 that makes them hard to work with is that there are two different units of analysis. Census data are fundamentally about individual people: names, ages, genders, and so on. This suggests arranging the data as a table where each item is an individual person and each field is one of the variables of interest.

Person

PID	First Name	Middle Name	Last Name	YOB	Sex	Profession	HID
34342	Emily	Lisa	Smith	1965	F	Mathematician	76
74523	George	Robert	Smith	1976	M	Primary School Teacher	76
92734	Felicia	Grace	Smith	1993	F	Student	76
98342	Nancy	Emily Grace	Smith	1998	F	Student	76
24973	Wilbur	Chase	Bucket	1956	M	Postal Worker	83
22534	Monique	Joelle	Floquet	1955	F	Painter	83
67342	Sandy		Leopard	1978	M	Systems Manager	14
66583	Carmela	Ghere	Ghimenti	1976	F	Chef	61
96743	Jeremy	Ralph	Prosit	1995	M	Student	61
86739	Phillip		Leopard	1993	M	Student	14

We have added two new fields to the original census data. The PID is a unique identifier for each person. HID is a unique identifier for each household. No PID value is repeated in the table, but HID will be the same for all the people in one household.

Within this table, queries can be answered easily. For instance, we want to know how many people there are of each gender. Use group/count on the Sex field:

Sex	Count
F	5
M	5

Want to examine the age distribution of students? Select on Profession matches student and project onto year-of-birth (YOB).

YOB
1993
1995
1998
1993

From YOB, age is easily calculated. The information in the resulting table can be used to make a histogram.

Census data are collected from households—a group of one or more individuals living together. The fact that a household is a unit of analysis for certain queries suggests storing each household in a table of households, which would also include the associated address and income information. Such data are naturally arranged as a table, where each item is a household.

Household

HID	Number	Street	Other	HouseType	PostalCode	Income
14	2483	Crown St.	Apt 9F	Apartment	10323	65,000
61	2483	Crown St.	Apt 4G	Apartment	10323	36,000
74	11	Elm St.		Single	10486	62,000
83	10	Elm St.		Single	10486	56,000

Again, some queries can be answered easily within the one table. For instance, to compute the fraction of households with incomes $\leq 40,000$, we perform group/count on income $\leq 40,000$:

Income ≤ 40,000	Count
T	1
F	3

Some queries require information in both tables. For example, to look at household income as a function of the number of children and number of adults, we would like to make a table with one item for each household. The table should look like this:

Income Analysis

HID	Income	# of Children	# of Adults
14	65,000	2	2
61	36,000	0	2
74	62,000	1	1
83	56,000	1	1

Once the table has been constructed using database tools, we can use other modeling tools (such as those in Chapter 17) to explore the relationship between income and household size.

It's tempting to fall back on the familiar single table operators. "If only we had planned ahead when designing the Household table and added a # of Children field and a # of Adults field to that table! This would make generating the Income Analysis table a trivial matter of projection."

But this add-a-field approach is ultimately limiting. The "# of Children" field would involve having a fixed definition of "child" that might not be appropriate for all purposes. For instance, we might want to consider both the age of the person and whether they are a student. Or, suppose we decided that we wanted to include the gender of single parents in our analysis of income. That would involve adding yet another field to Households. We can't possibly anticipate at the time that we set up the database all of the queries we would like to make. Or, seen another way, if we rely only on single-table operators, then we will limit the ways in which our data can be used.

What's more, the expanded Household table would be hard to maintain. Each time a new person entered a household, or a child aged to an adult, the household information would need to be updated. The expanded Household table would contain information with the same content as the Person table. Ensuring that the redundant information is consistent is a difficult maintenance problem.

Instead, the approach used in relational database is to generate tables like Number in Household using the relational operators on simple tables that do not contain redundant information.

The relational operators provide the means to combine the information in the various tables. In order for the operators to be effective, though, we have to incorporate enough information into each item so that the relevant data from other tables can be located. In the Households table, for example, each household has been assigned a unique identifier stored in the HID field. There is a similar unique identifier for each person in the Person table: the PID field. Using these identifiers enables one table to refer to another, but without creating any redundancy of information. This is why there is an HID column in the Person table.

To generate the Income Analysis table, we can follow these steps:

1. We'll define, for the sake of simplicity, a child to be a person younger than 18 years. Projecting the Person table onto HID and Age < 18,

Person

HID	Age < 18
76	F
76	F
76	T
76	T
83	F
83	F
14	F
61	F
61	T
14	T

2. Select from the Child table for T in the Age < 18 field. Then group/count on HID. This gives the number of children in each household.

Number of Children

HID	Age < 18
76	2
61	1
14	1

3. Select from the Child table for F in the Age < 18 field. Then group/count on HID. This gives the number of adults in each household.

Number of Adults

HID	Age > 18
76	2
83	2
14	1
61	1

4. Join the Number of Children and the Number of Adults tables based on HID. Since some households don't have kids, there are some HIDs that are not in both tables. If a natural join were used, any households with no kids or no adults would not appear in the output. Instead, by using an outer join, we ensure that any household with either children or adults is in the output table.

Number of Persons

HID	# of Children	# of Adults
76	2	2
83		2
14	1	1
61	1	1

Outer Join

Number of Persons

HID	# of Children	# of Adults
76	2	2
14	1	1
61	1	1

Natural Join

5. Project Households onto HID and income. Join this with the Number of Persons table. This gives us the Income Analysis table.

With the capabilities afforded by the relational operators in mind, we look for ways to avoid redundant information when designing tables. For example, in the Household table, we did not record the city or town of the address, only the postal code. Another table, Postal Code, can store the relationship between postal code and town name—a join suffices to associate a town name with each household. The following Postal Code table includes additional geographic information (latitude and longitude) that wasn't part of the census information presented in Chapter 4:

Postal Code

Code	Town	State	Latitude	Longitude
55116	St. Paul	MN	44.95	95.25
10323	Coventry	NY	41.26	74.33
94321	Tunitas	CA	37.21	122.45

An important reason to use the relational operators to generate new tables in response to specific queries is to allow us to incorporate new information easily. Suppose, for example, that a school district wants to examine the average family income for each teacher's class.

We can easily imagine that the census database was designed without the school district's needs in mind. The school district, let us suppose, has a table with information about each student's class.

School Class

Student ID	School	Grade	Teacher ID
96743	Mann	2	74523
84732	Mann	6	35323
98342	Expo	K	64323
92734	Expo	3	54523
86739	Mann	6	35323

It probably is unrealistic to expect that the school district uses the same ID number as the census bureau. Translating from the school district's ID into a census bureau ID might itself be a significant database task for preprocessing the school district information. But once this is done, the relational operators can be used to generate a table that has the average family income of the students in each class.

◀

11.5   Efficient Searching for Matches

When searching for a match, the function `selectFromTable` works through each item in the table. This can be inefficient; it's possible to search a table without looking at every item. To convince yourself of this, think about looking up a name in a real-world phone book. It takes about 30 seconds to do this despite the fact that such books typically have hundreds of thousands of items. Clearly, no human is capable of reading so many items in just 30 seconds. The key to the efficiency of the phone-book search is, of course, the alphabetical order of the items in the table.

Putting the telephone table in alphabetical order is not difficult. The `sort` function will take a cell array of strings and return the sorted array. Here, for instance, we sort on the last name:

```
≫   lastnames = sort(data(:,1))
     lastnames:  'Cummings'
                 'Cummings'
                 'Cummings'
                 'Cummings'
                 'Webster'
                 'Williams'
                 'Wilson'
```

Of course, we want to have the entire table, not just the names, in sorted last-name order. This means rearranging the rows of the table so that the last names are in order. `Sort` will provide a set of indices that, if applied to the table, puts it in order:

```
≫   [trash, indices] = sort(data(:,1));
≫   indices
     indices: 1
              3
              5
              6
              4
              7
              2
```

Now, the entire table can be ordered as desired:

```
≫  byLastName = data(indices,:)
   byLastName:
```

'Cummings'	'Elizabeth'	'231-6845'	'865 Bayard'
'Cummings'	'e.e.'	'642-2138'	'1209 Fairview'
'Cummings'	'James F.'	'642-5431'	'1701 Fairview'
'Cummings'	'Alice'	'853-6551'	'1701 Fairview'
'Webster'	'Miriam'	'433-4321'	'1812 Princeton'
'Williams'	'William Carlos'	'642-9123'	'1692 Princeton'
'Wilson'	'George'	'231-3211'	'392 Bayard'

We have seen `sort` being used previously with number vectors, which it sorts in ascending numerical order. When `sort` is given a cell array as an argument, it sorts alphabetically rather than numerically, but the alphabetical order is based on ASCII. This means that characters such as the space are used to determine order, and that all uppercase letters come before any lowercase letters. For example, the ASCII-sorted list of first names is not right:

```
>> sort(data(:,2))
   ans: 'Alice'
        'Elizabeth'
        'George'
        'James F.'
        'Miriam'
        'William Carlos'
        'e.e.'
```

The lowercase `e.e.` should, in standard alphabetical order, come after `Alice`, not at the end of the list.

With the table in sorted order, we can apply a search strategy that is efficient in the same way as the familiar telephone-book search. But neither page flipping nor finger dragging are capabilities of a computer; we need to build the algorithm on a different basis.

The general principle of divide and conquer offers an effective strategy for a search-for-match algorithm. The inputs are the sorted column of the table, which we call a *list*, and the term to be used for an exact match, which we call a *search term*. The steps are as follows:

Key Term

1. Bound the list with two points. For a telephone book, these points are the front and back covers of the book; for a computer cell-array list the points are the indices `front=0` and `back=length(list)+1`. Note that the bounds are outside of the list to be searched.
2. Look at the item in the middle of the two bounds. For a telephone book, this involves opening the book to a point halfway between the bounds: a very fast operation. For a cell-array list, this is the list item at the index `middle=floor((front+back)./2)`.
3. Compare the middle list item from step 2 to the search term. An alphabetical comparison has three possible outcomes:

   (a) The search term matches the middle list item. We have found the item we are looking for. Return `middle`, the index of the matching list item.
   (b) The search term is alphabetically before the middle list element. We haven't found a match, so we need to keep searching. But we need only to search that part of the list between `front` and `middle`; the rest of the list can be safely ignored. So update the bounds by setting `back=middle` and continue at step 2.

Iteration 1		Iteration 2		Iteration 3		Result
front						
	Cummings		Cummings		Cummings	
	Cummings		Cummings		Cummings	
	Cummings		Cummings		Cummings	
middle	**Cummings**	front	**Cummings**	front	**Cummings**	
	Webster		**Webster**	middle	**Webster**	**Webster**
	Williams	middle	**Williams**	back	**Williams**	
	Wilson		**Wilson**		Wilson	
back		back				

Figure 11.2. Binary search of the sorted last-name list for "Webster". The indices `front` and `back` define a subset of the list, shown in **boldface**. At each iteration, the element at `middle` is compared to the search term and `front` or `back` updated accordingly.

(c) The search term is alphabetically after the middle list element. We therefore need to search the second part of the list, which is accomplished by updating the bound by setting `front=middle` and continuing at step 2.

If the search term is in the list, this algorithm is bound to find it; eventually we will reach step 3a and terminate the algorithm. If the search term is not in the list, the front and back bounds will move closer together at each iteration of steps 2 and 3. Eventually, there will be nothing between `front` and `back` and so it will be clear that the search term is not in the list. This situation can be identified when `back-front==1` in step 2. When this situation holds, the algorithm should be stopped and the empty vector returned to signify that the search term was not found in the list. Figure 11.2 shows the progression of `front`, `back` and `middle`. The state of the algorithm at any step is given by the values of `front`, `back`, and `middle`. In iteration 1, the middle item, `'Cummings'`, is compared to the search term, `'Webster'`. Since `'Webster'` is after `'Cummings'` in the ASCII ordering, `'Cummings'` becomes the new front item at the start of iteration 2. In iteration 2, `'Webster'` is compared to `'Williams'`, and since `'Webster'` is before `'Williams'`, the middle item becomes the new back item. In iteration 3, there is a match between the middle item and the search term `'Williams'`; the algorithm terminates.

The function `findInSortedList` implements the preceding algorithm. In addition to taking as arguments the list and the search term, there is a third argument that specifies how to compare any two terms. The comparison function takes two terms as arguments and should return either 1, 2, or 3, depending on whether the first term matches the second or comes before or after it.

```
[2]  % findInSortedList(list, searchterm, compfun)
[3]  % Returns indices of items from <list> that
[4]  % match <searchitem>, according to <compfun>.
[5]  % <compfun> is a function of two arguments
[6]  % that returns:
[7]  % 1: if the two arguments match
[8]  % 2: if arg1 is before arg2
[9]  % 3: otherwise
[10] front = 0;
[11] back = length(list)+1;
[12] res = [];
[13] while back-front > 1
[14]     middle = floor((front+back)./2);
[15]     switch feval(compfun,searchterm,list{middle})
[16]     case 1
[17]         res = middle;
[18]         break;
[19]     case 2
[20]         back = middle;
[21]     case 3
[22]         front = middle;
[23]     end
[24] end
```

Item at `middle` matches [16]

We're done! [18]

It comes after search term [19]

It comes before search term [21]

An important part of the algorithm is the comparison between two terms done in step 15. This comparison is based on the output of the `compfun` function. We consider two such functions here: `alphabetcompare` and `asciicompare`. Both functions take arguments and return outputs in a standard format for compatibility with `findInSortedList`. To illustrate, consider `alphabetcompare`, which implements a comparison using the standard rules of alphabetization.

When the arguments match, 1 is returned:

```
>> alphabetcompare('orange', 'orange')
ans: 1
```

When the first argument is alphabetically first, return 2:

```
>> alphabetcompare('apple', 'pear')
ans: 2
```

otherwise return 3:

```
>> alphabetcompare('milk', 'honey')
ans: 3
```

Capital letters are the same as lowercase in the usual alphabetical order, so

```
≫  alphabetcompare('Milk', 'honey')
   ans: 3
```

But the asciicompare program uses the ASCII rules of order:

```
≫  asciicompare('Milk', 'honey')
   ans: 2
```

It is essential that the comparison function used in searching a sorted list be the same as that used in sorting it in the first place. For instance, if we use alphabetcompare to search a list sorted in ASCII order, the result will not be reliable. To illustrate, let's take the first names in ASCII order:

```
≫  firstnames = sort(data(:,2))
   firstnames: 'Alice'
               'Elizabeth'
               'George'
               'James F.'
               'Miriam'
               'William Carlos'
               'e.e.'
```

Searching using alphabetcompare won't always work:

```
≫  findInSortedList(firstnames, 'e.e.', 'alphabetcompare')
   ans: []
```

The returned value of [] suggests that the search term, 'e.e.', isn't in the list, but this is wrong.

Since sort uses the ASCII ordering of the characters, we need to use asciicompare to search a list arranged by sort:

```
≫  findInSortedList(firstnames, 'e.e.', 'asciicompare')
   ans: 7
```

There are good reasons why sort arranges things not in conventional alphabetical order but in the order dictated by ASCII. For one thing, this allows textually distinct strings to be kept distinct. For example, "greenhouse" and "green house" are different terms, but alphabetcompare doesn't recognize the difference, while asciicomp does:

```
≫  alphabetcompare('greenhouse', 'green house')
   ans: 1
≫  asciicomp('greenhouse', 'green house')
   ans: 3
```

So far as searching is concerned, it doesn't matter how the sorting is done, so long as the same ordinal relation between items is used in both sorting and searching.

The rapid divide-and-conquer search for a single matching item in findInSortedList is only part of the story. If a match is found, it is necessary to check whether there are other matches. But since the items are in sorted order, any other matches will be adjacent to the one found. Here is the remainder of findInSortedList that looks for additional matches:

```
[25]  if isempty(res)
[26]      return;
[27]  else
[28]
[29]      after = res;
[30]      for k=(res+1):min(length(list),back)
[31]          if 1==feval(compfun,searchterm,list{k});
[32]              after = k;
[33]          else
[34]              break;
[35]          end
[36]      end
[37]      before = res;
[38]      for k=(res-1):-1:max(1,front)
[39]          if 1==feval(compfun,searchterm,list{k});
[40]              before = k;
[41]          else
[42]              break;
[43]          end
[44]      end
[45]      res = before:after;
[46]  end
```

No matches → [26]

Are there neighbors that match? → [28]

Notice that since the table is sorted by last name, the first names are out of order. In a conventional telephone book, the first name is used to break ties in the last names. In this way, the subtable that is extracted based on a match for last name is already sorted by first name, so that a match by first name can be efficiently found. The sortrows program allows a table to be sorted in this tie-breaking way. It takes a list of columns, using the first to sort, the second to sort within ties, and so on. For instance,

```
≫ sortrows(data,[1,2])
  ans:
   'Cummings'   'Alice'              '853-6551'   '1701 Fairview'
   'Cummings'   'Elizabeth'          '231-6845'   '865 Bayard'
   'Cummings'   'James F.'           '642-5431'   '1701 Fairview'
   'Cummings'   'e.e.'               '642-2138'   '1209 Fairview'
   'Webster'    'Miriam'             '433-4321'   '1812 Princeton'
   'Williams'   'William Carlos'     '642-9123'   '1692 Princeton'
   'Wilson'     'George'             '231-3211'   '392 Bayard'
```

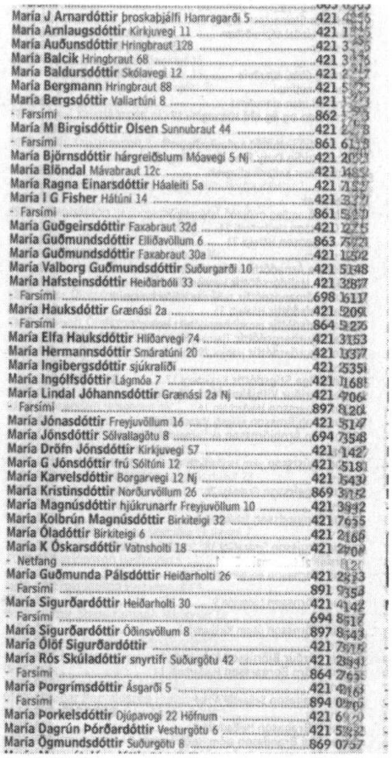

Figure 11.3. An excerpt from a phone book from Iceland. Since Icelandic last names are derived from the father's first name, there are many fewer last names than first names—only male names appear as last names, but there are different first names for males and females. Thus, it makes sense to sort by first name.

Sorting in this way, however, is no use if the search term is a first name rather than a last name. There are several reasons why conventional phone books are sorted by last name; for instance, the historical fact that telephones are associated with a household rather than an individual. But not all phone books are this way; see Figure 11.3.

The Benefits and Costs of Searching Sorted Lists

How much work is involved in searching a sorted list versus an unsorted list? In part, the answer depends on luck. When searching an unsorted list, if the search term happens to match the first one in the list, then the search can be extremely short; only a single comparison is needed. More typically, we expect to have to search through half of the list, on average, before finding a match. But if the search term is not in the list (this is the worst case), we have to check all of the items. So the worst-case cost for searching an unsorted list of length N is N comparisons.

When searching a sorted list, we don't need to compare to every item, so the search should be much faster. In the worst case, it takes approximately

$$\log_2 N$$

comparisons, since each comparison cuts in half the size of the part of the list remaining to be searched. After $\log_2 N$ cuts, the remaining list will have only one item, and searching it involves a single comparison.

This is a remarkable formula that explains why searching a large telephone book is not much harder than searching a small one. Suppose you have a small phone book with 50,000 entries. The formula for the number of comparisons needed for the binary search algorithm can be computed as

```
≫ log2( 50000 )
   ans: 15.6
```

which needs to be rounded up to 16. For a much larger city, the phone book might have 10 times as many entries. Searching such a phone book requires only 20 comparisons:

```
≫ ceil(log2( 500000 ) )
   ans: 19
```

The hypothetical phone book with one entry for each of the approximately 10^{10} people on earth would require 34 comparisons according to the formula: only about twice as much effort to search than the small-city phone book despite being 200,000 times larger. A computer that could do 1,000,000 comparisons per second would need more than three hours to search the whole-earth phone book if it were not sorted, but less than $\frac{1}{10000}$ of a second to search the sorted whole-earth book.

For large N, $\log_2 N$ is much smaller than N, and searching a sorted list is much more efficient than searching an unordered list. To put things in concrete terms, a search that takes $\log_2 N$ comparisons corresponds to using a telephone book to look up a number given a person's name. A search that takes N comparisons corresponds to being given a number and asked to find the person's name in the telephone book.

If searching a sorted list is so much more efficient than searching an unordered list, why not just sort every list before searching? The reason is that sorting itself is an operation that involves approximately $n\log_2 n$ comparisons. (See Exercise 11.9.) So when only a single search is to be done, the benefits of sorting are outweighed by the costs of sorting. However, if the sorted list will be used for many searches—as with a telephone book—then sorting is worthwhile. "Many searches" is not actually so many: It's worthwhile to sort if more than approximately $2\log_2 n$ searches will be performed.[2] For instance, if we anticipate using the small-town phone book more than 32 times, then sorting is worthwhile. The whole-earth phone book, if used more than 68 times, justifies sorting.

---

[2] Since it takes $n\log_2 n$ comparisons to sort, and $\log_2 n$ for each search of the sorted list, k searches of a sorted list takes $k\log_2 n + n\log_2 n$ comparisons. In contrast, k searches of an unsorted list takes, on average, $kn/2$ comparisons.

Algorithms for sorting are interesting in their own right, and interesting and difficult situations are presented when there is so much data that it cannot all fit into computer memory at one time. But for the moderate-sized datasets that we encounter here, less than, say, 100,000 items, the built-in `sort` function works well.

11.6   Simultaneous Collection and Access of Data

Consider the task of counting how many times each word appears in a body of text. This could be accomplished by setting up a table with two fields: one for the word and one for the count, as in Table 11.1. As each new word is encountered, we search the word field to see if the word is already listed. If so, increment the corresponding count. If not, add the word to the table and set the count to 1.

We need to be careful about using a sort-then-search strategy. Every time a new word is encountered, the table needs to be resorted. This could be dramatically inefficient.

There are a variety of approaches for dealing with this sort of situation where we need both to arrange a list to make searching efficient and to modify the list as new data arrive. The technique we examine here is called *hashing*.

Key Term

In Chapter 8, we looked at the frequency with which letters are used. We did this by setting up a vector array of counts with one element for each ASCII character. The numerical value of the ASCII code served as the index into the array. As each character of input was reached, we incremented the appropriate position in the count array.

Table 11.1   A Word Count Table for the Phrase "We cannot dedicate, we cannot consecrate, we cannot hallow this ground."

Word	Count
we	3
cannot	3
dedicate	1
consecrate	1
hallow	1
this	1
ground	1

This searching process was efficient because we could use indexing rather than sort-and-search to find the relevant count to update. This approach works because the items themselves (individual ASCII characters) served as unique identifying indices.

In theory, we could extend this approach to whole words. Rather than use a one-dimensional vector, we could use a multidimensional array with a dimension for each character in the word. If we limited ourselves to words of, say, five letters, then the counts array for any word w would be accessed as

```
counts(w(1), w(2), w(3), w(4), w(5) )
```

Of course, we also need to handle words with lengths other than five. Here's a possible trick. Recall from Chapter 8 that multidimensional arrays are an illusion—the multiple indices are reduced to a single integer, which is used to reference a one-dimensional vector. For our ASCII indices with 127 possibilities, this index could be computed as

```
index = w(1) + 128*(w(2) + 128*(w(3) + ... ))
```

This computation would work for a word of any length, as in the following function:

```
[1]  function ind = word2index(word)
[2]  %  Convert an ASCII string to a number
[3]  ind = 0;
[4]  for k=1:length(word)
[5]      ind = 128*ind + word(k);
[6]  end
```

Here's how it works:

```
>> word2index('we')
  ans: 15333
>> word2index('this')
  ans: 244987123
```

All we need to do in theory is use the output of word2index to access an array of counts. Unfortunately, the indices that word2index generates are big even for modest-sized words. Longer words, for instance,

```
>> word2index('consecrate')
  ans: 9.211746583542073e+020
```

would completely outstrip the memory of even the biggest computers. Indeed, the computation has lost precision because there are 21 digits in the index, but a double-precision floating point number can hold only about 16 of them.

But such large indices are not really needed. Table 11.1 has only eight items, so there's no reason why indices like 3396479809011 should be involved. For a source containing N words, we never need to have a table larger than N and, taking into account repeated uses of a word, the table might be much smaller.

In a hash table, we compute an index in a manner similar to `word2index` but arrange things so that the returned index is always smaller than a specified maximum `nmax`. This computation involves a *hash function*. Here is a suitable hash function for the word-index problem:

Key Term

```
[1]   function ind = wordhash(word,nmax)
[2]   % wordhash(word) --- a hash for words
[3]   ind = 1;
[4]   for k=1:length(word)
[5]       ind = 31*ind + word(k);
[6]       ind = rem(ind,nmax) + 1;
[7]   end
```

Put in the range 1 to nmax

This hash function is guaranteed to produce an output between 1 and `nmax` for any input word, so the output can be used safely as an index into an array of size `nmax`.

For the 11-word phrase in Table 11.1, we can use an array of length 11 to hold the counts. Here are the indices:

```
≫  wordhash('we', 11)
   ans: 9
≫  wordhash('cannot', 11)
   ans: 4
≫  wordhash('dedicate', 11)
   ans: 11
≫  wordhash('consecrate', 11)
   ans: 1
```

Regretably, things are not as simple as they might seem. Although the hash function produces small indices, there is no guarantee that different words will produce different indices. For example,

```
≫  wordhash('hallow', 11)
   ans: 9
```

which is the same as for `'we'`.

By making `nmax` large, we can reduce the probability of such coincidences but not entirely eliminate it. But clearly, it's not acceptable to have two different words sharing the same count in the table, so we need a way to deal with the coincidence problem. Here's one algorithm: If a coincidence is found, jump around the array until we find an empty slot or a matching word.

1. Set up a cell array of size nmax. Initialize each element to be empty. Initialize s to be the smallest integer that is relatively prime to nmax.
2. For the input word, compute the index using k=wordhash(word, nmax).
3. If array{k} matches word, return k as the index. Otherwise,
4. If array{k} is empty, set array{k}=word and return k as the index. Otherwise,
5. There is a different word occupying slot k, so let's find another slot. Do this by setting k = k+s. (If k > nmax, set k = 1.) Continue at step 3.

The choice of s as relatively prime to nmax will ensure that an empty slot will always be found unless the array is completely full. The possiblity of long searches for an empty slot can be reduced tremendously by making nmax larger than the number of items that will be inserted; a factor of, say, 25% will do the job nicely.

A simple hash table is implemented by the three functions makeHashTable, insertInHashTable, findInHashTable, and reportHashTable. Since the four functions need to communicate with one another, the hash table itself is a global variable. This means there can be only one hash table in use at a time: a rather burdensome restriction that is easily avoided in languages with variables that can be passed by reference.

To illustrate the use of these functions, here is a function to count how many times each word in a list appears:

```
[1]   function [words,cnts] = countWords(wordlist)
[2]   nmax = floor(1.5.*length(wordlist(:)));

[3]   h = makeHashTable(nmax);
[4]   counts = zeros(nmax,1);
[5]   for k=1:length(wordlist(:))
[6]       ind = insertInHashTable(wordlist{k});
[7]       counts(ind) = counts(ind) + 1;
[8]   end
[9]   [words,inds] = reportHashTable;
[10]  cnts = counts(inds);
```

nmax must be bigger for efficiency — [2]

Initialize counts — [4]

Slot for this word — [6]
Update count in that slot — [7]

Get rid of blanks — [9]

We use this function to count the uses of words in *Moby Dick*. First, read in the text from the file mobydick.txt, dividing it by words.

```
>> w = textread('mobydick.txt', '%s', 'delimiter', ' .=?-''"\n;');
```

Next, count the instances of the different words. We ignore capitalization for the present:

```
≫  [words,cnts] = countWords(lower(w));
```

The five most common words can be displayed this way:

```
≫  [trash,inds] = sort(-cnts);
≫  words(inds(1:5))
  ans: 'the'
       'and'
       'a'
       'of'
       'to'
```

with frequencies

```
≫  cnts(inds(1:5))
  ans: 1077
        594
        578
        529
        410
```

11.7 Exercises

Discussion 11.1:
What we call a grocery list is really a table with several fields. Translate the grocery list in the margin into a table with the necessary fields for effective shopping.

Discussion 11.2:
On page 273, there is an example of searching with a comparison function that doesn't match the sorted order of data.

```
≫  findInSortedList(firstnames, 'e.e.', 'alphabetcompare')
  ans: []
```

suggests that the search term `'e.e.'` isn't in the ASCII-sorted list `firstnames` even though it clearly is.

But sometimes, the search is successful, even when the wrong comparison function is used. For instance,

```
≫  findInSortedList(firstnames, 'George', 'alphabetcompare')
  ans: 3
```

Explain why the search result is correct in this case.

Note that a system that sometimes gives correct results is dangerous; it can be hard to find out that the system actually doesn't work.

Discussion 11.3:

Give the sequence of database operations needed to turn the Census and Postal Code tables in the text into a table giving family size versus latitude.

Exercise 11.1:

Generate a table with two fields, Teacher ID and Average Family Income. Describe the operators used. (*Answer:* Project Households onto PID and Income. Join this with People so that there is an income associated with each person. Join this with the Class Data [natural join] so there is a Teacher ID associated with each person in the School Classes table. Group/average over teacher ID to get the average income.)

Exercise 11.2:

Write the following criterion functions for use with the telephone dataset, and demonstrate that the functions work. Recall that a criterion function takes a single item—a row of a table—and returns a boolean.

- `streetPrinceton(item)`, telling whether the street address is on Princeton Street.
- `exchange231(item)`, telling whether the telephone number is in the 231 exchange. (The exchange is the first three digits of the telephone number.)

Exercise 11.3:

Writing criteria functions like `isCummings` is very easy. But it can be better to have general-purpose functions. Write a function `exactMatch` that takes a character string and a field index as arguments and returns an `inline` function. The returned function should implement a criterion function that checks for an exact match with the character string in the specified field. It should work like this:

```
≫  selectfromtable(data,exactMatch('Cummings',1))
   ans: 4
        5
```

Write similar general-purpose functions `onStreet(name)` and `inExchange(exchange)` to implement the functionality of the criterion functions in Exercise 11.2.

Exercise 11.4:

The `compsort` program takes a list and a comparison function and sorts accordingly. For example:

```
≫  compsort('Fred', 'fine', 'finish', 'alphabetCompare')
   ans: fine finish Fred
```

Write a comparison function that orders strings based on which one is longer. Use this, with `compsort`, to sort a cell array of strings into order of increasing length.

Exercise 11.5:

Use union and intersect on the indices returned by selectFromTable and the appropriate criterion functions to create the following tables:

- People in the 642 exchange living on Fairview
- People whose last name begins with W and who live on Princeton
- People whose last name is Cummings and first name is Alice

Exercise 11.6:

Use the intersect, union, setdiff and setxor operators, together with selectFromTable and the appropriate criterion functions to create the following tables:

- People who live on Fairview and have Cummings as a last name.
- People who live on Fairview or have Cummings as a last name.
- People who either live on Fairview or have Cummings, but not both.
- People in the 642 exchange who do not live on Fairview.
- People whose last name is Cummings and first name is not Alice.

Exercise 11.7:

The compsort(list,comparefun) program sorts a cell-array list into an order dictated by its second argument, comparefun. For example, compsort(wordlist, 'alphabetcompare') will sort into an alphabetical order ignoring cases and spaces.

Write a comparison function that orders strings based only on the vowels in them, sorting the vowels in the usual alphabetical order. Use this, with compsort, to sort the word list.

Exercise 11.8:

Write a comparison function, spellMatch, that can be used to frame searches that are relatively insensitive to misspellings, plurals, possesives, and so on. For instance, "apple" and "apples" should be equivalent according to spellMatch. There are many ways to do this; be creative. Some simple ideas are as follows: Convert everything to lowercase, drop any "s" at the end of the string, drop all but leading vowels from the string, eliminate repeated vowels and consonants. Demonstrate the effectiveness of your program by searching intentionally misspelled words in a list of correctly spelled words.

Exercise 11.9:

Experimentally, confirm that the time it takes to sort a list is proportional to $n \log_2 n$. Do this by constructing word lists of various lengths n, selecting a subset of length n randomly, and measuring the time needed for compsort to sort them using tic and toc. Make a graph of the sorting time versus list length n.

Similarly, show that the time it takes to search a sorted list is, on average, proportional to $\log_2 n$. Make a graph of the searching time versus list length.

Exercise 11.10:

Construct a geneology database with ancestor information in one table and disease information (person ID, disease, fatal or not) in another table. Write a program to extract all of the ancestors and descendants of a given person and tell what diseases they had and how many people had them.

Exercise 11.11:

Using the relational operators, including the group/min operator, describe how you would calculate how much money you will spend to buy your grocery list if you bought all of the items at the cheapest store.

11.8 Project: A Bridge Database

Figure 11.4 shows a schematic diagram of a truss bridge. The truss consists of a number of "nodes" and "members" that are joined together at the nodes. Trusses are important structures in the real world: strong and economical. They are strong because they consist of interconnected triangles. A triangle is a strong configuration because its shape cannot be distorted without distorting the members themselves. In contrast, a rectangle is a weak configuration since it can be skewed or folded without altering the individual members.

The bridge shown in Figure 11.4 consists of seven nodes (numbered 1 through 7). The ground supports the bridge at node 1 both vertically and horizontally and vertically only at node 4. (You can imagine that node 4 sits on a roller that can exert no horizontal force.) The seven nodes are connected in a triangular web with eleven members. The members are drawn as thick lines if they are under compression and as thin lines if they are under tension. When constructing the bridge, members under tension could be guy wires instead of rigid beams. The amount of stress in each member is color coded. (*Stress* is a general term for force that can refer to either tension or compression.)

Key Term

Engineers need to perform a variety of computations on trusses (for example, calculating the stresses in the members for various load configurations). The issue for us here is how to represent the information that describes a truss and to integrate it with other information, such as the loads on the truss or the computed stresses in the beams.

Since the bridge is made up of members, it might make sense to organize the bridge information as a table, with each item referring to one member. To draw the bridge, we would need to know the *x*- and *y*-coordinates of each end of each member. This suggests storing the information in a table that looks like this:

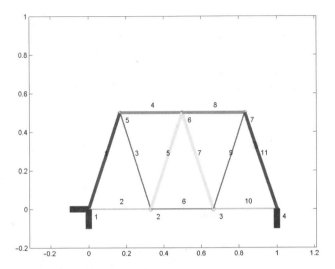

Figure 11.4.   A truss bridge with seven nodes and eleven members. The horizontal and vertical ground supports are shown as thick lines. If a member is in compression, it is drawn as a thick line; if it is in tension, it is drawn as a thin line. The magnitude of the compression or tension, that is, the stress, is used to shade the beam: Higher stress is darker. Each node is numbered, as is each beam. The database files representing this bridge are named `bridge7`.

Member Information

Member No.	x_1	y_1	x_2	y_2
1	0	0	0.1666	0.5
2	0	0	0.3333	0
3	0.1666	0.5	0.3333	0
4	0.1666	0.5	0.5	0.5
5	0.3333	0	0.5	0.5
6	0.3333	0	0.6666	0
7	0.6666	0	0.5	0.5
8	0.5	0.5	0.8333	0.5
9	0.6666	0	0.83333	0.5
10	0.6666	0	1	0
11	0.8333	0.5	1	0

We could store this table in a number of ways, perhaps just as an ordinary numerical matrix. Doing so would require that we keep track of the meaning of each column (for instance, that column 2 is the *x*-coordinate of the first node endpoint of the member). Perhaps it is clearer to use a structure type to give each column a name; for instance,

```
m.x1 = [0; 0; 0.1666; 0.1666; 0.3333; and so on ];
m.x2 = [0; 0; 0.5; 0.5; 0; 0; 0; 0.5; 0; 0; 0.5];
m.y1 = [0.1666; 0.3333; 0.5; 0.5; 0.6666; and so on];
m.y2 = [0.5; 0; 0; 0.5; 0.5; 0; 0.5; 0.5; 0.5; 0; 0];
```

With this scheme, drawing the truss is a matter of looping over all of the members and drawing a line for each. For example, for member j, the statement looks like

```
f = line([m.x1(j), m.x2(j)], [m.y1(j), m.y2(j)]);
```

The returned value, f, provides a handle to the line that can be used to set attributes of the line such as the color and thickness. The following two statements produce a thick, red line:

```
set(f,'linewidth', 20);
set(f,'color', [1, 0, 0]);
```

Although the preceding format for the bridge information is well suited to drawing, it has a number of deficiencies. Most fundamentally, the same information appears in several places. For example, members 4, 5, 7, and 8 all meet at one node, whose coordinates are (0.5,0.5). Suppose a design change were proposed, moving that node upward to (0.5,0.75) to give the bridge a more pyramidal shape? This would involve changing several of the y-coordinates in the table from 0.5 to 0.75. But which ones? Since some other nodes also have y-coordinate 0.5, care has to be taken to change only some of the entries and not others.

Another shortcoming is that it can be hard to add new information to the table. For instance, the ground supports are always positioned at nodes. They provide a support force to one end of the member, and the force is split up among the several members that meet at a node. Similarly, information about the loads on the bridge (weights of traffic, the weight of the bridge itself, wind, etc.) does not relate to individual members. The forces are usually modelled as acting at nodes, not on individual members.

Perhaps it would be better to arrange the information in terms of nodes: This would allow us to specify the position of each node and the support and loads without any redundancy. Unfortunately, then we face the problem of how to represent the members, since each member connects two nodes.

Here is a better scheme for representing the truss information: The information is in several tables. The first four tables contain what might be called "input information" that gives the geometry of the bridge:

- The (x, y) coordinates of each node. For the bridge shown in Figure 11.4, this table looks like

Node Locations

Nodes

Node ID	x-coord	y-coord
1	0	0
2	0.33	0
3	0.666	0
4	1	0
5	0.1666	0.5
6	0.5	0.5
7	0.8333	0.5

- The two nodes that are connected by each member. Rather than specifying the nodes in terms of their positions, we use a node ID. That way, there is no redundancy between this table and the one giving node positions.

Members

Member ID	Node1 ID	Node2 ID
1	1	5
2	1	2
3	5	2
4	5	6
5	2	6
6	2	3
7	3	6
8	6	7
9	3	7
10	3	4
11	7	4

- The horizontal and vertical loads at each node.

Loads

Load ID	Node ID	Horiz. Load	Vert. Load
1	1	0	−20
2	6	0	−5
3	4	0	−20
4	5	0	−5
5	2	0	−20
6	3	0	−20
7	7	0	−5

- The nodes that are supported by the ground. Vertical supports are encoded as 0, horizontal as 1.

Supports

Support ID	Node ID	Direction
1	4	0
2	1	0
3	1	1

From this input information, the stress in each member and the force on each support can be computed, as we'll see in Chapter 17. For the purposes of drawing, this information can be stored in two additional tables:

- The stress in each member. This table would have two fields, the Member ID and the Stress (in the member).
- The stress on each support. Again, this table has two fields, the Support ID and the Stress.

For the loads specified previously, these tables would look like this:

Member Stresses

Member ID	Stress
1	−29.1
2	9.2
3	23.8
4	−16.6
5	−2.7
6	17.4
7	−2.5
8	−16.6
9	23.6
10	9.1
11	−28.9

Support Forces

Support ID	Stress
1	47.5
2	47.5
3	0.0

Key Term

In most of the tables, there is one column that contains the *key* to the table. In the nodes table, the key is the node ID. In the beams table, the key is the beam ID. The rest of the table gives information about the items represented by the key. For instance, in the beams table is given the two nodes that are connected by the beam in question as well as the mass of the beam. In each table, there is only one instance of each key.

Key Term

Tables can also contain *foreign keys*. These are entries that refer to the key of another table. For example, in the beams table the node ID is a foreign key; the node IDs used in the beams table refers to the keys of the nodes tables.

Key Term

This arrangement of data into tables, using keys and perhaps foreign keys from other tables, is called a *relational database*. This is an extremely small database: Real-world databases can have millions of entries and thousands of tables and use extremely sophisticated software to maintain and update the database.

An advantage of using a relational database is that there is a minimum of redundant information. For example, to redesign the bridge to move node 6 upward, we need change only one entry in the node-position table. (Also, note that since node 6 has an explicit ID, it's easy and unambiguous to refer to node 6.) It's impossible to introduce inconsistency into the node positions: If an error is made in the location of the node, all of the members attached at that node will still be attached.

Another example: The structure in Figure 11.4 spans the gap between the two vertical ground supports. For other purposes, it might be desired to extend the structure as a cantilever by, for instance, moving the ground support from node 4 to node 2. This involves making a single change in the

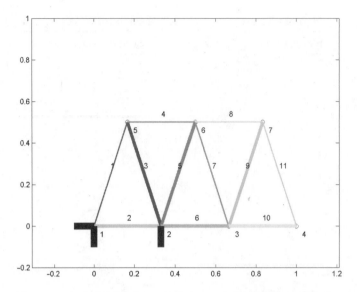

Figure 11.5. The same structure in Figure 11.4, but supported as a cantilever.

supports table, from which the new stresses can be computed. The cantilever is shown in Figure 11.5.

Exercise 11.12:

Design several trusses of your own and plot them out using the `readbridge` and `drawbridge` functions provided as p-code. Your design information will take the form of the four input tables described previously.

In a commercial relational database system, each of the truss information tables would be stored together with identifying information. Here, in our toy system, we will be more informal and store each table as a numerical matrix in an ASCII file. Since there are four input tables, there will be four files. To keep the four files organized, all four will have the same file name, but a different filename extension. For instance, if your truss is called `mybridge`, the four files should be `mybridge.nodes`, `mybridge.members`, `mybridge.supports`, and `mybridge.loads`. An example of the format of these files is given by the set called `bridge7` that contains the information for the bridge in Figure 11.4.

For a large project, we would want to use a database system to handle the bridge data. In this small case, it's reasonable simply to store the data as MATLAB structures and use `find` to locate specific items.

The function `readbridge` takes as an argument the name of the bridge files and returns a structure with several tables.

```
>> b = readbridge('bridge7');
```

The returned tables are `b.nodes`, `b.members`, `b.loads`, `b.supports`, `b.stress`, `b.supportforces`. Each of these components is arranged with subfields that correspond to the tables (e.g., `b.nodes.x`, `b.nodes.y`, and so on). Another function, `drawbridge`, will draw the truss, with appropriate labeling.

>> `drawbridge(b);`

A third program, `findtensions`, computes the stresses in the truss members. It returns a bridge structure that has all the same information as the structure given as an argument but has the `.stress` and `.supportforces` fields filled in.

>> `bt = findtensions(b);`

`Drawbridge` can use this information to color-code the drawing.

>> `drawbridge(bt);`

Exercise 11.13:
Write a program `memberCoords` that takes two arguments, a bridge structure (as constructed from files by `readbridge`) and a member number and returns two vectors, one giving the *x*-coordinates of the endpoints and the other the *y*-coordinates.

Write another function, `drawMembers`, that loops over all of the members and, using `memberCoords`, draws the whole bridge.

Exercise 11.14:
Using the output of `findtensions`, expand `drawMembers` to include graphical depictions of the tension information. In Chapter 17, we will show the techniques for writing `findtensions`.

Exercise 11.15:
Make a movie of the changing stresses on bridge elements as a load is moved across the bridge. Note that when a load is applied to the member between two nodes, the load is split between the two nodes. If the load is situated a fraction p of the distance from node A and node B, then node A carries fraction $1 - p$ of the load and node B carries fraction p. (*Hint:* Write a function that takes as an argument time t, computes the location of the load, and fills in the load's part of the database structure. Then use `findtensions` and `drawbridge` in the ordinary way.)

11.9 Project: Random Generation of Text

A simple model of text, proposed originally by the information theorist, Claude Shannon, looks at the probabilities of words. As we saw in Section 11.6, some words occur more frequently than others. By tabulating these frequencies, we capture some of the structure of the text. By generating text according to the frequencies, we can help to see whether this frequency-based structure indeed captures important aspects of the text.

To illustrate, we can read in a body of text, word by word. Here is the novel *Moby Dick:*

```
≫  w = textread('mobydick.txt', '%s', 'delimiter', ' .,:;=¿"!?');
≫  length(w)
   ans: 215743
```

To generate the random text, we can simply pull words at random out of this corpus. In so doing, we will follow statistically the frequency of words in the text. One way to sample randomly is to generate a set of random integers, each between 1 and the length of the text. For example, to sample 100 words at random, we generate 100 random integers:

```
≫  inds = ceil( length(w)*rand(100,1));
```

Here are the sampled words:

> **of ashore stood his of even I blackness heads that gin a sea take ladder that Mungo porcupine Fates soggy his again the the into in Jonah great town To all pitched look side ll months to wonder nothing mattress tell the by manliest picture sighs have and the the tapping where as prepares not do array his darkness opinions the belongs and as difficulty the his ship glides wide expostulations curious his previous shadows all part up no monstrous the with to in seemed of hand to I out has rather sees tempest place Thirty But**

A rough read! Randomly ordered words do not seem much like real text.

Shannon knew this. He proposed a model of text that takes into account the fact that the probability of a word depends on its context. For example, it's very unlikely to have pairs of words such as "the his" or "the the," but both of these pairs appear in the preceding text. In Shannon's model, one looks at the "conditional probability" of each word given the previous word.

Going through the list of words in w, we can pull out every word that occurs after the word "the." It turns out that there are 14,302 of them in *Moby Dick*. Here is part of the ASCII-ordered list:

```
21st act adulterer affrighted afternoon age ages
aggregated air air air ... world world world worm
wound wrinkled yawning year young
```

As anticipated, the phrase "the the" does not ever appear in the text. However, phrases like "the air" and "the world" do appear several times; perhaps this is to be expected in a novel involving a sailing ship.

The conditional probabilities contain additional information about the text. We can generate text that obeys these probabilities with the following algorithm. To generate N random words from the text,

1. Start with an initial word, called iword.
2. Construct a list of all the words in the text that follow iword.
3. Choose at random one of the words from this list, calling it rword.
4. Set iword = rword and repeat until N words are generated.

The sequence of N iwords constitutes the text.

Here is an example from *Moby Dick*, starting with "call" as an initial word. The superscripts indicate how many choices there were when each word was selected at random:

call^0 all^{55} the^{1477} fishery^{14302} in^{60} others^{4136} that^{38} strange^{2937} town^{96} some^{17} of^{614} oil^{6579} so^{78} day^{1041} grew^{158} out^{14} more^{521} and^{502} every^{6358} roll^{230} of^{17} it^{6579} or^{2375} the^{708} whales^{14302} these^{264} things^{397} such^{127} an^{366} african^{589} elephant^3 vast^{15} host^{73} would^{10} not^{427} love^{1131} steelkilt^{24} hailed^{35} him^{20} therefore^{1046} you^{66} be^{861} social^{1038} meeting^{12} a^{10} weaver's^{4630} loom^1 are^8 scattered^{595} people^9

This is much easier to read; all of the word-to-word transitions actually occur in the original text.

An even stronger model of text comes from defining the context for each new word to be the previous two words. To generate text according to this model, we start with two initial words, generate a third word, and then iterate. For example, the following text was generated starting with "Call" and "me." The third word, "an," happens to be one of three in the text that follows "Call me." The next word generation occurs in the context of "me Ishmael."

call^0 me^0 an^3 immortal^3 by^1 brevet^1 yes^1 there^1 is^1 again^{83} carried^1 forward^1 the^1 boat^3 and^{104} grievous^{28} loss^1 might^1 ensue^1 to^1 nantucket^1 and^3 come^{10} to^6 think^{34} of^{18} a^{34} baby's^{333} pulse^2 and^1 lightly^2 say^1 of^1 it^1 from^{95} an^{13} unquestionable^1 source^1 bear^1 with^1 me^2 i^{24} only^{23} wish^4 that^1 we^2 seemed^{28} to^1 drink^{57} a^7 bottle^1 of^2 bordeaux^1 he^1 added^{67} motioning^5 to^1 me^1 though^{89} indeed^2 i^9 might^3 proceed^9 with^1 a^2 globular^{247} brain^1 and^1 heart^5 and^3 hand^8 backed^1 by^1 a^2 heedful^{82} closely^1 calculating^1 attention^1 to^1 that^4 not^{53} an^{10} unfair^4 presumption^1 i^1 say^1 all^{70} of^2 ye^{25} nor^{23} can^1 whalemen^5 be^1 recognised^1 as^1 the^1 ship^{212} the^{193} ship^{17}

Compare this text to the original on page 100.

Exercise 11.16:

Write a function `textpairs(w)` that takes a list of words and generates, for each sequential pair of words in the original list, a list of all the words that follow that pair. You can do this with a function much like `countWords` (on page 280), but instead of incrementing the count, add the word that follows the pair onto the list. `Textpairs` should return a cell array: Each element of the array will contain the word list for the corresponding pair of sequential words.

(*Hint:* Keep in mind that the pair itself will be stored in the hash table. One way to hash on the pair of words is to concatenate them together.)

Exercise 11.17:

Write a function, `generateText(lists, one, two, n)`, that generates n words of text, starting from initial words `one` and `two`. You will have to use the function `findInHashTable` to get an index for the pair of words, then use this index into `lists` to pull out the set of words that follows the pair.

Exercise 11.18:

Rather than using `textread`, write your own parsing program that breaks a text file into a sequence of words and that treats each punctuation symbol as a word so that the text you generate can include punctuation.

12

Trees and Recursion

Great fleas have little fleas
upon their backs to bite 'em,
And little fleas have lesser fleas,
and so ad infinitum.
And great fleas themselves, in turn,
have greater fleas to go on;
While these again have greater still,
and greater still, and so on.
— Augustus DeMorgan

Most of us live in areas where trees are a common, even dominant feature of the landscape. In terms of a tree's production of food, the tree is a collection of leaves or needles. The trunk and branch structures of the tree arrange and support the leaves, keeping them separated so they can be exposed efficiently to light and air, but uniting them into a single entity.

The tree structure is also an effective way to organize information. We've seen how computer files can be arranged into directories and subdirectories, uniting them into a single file system while separately keeping accessible the contents of the different files. Many other sorts of data lend themselves to organization as trees and, as we shall see, many complicated-seeming computations become straightforward when they are arranged using a treelike branching structure.

From an algorithmic point of view, the most remarkable feature of trees is their self-similarity. Each branching point of a tree looks much like any other branching point. Cut off a large branch of a tree, and you have a structure that looks like an entire tree.[1] This self-similarity means that an algorithm can be designed for a small, simple tree, but then can operate without change on large, complex trees. As an example, consider the file system not of an individual computer but of the entire World Wide Web. The immense number of files on the WWW are organized as a large tree, much like the file system of a single computer, or even the system of files stemming from a single directory on that computer. The same principles of access are involved in the relative pathname book/chap11/programs accessible on my computer as are in the absolute pathname www.macalester.edu/~kaplan/cs121/chap11/programs accessible from anywhere on the Internet.

In this chapter, we explore ways of representing trees as computer data structures, ways to create trees, and ways to process the data in them. We use an important algorithmic style, recursion, whereby algorithms are described in terms of themselves.

12.1   Simple Recursion

We will start with an example where recursion offers little advantage, simply to contrast the recursive approach to the looping approach. Recall that the factorial of an integer n, denoted $n!$, is

$$1 \times 2 \times 3 \times \ldots \times n = \prod_{k=1}^{n} k$$

In Chapter 8, the function productToN performed this computation using loops. For reference, we reprint it here:

Compute $\prod_{k=1}^{n} k$ [1]   `function res = productToN(n)`
Set up the accumulator   [2]   `res = 1;`
   [3]   `for k=1:n`
Update the accumulator on   [4]   `res = res .* k;`
each iteration
   [5]   `end`

Here's the computation written in a recursive style:

   [1]   `function res = factorialrecursive(n)`
   [2]   `if n == 1`
Base case: $1! = 1$ [3]   `res = 1;`
   [4]   `else`
Simplification and delegation   [5]   `res = n*factorialrecursive(n-1);`
   [6]   `end`

---

[1]The self-similarity of tree geometry can be exploited to make simple but realistic descriptions of trees for computer rendering of graphics. See [25].

Note that `factorialrecursive` contains no loop. Instead, it relies on two properties of the mathematical factorial function:

Base case: We know the answer for some simple values. In particular, $1! = 1$.

Simplification: When $n > 1$ we can break down $n!$ into $n \times (n-1)!$ [Proof: $n! = n \times (n-1) \times \ldots 2 \times 1$, which is n times $(n-1) \times \ldots 2 \times 1$, which is $n \times (n-1)!$]

All the program `factorialrecursive` needs to do is test whether the argument n corresponds to the base case and, if so, return the appropriate answer or, if not, simplify the factorial computation and request, via a function call, that the simplified computation be carried out.

It might help to think of the chain of events that is invoked by a statement like

≫ `factorialrecursive(10)`

in terms of the workings of a large corporation. Imagine yourself employed as a corporate executive who has just been asked by your boss to compute 10! Perhaps this request has taken the form of a memorandum sent to you asking for the value of `factorialrecursive(10)`. This is too tedious a computation for an important person such as yourself, but you are not so important that you don't have to give your boss an answer. So you decide to delegate the task to a subordinate. You know, though, that the corporation consists of people who follow your example. If you asked your subordinate to compute `factorialrecursive(10)`, he would just pass the request along to someone else who would do the same, and so on, ad infinitum. The result would be that you would never get an answer back to report to the president.

Instead, you simplify the problem a little bit, asking your subordinate to compute `factorialrecursive(9)`. When you get the answer back from your subordinate, you will multiply the answer by 10 to get the value to give to your boss. Your subordinate will follow the same strategy: He will ask his own subordinate to compute `factorialrecursive(8)`. When he gets the answer back, he'll multiply it by 9 to get the value to return to you. But before this happens, your subordinate's subordinate will also simplify and delegate, asking yet another corporate employee for `factorialrecursive(7)`. And so on down the corporate line until someone receives the task of computing `factorialrecursive(1)`. This problem is so simple that there is no need to delegate further.

The program `factorialrecursive` constitutes the corporate "standard operating procedure" that each corporate automaton will use to handle the problem that he receives. The program tells how to identify and handle the base case as well as how to simplify and delegate nonbase cases. The same procedure can be used at every level of the computation.

The following diagram illustrates what computations and delegations are being done in the recursive calculation of `factorial(10)`:

```
factorial(10)
- - - - - - - - - - - -
|
10 * factorial(9)
     - - - - - - - - - -
     |
     9 * factorial(8)
         - - - - - - - - - -
         |
         8 * factorial(7)
             - - - - - - - - - -
             |
             7 * factorial(6)
                 - - - - - - - - - -
                 |
                 and so on until the base case
                             factorial(1)
                             - - - - - - - - - -
                             |
                             1
```

Admittedly, the preceding diagram doesn't look like any sort of natural tree. First, the base of the tree is drawn at the top to honor the ancient convention of reading from the top down. Second, the branches aren't splits in the structure. Perhaps we can think of it as a tree where only one child branch comes off of each parent branch—more like the segments in bamboo than a tree. In the next section, we'll see trees with multiple children at each level.

Still, there is more going on here. What plays the role of the corporation's mail office that delivers the memoranda delegating computations and returns the results to the appropriate person? While you are waiting for the answer from your subordinate (which might take quite some time since the subordinate has a lot of delegating to do himself) how do you remember that you will need to multiply his answer by 10 to produce your own result? But this complexity is hidden from the programmer, being contained entirely in the language's function-calling mechanism. The function-calling mechanism automatically keeps track of the pending calculations until the value of the delegated computations are returned; it makes sure that each of the delegated computations are returned to the appropriate superior.

Recursion is by no means limited to mathematical functions like the factorial. For example, any algorithm that uses a loop to consider each element of a list or vector can be written recursively. The simplification consists of removing one element, typically the first, from the list, and the delegation is the processing of the shorter-by-one list.

To illustrate, consider the algorithm for finding the sum of all of the values in a vector:

Base case: If the vector has length 1, return the single element as the sum.

Simplification and recursion: If the vector has more than one element, return the value found by adding the first number in the vector to the number returned by a delegated task: the sum of all of the numbers in the rest of the vector.

The function sumrec implements this algorithm:

```
[1]   function res = sumrec(vec)
[2]   if length(vec) == 1
[3]       res = vec(1);
[4]   else
[5]       res = vec(1)+sumrec(vec(2:end));
[6]   end
```

The base case — line [3]

Simplification and recursion — line [5]

Tracking the Recursion

The debugging tools introduced in Chapter 9 can be used to help visualize what is going on during the recursion. To illustrate, let's put a debugging breakpoint at the first executable line of factorialrecursive

```
≫   dbstop factorialrecursive
```

Now we run the function

```
≫   factorialrecursive(10)
```

The debugger will cause execution to stop at the breakpoint. At this point, the editor may pop up to allow you to look at the function, and you may prefer to use the graphical user-interface to run the debugger. But we will use the command-line interface so that commands and their results can be printed here.

First, we verify that we are indeed where we expect to be:

```
K≫   dbstack
    ans: > In c:/book/factorialrecursive.m at line 2
```

The argument to factorialrecursive is named n, and we see it has the correct value:

```
K≫   n
    ans: 10
```

Now, continue with the recursion.

```
K≫   dbcont
```

The execution doesn't continue for very long. It stops again almost immediately because `factorialrecursive(10)` invokes `factorialrecursive(9)` on line 5. We see that n has the anticipated value

```
K≫  n
  ans: 9
```

Examining the stack

```
K≫  dbstack
  ans: > In c:/book/factorialrecursive.m at line 2
        In c:/book/factorialrecursive.m at line 5
```

reveals that we are once again at line 2 and that we got there because of the recursive invocation on line 5.

Continue execution again and examine the stack when execution stops:

```
K≫  dbcont
K≫  dbstack
  ans: > In c:/book/factorialrecursive.m at line 2
        In c:/book/factorialrecursive.m at line 5
        In c:/book/factorialrecursive.m at line 5
```

`factorialrecursive(9)` invoked `factorialrecursive(8)` when it got to line 5. We are now at line 2 and n is 8:

```
K≫  n
  ans: 8
```

We could continue this process of stopping the execution as the recursive calculation goes on. We would see n count down steadily and the depth of the stack increase accordingly. Only when n reaches 1 will the recursive invocation on line 5 not be reached.

Keep in mind that the environments that have been created with each of the recursive function invocations are being maintained in memory. We can access them with the dbup and dbdown commands. At the point where we've stopped in this example, n is 8. In the environment where the invocation `factorialrecursive(8)` was made, n had the value of 9. We can see this directly:

```
K≫  dbup
K≫  n
  ans: 9
```

The dbstack command displays the whole stack but indicates which environment we are visiting with a > marker.

```
K» dbstack
  ans: In c:/book/factorialrecursive.m at line 2
      > In c:/book/factorialrecursive.m at line 5
        In c:/book/factorialrecursive.m at line 5
```

Let's go up one more level:

```
K» dbup
K» n
  ans: 10
K» dbstack
  ans: In c:/book/factorialrecursive.m at line 2
        In c:/book/factorialrecursive.m at line 5
      > In c:/book/factorialrecursive.m at line 5
```

One more hop up and we are at the command-line environment:

```
K» dbup
  ans: In base workspace.
K» n
  ans: ??? Undefined function or variable 'n'.
```

Although we need to use the debugging commands to examine the stack, the stack exists even when the debugger is not in use. The stack is the means by which MATLAB keeps track of the various environments that are in existence. Usually the stack is invisible to us. One situation when it is not is when the stack, and the environments associated with each of the function invocations represented on the stack, becomes large. This may consume so much memory that MATLAB cannot operate properly, and a stack error is generated. By default, MATLAB is conservative with the memory allocated for the stack and a stack error may be generated well before memory is exhausted. If you encounter a stack error, you may choose to allocate more memory to the stack (with the set(0,'RecursionLimit',N) command), or to switch to a nonrecursive technique.

12.2 Multiway Recursion

Key Term

In all of the examples thus far, the simplify-and-delegate component of recursion has involved the delegation of only one task. *Multiway recursion* refers to the situation where more than one task is delegated. Later in this chapter, we shall see some important practical uses of this to implement divide-and-conquer algorithms, but to illustrate the basic concept, we focus on a historical example that long predates electronic computers.

Recall from Chapters 2 and 8 the Fibonacci numbers, defined recursively as

```
fib(n) = fib(n-1) + fib(n-2)
```

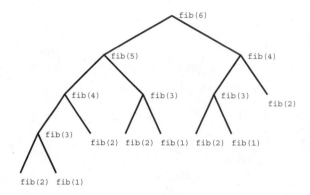

Figure 12.1. The computational tree for a two-way recursive computation of `fib(6)`.

The base case is really two base cases: `fib(1)` → 1 and `fib(2)` → 1. Altogether, the computation of `fib(n)` can be implemented thus:

```
                          [1]  function res = fib(n)
                          [2]  if n == 1
Base case: n = 1          [3]      res = 0;
                          [4]  elseif n == 2
Base case: n = 2          [5]      res = 1;
                          [6]  else
Simplify and delegate     [7]      res = fib(n-1) + fib(n-2);
                          [8]  end
```

Somewhat amazingly, this is a functional program:

```
≫  fib(9)
  ans: 34
≫  fib(10)
  ans: 55
```

The use of the word "simplify" in the recursive step arguably is misleading. Although each of the delegated tasks involves a smaller argument that is closer to the base case than the original argument, there are two delegated tasks, each of which itself can lead to two more delegated tasks. Figure 12.1 shows the tasks delegated to compute `fib(5)`.

The result is a computational chain reaction that manifests itself as `fib(n)` being very slow for values greater than, say, 20. Notice that some identical computations are being repeated several times: `fib(3)` is done twice, `fib(2)` is done three times, `fib(1)` is done five times. The recursive method can be sped up fantastically by a simple approach: Store the results of previous calculations. This is called *memoization*. (See Exercise 12.4.)

Key Term

12.3   Accumulators

The recursive style presented in the programs `factorialrecursive`, `fib`, `maxrec`, and `sumrec` is one where the final output is computed only when the delegated tasks return their partial answer to the boss. This is not the only way to structure the computation. Another recursive style involves accumulators that are explicitly formed: The boss hands down the incomplete, accumulated answer so far and each delegated task adds on to that answer. When the base case is reached, the complete answer can be formed.

Key Term   This is called *tail recursion*.

In the tail-recursive algorithm for recursively adding all the values in a vector, at each step the accumulator is updated by adding the first number in the vector to the accumulator, and the updated accumulator and the rest of the vector is handed off recursively. The accumulator needs to be initialized

Key Term   to zero at the beginning. This typically is done by a nonrecursive *wrapper function* that initializes the accumulator and then calls a recursive helper function. Here is the program for adding up the numbers in a vector:

Wrapper function	[1]	`function res = sumaccum(vec)`
Initialize the accumulator to 0	[2]	`res = helper(vec,0);`
	[3]	`%%%%%%%%%%`
Helper subfunction	[4]	`function res = helper(vec,accum)`
	[5]	`if isempty(vec)`
Done. Return accumulator	[6]	`res = accum;`
Update the accumulator	[7]	`else`
and recurse	[8]	`res = helper(vec(2:end), accum+vec(1));`
	[9]	`end`

As a more complex example of accumulation, consider the problem of constructing all of the ways of taking n items out of a set of k items. For instance, there are three different ways of taking two items out of the set $(1,2,3)$; they are $(1,2)$, $(2,3)$, and $(1,3)$. We are considering all of the permutations of the set to be equivalent, so that $(1,2)$ and $(2,1)$ are the same thing.

There is a deceptively simple recursive algorithm for finding all of the unique subsets with k items. It is based on the trivial observation that there are two different types of subsets with n items selected out of a set of k members: those subsets that include the first member of the set and those that do not. For instance, $(1,2)$ and $(1,3)$ both have the first member of $(1,2,3)$, whereas $(2,3)$ does not.

Let's consider the two different types of subsets separately. If we know that the first member of the set is in the combination, then there are $n-1$ items left to be taken from the remaining members of the set. This suggests a recursive approach: Construct all of the subsets that have $n-1$ items selected from all of the members of the set except the first one. Or, using an accumulator style, we can set the accumulator to hold those elements that are already known to be in the subsets and allow the recursive computation to add in the remaining elements.

Those subsets that do not include the first member of the set can be generated by selecting n items from all of the members of the set except the first one.

There are several base cases:

- There are $k = n$ items in the set from which we are to take n items. In this case, there is only one combination, and that is the entire set.
- There are $k < n$ items in the set from which we are to take n items. This are no possible subsets of size n, so the empty set is the answer.
- There are no items left to be taken; that is, $n = 0$. In this case, we are done assembling the subset, so return the subset already assembled. That is, return the accumulator.

The program `combinations` implements this recursion. Rather than taking any set at all, it works on the set of integers `1:k`. This is quite general, since we can use these integers as indices into any desired set. The function returns a matrix where each row is one of the possible ways of picking n items from the set `1:k`. (The matrix has n columns.) A wrapper function creates the set `1:k` and initializes the accumulator to be the empty vector.

The wrapper function

```
[1]  % combinations(k, n)
[2]  % All the ways of taking n items
[3]  % out of a set of k items
[4]  function res = combinations(k, n)
[5]  res = helper(1:k,n,[]);
```

The wrapper function calls a recursive helper function:

Recursive helper function
Base case: can't find n items in set
Base case: have already found n items
Base case: the whole set is needed
The two recursive branches
Those subsets with `set(1)`
Those subsets without `set(1)`
Collect together the two types of subsets

```
[6]  % Helper function
[7]  function res = helper(set,n,accum)
[8]  if n > length(set)
[9]      res = [];
[10] elseif n==0
[11]     res = accum;
[12] elseif n == length(set)
[13]     res = [accum, set];
[14] else
[15]     a = helper(set(2:end),n-1,[accum,set(1)]);
[16]     b = helper(set(2:end),n,  accum);
[17]     res = [a; b];
[18] end
```

The accumulator `accum` keeps track of those elements of the set that have already been selected for the subset. Here are all of the ways of picking three items out of a set of five; each row is one of the ways.

```
>> combinations(5,3)
  ans:  1 2 3
        1 2 4
        1 2 5
        1 3 4
        1 3 5
        1 4 5
        2 3 4
        2 3 5
        2 4 5
        3 4 5
```

Note that the first several rows all contain 1 as the first member—these are the subsets generated by the first type of recursion. The remaining rows do not have 1 in them at all—these are the subsets generated by the second type of recursion.

12.4   Example: Optimal Matching (Recursive)

In Section 8.4, we examined how best to match up a set of people to a set of jobs. The objective is to find the assignment of people to jobs that produces the maximum quality; it is assumed that we know how good a job each person would do on each job.

In the program `matchpeople` on page 158, we used nested loops: one loop for each person. Clearly, the number of nested loops depends on the number of people to be matched, and so, a different program is needed for each different-size problem. There was also a rather complicated logic to make sure that two people aren't assigned to the same job or two jobs to a single person.

The problem takes on a rather simpler form if we arrange it recursively. The recursion, as always, has three logical parts:

Base case: If there is just one person and one job, then match the person to the job.

Simplification: Take the first person and, on a trial basis, match that person to the first job. With that person and that job eliminated from the problem, find the best match between the remaining people and the remaining jobs. This problem is simpler than the original because there is one fewer person and one fewer job. The quality for the trial assignment is the sum of the quality of matching the first person to his or her job plus the total quality for the best match of the remaining people to the remaining jobs.

Repeat the trial assignment of the first person to each of the jobs in turn, choosing the assignment that gives the largest total quality.

Combination: The overall answer to the problem is constructed by combining the best trial assignment of the first person to a job with the answer to the recursive problem.

The basic idea is to try all of the possibilities and pick the best one. The recursive structuring of the problem enables us to consider only the job possibilities for the first person, delegating the solution to the rest of the problem to the recursion.

To illustrate, consider the same problem solved in Example 8.4: four people to be matched to four jobs with the quality of each person's performance of each job given by the table on page 156. Making a trial assignment of the first person to job 1, we are left with the problem of the best match of persons 2, 3, and 4 to jobs 2, 3, and 4. Given the answer to this simplified problem, we can move on to another trial assignment and see what happens when the first person is assigned to job 2, leaving persons 2, 3, and 4 to be assigned to jobs 1, 3, and 4. There are, altogether, four possible trial job assignments for person 1. To assess each of these four possibilities, we need to solve the problem of assigning the remaining jobs to persons 2, 3, and 4. We can then pick the best of the four possibilities as our final answer.

The subproblem of finding the best assignment of persons 2, 3, and 4 to the three remaining jobs works the same way: Make a trial assignment of person 2 to each of the three jobs in turn, and for each trial assignment, calculate the best assignment of persons 3 and 4 to the remaining two jobs. This sub-subproblem is then handled in the same way, assigning person 3 to each of the two remaining jobs, and finding the best assignment of person 4 to the single remaining job. Of course, the assignment of a single person to a single job is completely trivial; this is the base case.

Figure 12.2 shows the various possible job assignments for persons 1, 2 and 3. There are four possible trial assignments for person 1. Given that assignment, person 2 has three possible trial assignments. Given the assignments of persons 1 and 2, person 3 has only two possible assignments. Since there are only four jobs, person 4's assignment is set by the assignments of the first three people. There are 24 different possibilities to be considered arranged as a tree. The recursive formulation of the problem loops over the possibilities at the highest level in the tree and recursively delegates the solution to the problem of each of the subtrees.

A crucial step in writing a recursive program is deciding what will be the inputs and outputs of the recursive function. The inputs, in addition to the cost matrix that defines the specific problem, need to include the list of remaining people and remaining jobs, that is, those that have not already been assigned on a trial basis. The important output information is the best

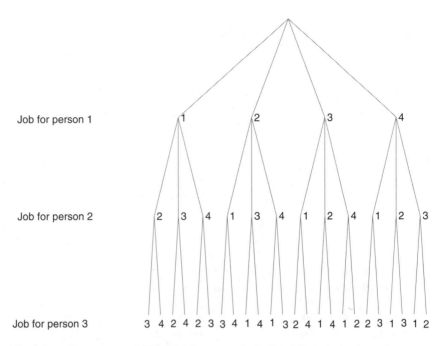

Job for person 1

Job for person 2

Job for person 3

Figure 12.2.   The job assignment possibilities for four people to four jobs, arranged as a tree.

match between people and jobs; this can also be represented by the list of people and a list of corresponding jobs. Since the cost of this matching needs to be considered at the higher level in the recursion, we need to return this as well.

Here is a matching program written in the recursive style. A wrapper function sets up the problem:

```
[1] %  bestAssign(qual): Assign people to jobs
[2] %  to maximize quality
[3] %  qual is a square matrix:
[4] %  each person is one row each job
[5] %  is one column
[6] function jobs = bestAssign(qual)

[7] [r,c] = size(qual);
[8] [qual, people, jobs] = helper(1:r,1:r,qual);
```

A wrapper to set up the recursion

Set up the lists of people and jobs and start the recursion

The actual assignment is done with a recursive helper function. The basic strategy is to use a loop to find the best job assignment for the first person in the list. Of course, the job assigned to this first person affects the possible job assignments for the rest of the people; the optimal job assignment for the rest of the people is computed recursively on line 23.

Recursive helper function returns
the best found quality bq, and
the corresponding best people
bp and job bj assignments

Base case: one person and one job

The person in the trial assignment
Remaining people
The best found so far
Try all the job possibilities
for the first guy
Remaining jobs
Recursive solution of the
simplified problem
c carries the result of the recursive
problem to the solution of this one
Is this new one the best so far?

```
[9]  function [bq,bp,bj]=helper(plist, jlist, qual)

[10]   np = length(plist);
[11]   nj = length(jlist);
[12]   if np==1
[13]      [bq,ind] = max(qual(plist, jlist));
[14]      bp = plist;
[15]      bj = jlist(ind);
[16]   else
[17]      firstguy = plist(1);
[18]      rpeople = plist(2:end);
[19]      bq = -Inf;
[20]      for k=1:nj
[21]         rjobs = jlist(k ~= (1:nj));
[22]         [c,p,jbs]=helper(rpeople, rjobs, qual);
[23]         thisqual=qual(firstguy,jlist(k))+c;
[24]         if thisqual > bq
[25]            bq = thisqual;
[26]            bp = [firstguy, p];
[27]            bj = [jlist(k), jbs];
[28]         end
[29]      end
[30]   end
```

The recursive function invocation is modestly situated on line 22. We can apply this program to the quality table on page 156:

```
>> qual = [7 4 2 4; 6 8 5 2; 4 7 1 3; 6 5 2 1];
>> bestassign(qual)
   ans: 4 3 2 1
```

As expected, this gives the same answer as the nested-loops approach to the problem. But since there is just one loop that goes over the possible assignments at the top level of the assignment tree, the same program can be applied to problems of different sizes. (See Exercise 12.1.)

Although the recursive formulation of bestAssign enables it to deal with problems of various sizes, it does not eliminate the essential mathematical difficulty of the problem. For n people there are altogether $n!$ possible job assignments. This is a huge number for even moderate n, for $n = 15$ there are more than 10^{12} possible assignments to be considered.

In order to handle such situations, a different strategy is called for. Since it is impossible to examine each of the possible assignments, we need to examine just a subset of them. Two approaches to doing this, beyond the scope of this book, are called "branch and bound" and "simulated annealing" [17].

12.5   Trees as Data Structures

In Chapter 5, we saw that computers generally store files in a treelike manner: Each directory can hold files as well as other directories. We've also seen that recursive functions can be thought of as organizing a computation like a tree. In the next sections, we will combine the two approaches and consider computations that generate and process treelike data.

Before we can do this, though, we need to have a MATLAB data type that is suitable for storing data organized as trees. For the sake of definiteness, let's consider some data comprising the phylogenetic tree shown in Figure 12.3. The root of the tree is a node entitled "Trees in general." From this node emerge two smaller trees, one labeled "Deciduous" and the other labeled "Evergreen." The subdivision of types continues recursively.

Taken as a whole, the tree in Figure 12.3 is complicated. But keeping in the spirit of recursion, where we simplify things by delegating out the details, let's consider divide the tree into nodes. The root node of the tree, the

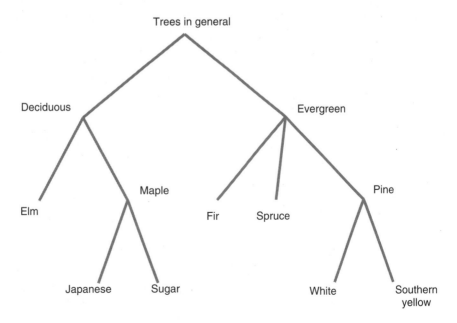

Figure 12.3.   The phylogenetic relation of some common trees.

one labeled "Trees in general" has three pieces of data associated with it: (1) the label of the node ("Trees in general"); (2) the right branch (the tree that has "Evergreen" at its root node); and (3) the left branch (the "Deciduous" tree). One way to organize this information is as a structure type, perhaps with three fields: We can store the label in a field called `value`, the left-hand tree in a field called `leftbranch`, and the right-hand tree in `rightbranch`.

This strategy would work well if all nodes had two branches coming from them, but in Figure 12.3 we see one node, "Evergreen," that has three branches. There are several nodes—"Elm," "Japanese," "Sugar," and so on—that have no branches coming off of them. These are *terminal branches*, also known as "leaf nodes."

Key Term

This diversity of node types suggests a simple generalization: Store each node's information as a structure with two fields: a `value` field to store any information associated with the node itself, and a field named `branches` that will hold information about the branches. Of course, we could use names other than `value` and `branches`. The names themselves hardly matter, but by standardizing on some fixed names, we will make matters easier when it comes time to write functions that operate on trees.

The `branches` field needs to hold nodes. For instance, the `branches` field for the "Deciduous" node needs to hold the "Maple" node and the "Elm" leaf node. Since the node itself is a structure, holding a set of nodes involves an array of nonnumerical data; this suggests a cell array. For a leaf node, the `branches` array will be empty since there are no branches.

Here's a function that collects the information needed to for a node and formats it as a structure with the `value` and `branches` fields that I have standardized on.

```
[1]  function res = makeNode(value, varargin)
[2]  %  Collect a value and some other nodes into
[3]  %  a structure with fields
[4]  %  <value> and <branches>
[5]  res.value = value;
[6]  if length(varargin)==1 & iscell(varargin{ 1} )
[7]     res.branches = varargin{ 1} ;

[8]  else
[9]     res.branches = varargin;

[10] end
```

The value field — `[5]`

Branches given as a single cell-array argument — `[7]`

Branches as multiple arguments — `[9]`

The value of the node is the first argument to `makeNode`. Any branch nodes should be given as additional arguments, either collected by the cell-array brackets or as separate arguments. The `varargin` construction is a special bit of syntax that instructs the interpreter to collect all of the additional arguments into a cell array named `varargin`.

The `makeNode` function potentially is confusing because it is so simple: All it does is collect information and arrange it into a simple structured data type. It's easiest to understand with an example. Here's how to make the leaf node labelled "Sugar" in Figure 12.3:

```
≫  a = makeNode('Sugar')
  ans: value: 'Sugar'
       branches:
```

Now that the Sugar node is created, we can use it as an argument when creating another node that has Sugar as a branch. For instance, the Maple node has Sugar and Japanese as branches, so if we make a Japanese node

```
≫  b = makeNode('Japanese')
  ans: value: 'Japanese'
       branches:
```

then we can specify the two nodes as the branches to Maple:

```
≫  c = makeNode('Maple', a, b)
  ans: value: 'Maple'
       branches: [1x1 struct] [1x1 struct]
```

Note that the `branches` field of c has two entries; these are the nodes Japanese and Sugar.

The `makeNode` function also allows you to prepackage the various branches as a cell array; for instance,

```
≫  a = makeNode('Maple', a,b);
```

This will simplify things later when we write programs that grow trees.

It's not strictly necessary to give names like a, b, and c to the nodes as they are created. We could have created the Maple tree with the expression:

```
≫  makeNode('Maple', makeNode('Sugar'), makeNode('Japanese'))
```

The entire "Trees in General" tree was constructed with the statement

```
a = makeNode('Trees in General', ...
    makeNode('Deciduous', makeNode('Elm'), ...
    makeNode('Maple', makeNode('Japanese'), makeNode('Sugar'))), ...
    makeNode('Evergreen', makeNode('Fir'), makeNode('Spruce'), ...
    makeNode('Pine', makeNode('White'), makeNode('Southern Yellow')))));
```

and drawn using

```
≫  drawtree(a)
```

In our standard notation for a node, a leaf is identified because it has no branches. This can easily be tested with a statement like

```
length(node.branches) == 0
```

Somewhat clearer is to have a function `isleaf` that returns a boolean indicating whether a node is a leaf or a tree.

```
[1]  function res = isleaf(node)
[2]  %  Test whether a node is a leaf
[3]  res = ~isfield(node,'branches') |...
     isempty(node.branches);
```

To handle cases where the node is formed improperly, test whether `branches` is a fieldname of `node`

12.6   Processing Trees

The fact that `makeNode` makes trees whose nodes all have a consistent format makes it easy to write general-purpose functions to operate on the trees. For instance, the program `drawtree` will draw a graph representing a tree, marking each node or leaf with text reflecting the `value` field of the node. `Drawtree` is somewhat complicated, since in order to figure out nice locations for each node it is necessary to know where the other nodes are. But other tree-processing functions have a simple recursive format.

Consider the problem of finding the maximum depth of a tree. This problem has a recursive formulation: The depth of a tree is 1 plus the depth of the deepest branch. The base case is a leaf; a leaf has 0 depth. For instance, the "Trees in general" tree has two branches, "Deciduous" and "Evergreen." Each of these branches has a depth of 2, so the depth of the entire tree is 3. The program `treedepth` merely applies itself recursively to each of the branches in the tree, finding the branch with the maximum depth. If there are no branches (that is, the tree under consideration is a leaf), then the base-case situation applies: The depth is zero.

```
[1]  function res = treedepth(tree)
[2]  % treedepth(tree)- maximum depth of tree
[3]  if isleaf(tree)
[4]      res = 0;
[5]  else
[6]      bdepths=zeros(length(tree.branches),1);
[7]      for k=1:length(tree.branches)
[8]          bdepths(k)=treedepth(tree.branches{k});
[9]      end
[10]     res = 1 + max(bdepths);
[11] end
```

A node has no depth

Find the depths of the branches

Total depth: 1 + the maximum branch's depth

A function to collect the leaves in a tree has a similar recursive structure: If the tree is a leaf, then return the value of that leaf. Otherwise, assemble together the leaves from each of the branches. Since we don't know ahead of time what type of variable the value of the leaves will be, it's best to collect the leaves as a cell array.

```
[1]  % collectLeaves(tree): put leaves in array
[2]  function res = collectLeaves(tree)
[3]  if isleaf(tree)
[4]      res = {tree.value};
[5]  else
[6]      res = {};
[7]      for k=1:length(tree.branches)
[8]          foo = collectleaves(tree.branches{k});
[9]          res = {res{:}, foo{:}};
[10]     end
[11] end
```

Base case — [3]

Gather leaves for this branch and add to the overall collection: cell-array concatenation — [8], [9]

```
≫  collectLeaves(a)
ans: 'Elm' 'Japanese' 'Sugar' 'Fir' 'Spruce'
     'White' 'Southern Yellow'
```

12.7   Example: Huffman Encoding of Information

The Shannon information provides a theoretical measure of information that takes into account the fact that some letters are frequently used and others not. In Chapter 8, we analyzed the frequency of letters in the text of the novel *Moby Dick* and found a Shannon information of 4.3967 bits/character. To achieve this information rate in practice requires the use of a variable-length code.

There are two problems that need to be solved to use variable-length codes effectively: how to generate an efficient encoding for a given character set and its corresponding relative frequencies, and how to decode a bit string according to the encoding. The solutions to both of these problems involve trees.

The problem of constructing an efficient encoding was solved by David Huffman (1925–1999). The information needed is a list of symbols and a list of frequencies: how often each symbol is used. To illustrate, consider the E–Z code from Chapter 8. Character E is used 50% of the time, Z and – are each used 25% of the time. One method to construct a Huffman encoding is to find the two least-used symbols (Z and – in this example). These two symbols are then merged as two branches of a single tree. This tree is then treated itself as a symbol whose frequency is taken to be the sum of the frequencies of the two branches. Then the two merged symbols are discarded,

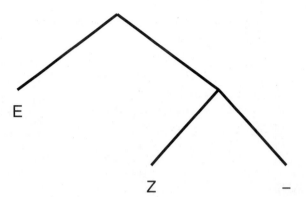

Figure 12.4. The Huffman tree for the E–Z code. The Huffman code is described implicitly by this tree by assigning a 0 to each left branch and a 1 to each right branch. In this tree, E is coded by 0, Z by 10, and – by 11.

keeping in their place the tree that is composed of the two symbols. Since two symbols have been replaced by one—the newly created tree—the list of symbols is smaller by one. This process is repeated until there is only one symbol left. For instance, the Z-tree will be merged with the E symbol to produce a single tree, as in Figure 12.4.

The program huffmantree implements this process.

```
[1] function res = huffmantree(symbols, freqs)
[2] % huffmanTree(symbols, freqs) make Huffman code
[3] % symbols: char or numeric vector
[4] % freqs: counts or relative frequencies
[5] treelist = cell(length(symbols),1);
[6] for k=1:length(symbols)
[7]     treelist{k} = makenode(symbols(k));
[8] end
[9] while length(treelist) > 1
[10]    [treelist,freqs]=reduceByOne(treelist,freqs);
[11] end
[12] res = treelist{1};
```

Set up each symbol as a tree [5]

Merge until only one is left [9]

This program sets up a list of trees, each of which is a single node containing a symbol with no branches. The program loops, each iteration combining the two least frequent symbols, until only one tree is left.

The hard work is being done by the helper function reduceByOne, which has been defined as a subfunction to huffmantree.

Helper function	[13]	```function [t,f] = reduceByOne(t,f)```
	[14]	```if length(t) == 1```
	[15]	``` return; % done```
Merge two into one tree	[16]	```else```
	[17]	``` [trash,inds] = sort(f);```
Replace one with the new tree	[18]	``` t{inds(1)} = makenode(' ', ...```
	[19]	``` t{inds(1)},t{inds(2)});```
	[20]	``` f(inds(1)) = f(inds(1)) + f(inds(2));```
Replace the other with the last item	[21]	``` t{inds(2)} = t{end};```
	[22]	``` f(inds(2)) = f(end);```
Shorten the list	[23]	``` t = t(1:(end-1));```
	[24]	``` f = f(1:(end-1));```
	[25]	```end```

The sort operator has been used to identify the two least frequently used symbols; these will be the symbols whose indices are inds(1) and inds(2). On line 18, these are merged together into a new tree and, on line 19, the frequencies of the two merged symbols are added together. Note that the new tree and its frequency are inserted into the place in the lists t and f that were previously filled by the first of the merged symbols. We also need to eliminate the second of the merged symbols from the lists. A simple way to do this is to copy over the element at the end of the list into the slot occupied by the second merged symbol. This is done on lines 20 and 21. Now the last element is redundant: It appears in two places in the list. We eliminate the last element in lines 22 and 23, thereby shortening the list by one.

To use huffmantree, we need to specify the set of characters and their frequencies of use. As in Chapter 8, we'll draw from *Moby Dick*.

```
>> [cnt,letters] = countCharsInFiles({'mobydick-excerpt.txt'});
```

The program countCharsInFiles used the entire set of 256 possibilities for bytes; most of these never appear in *Moby Dick* and so don't need to be included in a code:

```
>> keep = cnt>0;
>> hcode = huffmanTree(letters(keep), cnt(keep));
```

The tree produced is shown in Figure 12.5. The left branches are encoded with a 0, the right branches with a 1. Some of the codes that can be read off the tree are as follows:

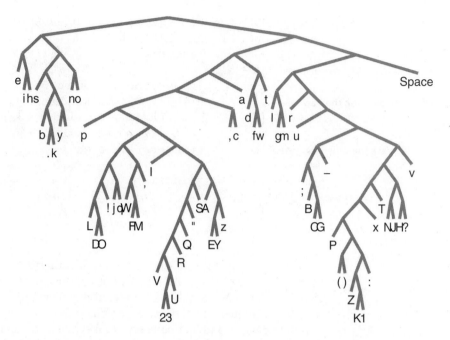

Figure 12.5. The Huffman code tree for the character frequencies from *Moby Dick*.

space	111	c	100011
C	11011100110	e	000
a	1001	s	0100
l	11000	h	1011
m	110011	period	0101010

The common characters, such as the space character and e, are encoded with short patterns, while the long patterns are used for less common characters, such as C.

Using this code, *Moby Dick* can be encoded. Here are the first few characters in plain form (where the spaces are, of course, part of the text)

        Call me Ishmael.

and in Huffman-code form

110111001101001110001100011111001100011100001100100001110011100100011000010101010

The entire 101,412 characters of Chapter 1 of *Moby Dick* need 449,351 bits, or 4.431 bits per character: close to the Shannon information of 4.40 estimated in Chapter 8.

The Huffman code enables us to find a practical encoding for an alphabet that closely approximates the Shannon information. Just as the Shannon information depends on the relative frequencies of the letters in the alphabet, the Huffman code is based on the relative frequencies. For example, since English text involves a different set of relative frequencies than French text, Huffman codes for text in the two languages would be different.

Computing a Huffman code is only part of the story for compressing data. Compressing *Moby Dick* with a file compression program like `gzip` results in a file that is only 42,293 bytes, or 3.336 bits per character. This compression program beats the Shannon information not because the Shannon information is wrong, but because we calculated the Shannon information assuming a certain kind of alphabet: the ASCII characters. The correspondence between ASCII and the ordinary A–Z alphabet makes it seem natural to use these when encoding text files, but there is no reason why the alphabet used for encoding can't have sequences of letters, or even words or phrases, as its alphabet. For example, our alphabet for encoding might consist of `'en'`, `'the '`, `'ing'`, and so on. Compression programs such as `gzip` use algorithms for finding efficient alphabets; the alphabet might be unique to the file being compressed. Once the alphabet is found, the Shannon information puts a lower bound on the number of bits needed.

Constructed alphabets like this aren't restricted to text. Formats such as JPEG, for storing photographs efficiently, use a fixed alphabet consisting of very small striped and dotted subimages. Bitmapped pictures are often transmitted as `png` or `gif` files. These are compressed using a scheme that involves an alphabet with entries like this: "seventeen 0s in sequence" or "nine repetitions of 001."

High-efficiency compression systems, such as `jpeg` or `mp3`, exploit an additional issue not considered in the formulation of the Shannon information: The message transmitted depends on the interpretation by the recipient. The listener to music or viewer of a photograph does not in fact make use of every bit of information in the original. By throwing away information that is not used by the human viewer or listener, these compression programs save bits. This *lossy compression* works for images and music, where much of the information is redundant. It evn wrks fr txt t a sm xtnt, but the risk of losing critical information that might appear in even a single word or letter in text is so great that it is unusual to use lossy compression for text.

12.8 Exercises

Discussion 12.1:
For the job-matching problem in Section 12.4, measure the time taken to run the program for different problem sizes; that is, different sizes of the `qual` matrix. You can generate random matrices of size n using `rand(n)`.

Compare your experimental results to the theoretical answer that there are $n!$ ways of assigning n people to n jobs, each of which is considered in bestAssign.

Discussion 12.2:

Use makeNode and drawtree to make a tree like that shown in Figure 12.1 but for fib(7). You can start by defining a few nodes:

```
≫  one = makeNode('fib(1)');
≫  two = makeNode('fib(2)');
```

Give three the title 'fib(3)' and, as branches, one and two.

Following this pattern, you can draw the tree with

```
≫  drawtree(seven)
```

Why did you only need seven nodes to describe the whole tree, which obviously has many more than seven nodes?

Exercise 12.1:

Write a recursive program, maxrec(vec), for finding the largest number in a vector. Here's the recursive logic:

Base case: If the vector has a length of 1, return the single element as the largest.

Simplification and recursion: If the vector has more than one element, return the larger of these two numbers: the first number in the vector and the largest number in the rest of the vector.

Exercise 12.2:

There are several values of n for which the programs fib and factorialrecursive will fail, resulting in a theoretically infinite recursion. Find these ways and rewrite the functions so that they intercept them and generate a helpful error message.

Exercise 12.3:

The recursive fib program is an effective way of computing Fibonacci numbers only for small n. Even when $n = 30$, fib takes a minute or so to run. You can time it on your computer with a sequence of statments:

```
≫  tic; fib(30); t=toc
   t: 85.3000
```

The reason for the long run time is that fib makes redundant computations, computing over and over again the same values, as shown in Figure 12.1.

Show theoretically that the number of recursive invocations involved in computing fib(n) is the nth Fibonacci number.

Demonstrate the relationship experimentally by measuring the time of running fib(n) for various *n* and plotting each of these versus the corresponding theoretical result.

Watch out, though. According to the theoretical result, fib(40) will take 123 times as along as fib(30). Using the theory and your experimental results, estimate how long it will take to compute fib(50) and fib(60). You will need a practical way to compute the fiftieth and sixtieth Fibonacci numbers; you can use either the program in Exercise 12.4 or the looping program in Chapter 8. (If you accidentally get stuck in a computation that is taking too long, you can terminate it by pressing CNTL-c in the command window.)

Exercise 12.4:

The problem with the recursive formulation of fib is that it wastes time: It recomputes values that have already been computed. Here we use a technique "memoization," that speeds things up enormously by exploiting the fact that many of the delegated tasks already have been performed at a higher level.

Write a program, fibmemo(n), that remembers the results of previous calculations. You can do this by setting up a persistent variable, a vector whose *n*th element contains the previously computed results (or, initially, zero). For $n \geq 2$, fibmemo should check to see if the persistent array has a nonzero entry at *n*. If so, simply return this value. If not, call fibmemo recursively, but before returning the result, save it in the array.

This simple modification to the recursive algorithm allows numbers like fib(100) to be computed rapidly. (*Caution:* If you try large *n*, you may get a RecursionLimit error. Build up *n* gradually, in steps of no more than a few hundred, to avoid this problem.)

Exercise 12.5:

Write a function like collectLeaves on page 313 that collects the leaves but also returns a second value: a vector telling how deep in the tree each leaf is.

Exercise 12.6:

Write a program, huffmanDecode(bits, codeTree), that takes as arguments a vector of 1s and 0s and a code tree (as produced by huffmanTree) and decodes the first argument. An algorithm for decoding the bit string involves keeping a marker pos for the position in the code tree, looping over the bits, and appropriately modifying the position:

0. Set pos to be the root node, the node at the top of codeTree; that is, pos = codeTree;. Loop over the elements of bits.
1. If pos contains a letter, emit that letter and reset pos to be the root node.
2. If the bit under consideration is 0, go down the left branch of the tree; that is, pos = pos.branches{1};. If the bit is 1, go down the other branch.

Exercise 12.7:

Write a program, [vals,codes] = getCodes(htree), that takes a Huffman code tree as an argument and returns two cell arrays. The first cell array contains the values of the leaves of the tree; the second contains the corresponding codes represented as a vector of 1s and 0s. You will have to assemble those vectors as you traverse the tree toward the leaves.

With vals and codes, it is relatively easy to write a program to encode text using htree. Do so, producing a program bits = encode(text, htree).

12.9 Project: Clustering of Data

An iris.

In clustering, the problem is to take individual items and decide how to group them; one wants each group to consist of similar items. The problem is to define "similar." Often, we just have some measurements made on the items without any clear definition of when two measurements are similar and when they are different. For the sake of definiteness, consider some famous data on the shapes of various species of irises. (These data were collected by Edgar Anderson [20].) Here is a small subset of the data; the complete set is in the file iris.dat.

Item ID	Sepal Length	Sepal Width	Petal Length	Petal Width
1	5.1	3.5	1.4	0.2
2	4.9	3.0	1.4	0.2
3	4.7	3.2	1.3	0.2
4	4.6	3.1	1.5	0.2
5	5.0	3.6	1.4	0.2
⋮	⋮	⋮	⋮	⋮
149	6.2	3.4	5.4	2.3
150	5.9	3.0	5.1	1.8

In tabular format, the data are not particularly informative. Since there are four measurements on each flower, it's difficult to plot all of the data at once, but we can look at scatterplots of pairs of measurements. An effective technique is to look at all of the pairs of variables. Since there are four measurements, there are six possible pairs of them:

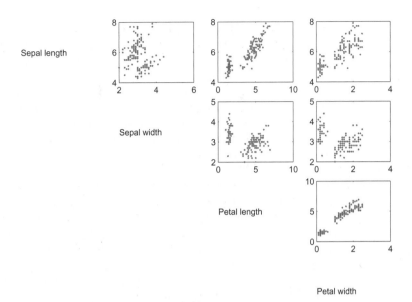

Figure 12.6.   Pairwise scatterplots of the Anderson iris data.

```
≫ combinations(4,2)
  ans:  1 2
        1 3
        1 4
        2 3
        2 4
        3 4
```

Figure 12.6 suggests that there are two distinct types of irises in the data. There is clearly a gap between two main clusters of points. What's not clear is whether there are meaningful divisions within the main clusters. Without an objective definition of what constitutes a cluster, it is difficult to know how to divide the clouds of data points.

A problem such as this, where we seek to put the individual data points into groups without any prior definition of what constitutes a group, is called *unsupervised learning*. The alternative sort of problem, called *supervised learning*, is when each data point comes with a group label, and we seek to find a definition for the groups in terms of the measurements. Supervised and unsupervised learning are problems at the interface of computer science and statistics. In computer science, the problems are one aspect of "artificial intelligence."

There are many approaches to solving the unsupervised clustering problem. A simple but effective method involves the construction of a

Key Terms

Key Term

dendrogram. In a dendrogram, each data point is the leaf of a tree. The dendrogram collects the leaves together into nodes, always two at a time. Figure 12.7 shows a dendrogram for a subset of the Anderson iris data, points `1:5:150`.

To understand how the decision is made about which nodes to group together, consider each of the points in the data set in terms of one of the scatterplots shown in Figure 12.6 (for example, sepal length versus petal width: the plot in the upper-right corner of the figure). We start with each of the data points as a node with no branches: a leaf. The position of any particular node is the position of that node's data in the scatterplot. Then we calculate the interpoint distance from each node to every other node; that is, the distance between every pair of nodes. One of these pairs will have the shortest interpoint distance (although there might be a tie). Take the pair of nodes that have this shortest distance (breaking ties in any convenient way), and group them together as the branches of a new node. Treat this node as a new point that has its position at the average of the positions of its branch nodes. Take as the value of the node the distance between the two nodes that have been brought together. Repeat this process. Every time a new node is formed, two previous nodes are grouped together. The func-

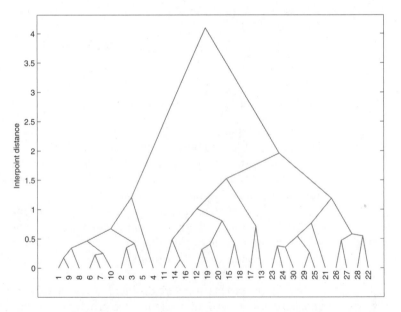

Figure 12.7. A dendrogram of points 1, 6, 11, 16, ... of the Anderson iris data. The dendrogram is based on the distances between points in the four-dimensional space—the position of each point is given by the four numbers: sepal width, sepal length, petal width, and petal length.

tion that does the grouping into trees is dendro: It takes as an argument a numerical matrix giving the position of each data point.

If there are N data points, then after N passes of this algorithm all of the points will be grouped together into a tree. When the tree is plotted out, each node is graphed at a vertical position that represents the distance between the two branch nodes.

To interpret the dendrogram, we look for groups of points where the points within a group are close to one another, but the groups themselves are far from other groups. One way to do this is to pick a clustering distance and divide the points into groups based on the number of branches at that distance. For example, if we believed there were three groups then, with reference to Figure 12.7, we would pick a distance of about 1.7 because at that distance there are three groups that have not yet been brought together. Then given the distance, we can see which group each individual data point belongs to.

The problem with unsupervised learning is that it's hard to be sure how good our answer is. In this case, we can check that things make sense because, in reality, the species of each of the flowers in the data set is known. The division of this dendrogram into three groups corresponds exactly to the species information. If we didn't know this information, we might examine the *robustness* of our classification by making a dendrogram with a slightly different set of data, perhaps different measurements, or by selecting a different subset of the data.

Exercise 12.8:
Write a program to cluster data using the preceding algorithm. Note that the overall algorithm is similar to that used in huffmanTree, except that the freqs data will not be a vector but the matrix of data, with one row for each item. You will need to supply your own logic for finding the two rows whose data are closest and modify the updating rule as the for the row corresponding to a newly merged tree. Note that the drawtree program looks for a node component with a fieldname of vertical, which, if present, tells drawtree what vertical level to use for drawing that node. Setting each node's vertical field to the distance between the merged trees will allow the tree to be plotted as a dendrogram.

Use your program and drawtree to make a dendrogram for the iris data.

CHAPTER 13

Sounds and Signals

Much of the information that we manipulate on the computer comes from measurements of physical quantities. In many cases, these quantities can be represented by a single number. For example, the body temperature of a person with the flu might be 39 °C (or, equivalently, 102.2 °F). In the case of body temperature, as useful as the single temperature reading is in directing treatment, we recognize that the quantity is changing continuously in time, varying not just in sickness and health but also over the 24-hour cycle of sleep and wakefulness. We take repeated measurements of temperature in order to track these changes.

For some physical quantities, a single measurement at a single point in time is useless. For instance, consider the air pressure involved in sound waves. The important phenomenon, the one that gives sound its character and meaning, is the continuous fluctuation of the quantity. In this chapter, we consider how continuous physical quantities can be represented and manipulated on the computer. A few examples: the minute fluctuations in electrical potential on the body surface that give rise to the electrocardiogram (ECG); air temperature, wind velocity and direction measured at a weather station; seismological recordings of movements in an earthquake; brightness recordings from stars at different frequencies of the electromagnetic spectrum.

In all of these examples, the physical quantity varies in a continuous way; it is a relationship between two variables: the dependent variable (e.g., voltage, temperature, pressure, or velocity) and an independent variable (e.g., time, space, or frequency).

In the days before computers became so widespread, the technology for recording continuous quantities was intuitive. In most cases, the independent variable is time (or can be converted to time) and a roll of paper would unscroll past a pen or series of pens. The pens would rest continually on the paper, moving from side to side proportionately to the quantity being measured and thereby tracing out in a continuous stream of ink the quantity versus time. Such chart recorders are still occasionally to be seen in research laboratories, but they are vanishing fast. Perhaps the best place to look for them now is in crime movies from the last century where they are a central part of lie-detection equipment.

In the digital era, collection, transmission, and recording of data tends to be done by computer. These might be expensive laboratory instruments capable of handling vast amounts of data from many sources or a palm-sized mobile telephone.

In order to be concrete, we focus on computation with sounds. But the principles that we need to understand how computers store and manipulate sound are equally applicable to a large variety of types of data collected in diverse fields.

13.1   Basics of Computer Sound

With developments in computer hardware and multimedia software, sounds now seem to be an almost natural part of the computer. We use computers to play recorded music, to generate realistic sounding musical compositions, to translate sound to speech automatically, and even to recognize spoken sound and translate it into text. There is an astounding amount of scientific and technological knowledge that underlies these modern capabilities.

To understand the basics of the acquisition, storage, and manipulation of sound, we take a step back into the past. The file fh.wav contains part of Winston Churchill's famous 1941 speech after the Battle of Britain: "This was their finest hour." (A longer excerpt of the speech is in finehour.wav.) Such files are usually read directly into sound playing and editing software, but we read it into MATLAB:

```
>> [x,fs,bits] = wavread('fh.wav');
```

We can play the sound on the computer speakers with the sound operator:

```
>> sound(x,fs)
```

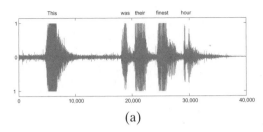

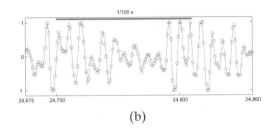

(a)   (b)

Figure 13.1.   The waveform of Winston Churchill speaking, "This was their finest hour." (a) The entire waveform. (b) A close-up of the part of the waveform from the "i" in *finest*, samples 24,675 to 24,860.

Wavread returns three items of information.

x   contains the sound waveform. This is plotted in the left panel of Figure 13.1. The waveform oscillates up and down very rapidly, but only the overall envelope of the waveform can be seen when the whole waveform is plotted. Plotting a small segment of the waveform—just over 1/100 second as shown in Figure 13.1(b)—shows that there are oscillations that take approximately 1 millisecond and that the waveform has been sampled at discrete times.

Key Term   fs   is the *sampling frequency*. This describes how many discrete samples of the waveform have been taken per second. In the fh.wav file, the sampling frequency is 11,025 samples per second.

bits   describes the number of bits used to represent each individual sample in the recording. In fh.wav, each sample was stored as 8 bits but is translated into a floating point number when read into MATLAB.

Looking at the sound waveform, and listening to it, prompts some questions:

- Why does the sound waveform range in amplitude from −1 to 1?
- What role does the sampling frequency play in the quality of the sound?
- What happens if we play back the sound at a different sampling frequency? For example,

    ≫ sound(x, fs/2)

  or

    ≫ sound(x, 2*fs)

- What is it about the waveform that gives its vocal qualities (for example, the "i" sound or the low pitch of Churchill's voice)?

13.2 Perception and Generation of Sound

Sound is a perception. The physical reality that underlies the perception is small, rapid vibrations: up-and-down fluctuations in air pressure. It is these fluctuations—pressure as a function of time, $P(t)$—that are recorded in a sound waveform.

A simple mathematical model of fluctuations is a sine wave:

$$P(t) = A\sin(2\pi ft + \phi)$$

There are three parameters in this model: the amplitude A, the frequency f, and the phase ϕ. Figure 13.2 shows how these parameters relate to the shape of the sine wave.

A single sine wave models a pure musical tone. The amplitude of the sound sets the perceived loudness. The frequency governs the tone of the sound. Frequency is measured in a unit called Hertz (Hz). One Hertz is one up-and-down cycle per second; 100 Hz is 100 cycles per second. Human ears are sensitive to sounds that involve vibrations faster than about 50 Hz and slower than about 15,000 Hz. There are, of course, sounds whose frequencies are outside the range of perception of the human ear. Whales can hear lower-frequency sounds than humans; bats navigate using higher-frequency sounds, called ultrasound.

The frequency is what enables us to distinguish one pure musical tone from another. Most sounds are complicated mixtures of many frequencies. The way the sound is mixed gives musical instruments their timbre and distinguishes the various vocal sounds of speech.

To illustrate, consider the waveform stored in `whistle.wav`, which records the author's attempt to whistle a musical scale.

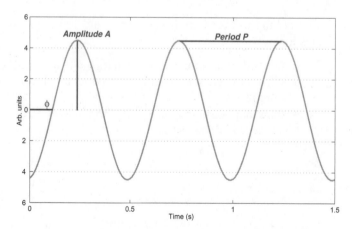

Figure 13.2. A sine wave signal. The amplitude is $A = 4.5$. The period is $T = 0.5$ seconds corresponding to a frequency of $f = \frac{1}{0.5} = 2$ Hz. The phase is $\phi = -1.4$ radians.

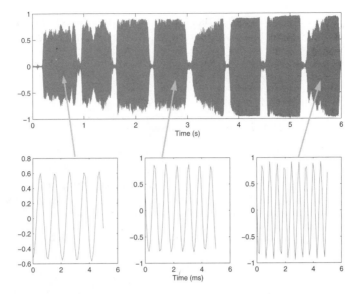

Figure 13.3.   The waveform of whistling a musical scale. The overall recording is six seconds long. Each of the small panels shows 5 ms (i.e., 0.005 s) of the signal.

```
>> [w,fs,bits] = wavread('whistle.wav');
>> sound(w,fs);
```

You will hear the eight ascending notes of the scale. Figure 13.3 plots out the entire waveform, as well as close-ups of three of the notes.

Seen up close, each of the notes closely resembles a sine wave. We can calculate the frequency of the sine waves by counting the number of cycles per second. The leftmost panel of Figure 13.3 contains approximately 5 cycles. Since the duration of the panel is 5 ms, the frequency is $\frac{5\,\text{cycles}}{.005\,\text{s}} =$ 1000 Hz. Similarly, the middle panel has about 6 cycles, giving a frequency of 1200 Hz, and the rightmost panel has about 10 cycles, for a frequency of 2000 Hz. (Note that the first and last notes in a musical scale are separated by one octave. An octave corresponds to a factor of two in frequency: The tones at 1000 Hz and 2000 Hz are an octave apart.)

The function puretone generates a pure sine waveform. It takes as arguments the frequency and duration of the desired sine wave. Here is a pure tone at 1000 Hz lasting for 0.5 seconds:

```
>> [tone1,fs] = puretone(1000, 0.5);
```

This can be played in the ordinary way:

```
>> sound(tone1,fs)
```

Puretone returns a vector containing the samples of sound and a number giving the sampling frequency. The length of the vector

```
≫ size(tone1)
  ans: 1 4098
```

is the sound duration in seconds times the sampling frequency:

```
≫ fs.*0.5
  ans: 4098
```

The amplitude of the waveform generated by puretone is 1. We can use ordinary multiplication to transform the amplitude to whatever we want: Multiplication is amplification. To play the sound at a lower volume, multiply it by a fraction less than 1:

```
≫ sound(0.1.*tone1, fs);
```

and an even lower volume:

```
≫ sound(0.01.*tone2, fs);
```

It's possible that you could not hear this last sound.

Multiplying the waveform by a number greater than 1 makes the sound louder, but since the sound generation hardware has limited capabilities, any values outside the range −1 to 1 are truncated, or clipped, to that range. This *clipping* distorts the waveform; you can hear it if it is severe enough. For instance,

Key Term

```
≫ sound(10.*tone1, fs)
```

When we add two sounds together, we hear (more or less) the sum of the two sounds. To illustrate, let's make another tone at a frequency that is 50% higher than the first and with the same duration and sampling frequency:

```
≫ tone2 = puretone(1.5.*1000, 0.5);
```

Adding together the two tones (and scaling by 0.5 to make sure that we don't exceed the clipping threshold of 1),

```
≫ pair = 0.5.*(tone + tone2);
≫ sound(pair, fs)
```

The result should sound somewhat pleasant or *consonant*. But if the two tones have frequencies that differ by only, say, 10%, the result is *dissonant*. (See Exercise 13.4.) Of course, in order to add together two sounds, they need to have the same number of samples, and they will be played back at the same sampling frequency. You can, of course, add together more than two tones to synthesize more complicated sounds; we do this in Section 13.3.

When addition is used to combine sounds, the result is that the sounds are played at the same time. To play the sounds in sequence, concatenation is used. Since `tone1` and `tone2` are row vectors, we need to use horizontal concatenation:

```
>> sound([tone1, tone2], fs)
```

Long sounds can be assembled in this way. Column vectors could also be used, in which case vertical concatenation would place the sounds in sequence. If two columns are given, for instance,

```
>> sound([tone1', tone2']
```

the sound will be played in stereo with one column sent to each speaker.

It would seem reasonable enough to believe that such long sounds can also be played by using multiple invocations of the `sound` command, as in

```
>> sound(tone1, fs); sound(tone2, fs);
```

But this does not work because of the way that MATLAB tries to communicate with the sound generation hardware; in the first invocation, MATLAB hands off the samples for the speaker to deal with and then immediately sets to work on the second invocation. Unfortunately, the speaker hardware is still busy with the first set of samples and isn't set up to accept the second set. An error message results.

13.3 Synthesizing Complex Sounds

The `puretone` program generates a single sine wave: not a very natural or musical sound. Musical instruments produce much more pleasant sounds that have a well-defined pitch but are not pure sine waves. These sounds are a complicated sum of simpler sine waves. Realistic sounding music is a sequence of such sounds, often adding together multiple instrumental tones at the same time.

In order to synthesize music on the computer, we need a way to represent the sequence, frequency, and duration of musical notes. There are many ways to do this. We'll take a particularly simple approach: a flat file in which we store the starting time, duration, frequency and amplitude of each note in a piece. Figure 13.4 shows the file `scale.csv`, which describes a musical scale.

The function `playmusic` takes this information, synthesizes the notes, and adds them together, doing everything at a sampling frequency of 48,000 Hz. For example,

```
>> a = playmusic('scale.csv');
```

constructs a sound waveform of a musical scale, using the note sequence given in `scale.csv`. This can be played in the normal way:

```
>> sound(a,48000)
```

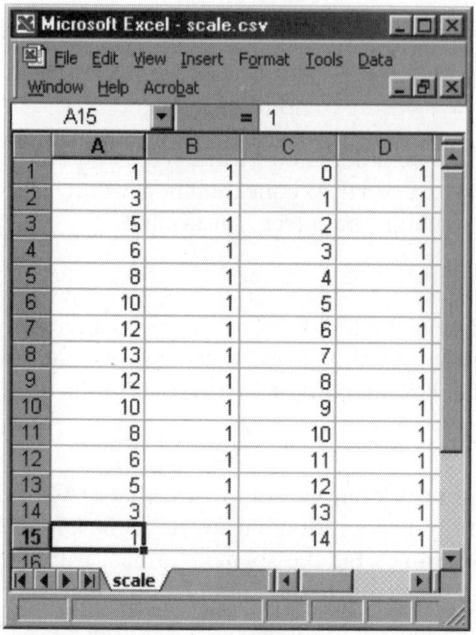

	A	B	C	D
1	1	1	0	1
2	3	1	1	1
3	5	1	2	1
4	6	1	3	1
5	8	1	4	1
6	10	1	5	1
7	12	1	6	1
8	13	1	7	1
9	12	1	8	1
10	10	1	9	1
11	8	1	10	1
12	6	1	11	1
13	5	1	12	1
14	3	1	13	1
15	1	1	14	1

Figure 13.4. A spreadsheet file containing the notes of a musical scale.

This has a somewhat artificial sound, because only a single tone is being played at one time. For a richer sound, multiple tones can be played simultaneously. The file `hushlittlebaby.csv` is a familiar song that involves multiple simultaneous notes.

You can write your own tunes in this format, or, even better, write a program to translate another form of musical notation into this format. For reference, here is the meaning of each of the columns in the file format read by `playmusic`.

The input file should be a spreadsheet consisting of four columns:

1. A number indicating the note to be played.

middle C	1
D	3
E	5
F	6
G	8
A	10
B	12
C	13
and so on	

Notes below middle C are negative numbers. Fractional numbers are allowed.

2. The duration of the note (1 is a quarter note).
3. The starting time of the note (in quarter notes).
4. The amplitude of the note (1 is typical).

By arranging things in this way, multiple notes can be played simultaneously for chords, harmonies, and so on.

By default, `playmusic` uses the `simpletone` program to generate the individual notes. But you can instruct `playmusic` to generate different sorts of notes by writing your own programs. You do this by giving a second argument that is the name of the program you wrote; for example,

```
>> vec = playmusic('scale.xls', 'pipetone');
>> sound(vec,48000)
```

Exercise 13.8 gives directions for writing such programs and particularly, for taking a sound recorded from a real musical instrument or other source and modifying it to change its pitch.

13.4 Transducing and Recording Sound

The ear (the entire system from external pinna to auditory nerve and the brain) is a complicated transducer that doesn't merely transform vibrations in the air into activity in the auditory nerve; the ear also performs computations that transform and reformat the information in the sound vibrations.

Sound recording devices have a much simpler task: Transduce the sound into a form suitable for storage, store it, and enable it to be played back as sound vibrations that can then be processed by the ear. The earliest sound recorder, the phonograph, transduced sound's up-and-down fluctuations in air pressure into waves cut by a needle into a spinning wax cylinder. The air pressure was funneled into a large horn, where it moved a small diaphragm, displacing a needle up and down in the wax and cutting a wavy groove as the wax moved. The wax was permanently deformed by the needle, effectively storing the sound. Playing the sound back involved reversing the process; tracking the needle along the wavy, waxy groove and moving the attached diaphragm to vibrate the air. This sort of recording is called an

Key Term

analog recording because the continuous-in-time fluctuations in the shape of the groove are analogous to the continuous-in-time fluctuations in air pressure that are being recorded.

Microphones ("phone" for sound, "micro" for small, so called because they are much smaller than the horn used in the early phonograph) are generally arranged to transduce sound pressure into electrical voltage,

Edwards

producing a voltage whose fluctuations are like those in sound.[1] This is also an analog transduction.

The fluctuating, analog voltage output of a microphone is useful because it can be used as an input to many different devices: It can be amplified and used to drive loudspeakers that reproduce the sound captured by the microphone; it can be transduced into radio waves and broadcast; it can drive the electromagnetic heads of tape recorders that transduce the voltage into magnetic fields that can be stored permanently on tape.

Key Term

Computers store information in a *digital recording* format: a series of bits. To store sound on a computer, the analog voltage output of the microphone needs to be converted to bits. The most common way of doing this involves measuring the voltage; producing a number. This number can then be stored in bit form. To capture the fluctuating voltage in a digital recording, we need to measure the voltage at many different instants; storing the sequence of fluctuating numbers. It's impossible to measure the voltage continuously in time; doing so would produce an infinite sequence of numbers. Instead, the fluctuating voltage is sampled at discrete times.

Key Term

The process of converting a continuous-in-time analog signal into a sampled digital output is called *analog-to-digital conversion* ("A/D conversion" for short) or "digital sampling."

There is also a reverse process; converting a digital record into an analog form suitable for, say, playing over a loudspeaker. This process is called *digital-to-analog conversion* ("D/A conversion" for short).

Key Term

Digital Sampling

Key Term

Key Term

The standard procedure for digital sampling is to measure the analog signal at fixed, repeated, evenly spaced intervals in time. The time between samples is called the *sampling interval*. Another way to describe the sampling interval is in terms of the number of samples per second; this is called the *sampling frequency*. Numerically, the sampling interval and the sampling frequency are reciprocals of one another; a sampling interval of $\frac{1}{100}$ second corresponds to a sampling frequency of 100 Hz.

The two main issues in digital sampling are these:

- How to set the sampling interval; that is, how often to measure the analog signal.
- How precisely to measure the analog signal at any single instant; that is, how many bits to use in storing each number and how to represent this number.

The second issue is generally much more straightforward, so we will handle it first.

---

[1]Depending on how the microphone is designed, the voltage output may not measure the pressure directly but might measure some function of the pressure, such as its time derivative.

Quantization

Key Term

It seems natural enough to measure instantaneous amplitude of the analog signal using a number. This decided, we need to think about the number of bits—called the *resolution* of the measurement—and format of the number. For general arithmetical purposes, MATLAB stores numbers as double-precision floating point numbers. These have 64 bits, separated into a mantissa and an exponent. But there is nothing magical about the 64-bit size of a double-precision number. This size was established as a compromise between the needs for precision in mathematical operations versus the demands of space for storage and of speed of processing. The best compromise changes, depending on the requirements and costs: The term *double-precision* testifies to a change from 32 bits to 64 bits that occurred as storage became bigger and processing faster.

In the case of a sampled analog signal, the issues are different from those involved in numbers used for mathematical operations. Analog signals almost invariably involve amounts of random noise that are introduced by mechanical, electrical, or other interference or even by fundamental thermodynamic or statistical fluctuations; there is little point in measuring a quantity to a precision that is much finer than the noise involved in the quantity itself. Additional factors motivating a low-precision measurement are that the time and expense of making a measurement increase with the required precision, as does the space needed to store the measurement.

The best compromise among these factors depends on the details of the signal being recorded and the transduction method, but for most purposes, such as sound recording, it has been found that a resolution of 16 bits offers a very high quality, that 12 bits is generally adequate, and that even 8 bits can do a serviceable job.[2] An inexpensive computer system or a telephone generally uses 8 bits of resolution; laboratory A/D boards often use more.

Recall that 16 bits allows the storage of $2^{16} = 65,536$ different patterns, 12 bits allows $2^{12} = 4096$ patterns, and 8 bits allows $2^8 = 256$ patterns. For most purposes, these patterns are used to store the A/D converted measurement as an integer number on a linear scale. For example, 8-bit measurements are used to represent the 256 integer values from -128 to 127; 12-bit measurements represent -2048 to 2047.

The precision of a 12-bit, linear-scale measurement is 1 part in 4096. But, you ask, 1 part in 4096 of what? The A/D hardware is set up to measure voltage, and it does so over a specific range. In the following, we'll take that range to be -5 V to 5 V, which is typical but not universal. Each digital value

---

[2]There are, of course, exceptions, particularly when the recorded signal contains the sum of several different signals of different amplitudes. This happens, for instance, in seismology, where a single seismograph responds to a mixture of vibrations from earthquakes and other activities from all over the world. Events from nearby sources will tend to be high in amplitude, while distant sources have low amplitude.

corresponds to one specific voltage in this interval. For instance, if a digital value of 0 corresponds to 0 V, and a digital value of 2047 corresponds to 5 V, then a digital value of 1 will correspond to 0.0024 V, a digital value of -1 corresponds to -0.0024 V, a digital value of 2 corresponds to 0.0048 V, and so on.

The analog output of the transducer—a microphone in the case of sound recordings—is connected to an analog amplifier. This amplifier is adjusted, typically by human hands, so that the output range of the amplifier in response to the analog input signal will cover the complete range of operating voltage of the A/D converter hardware.[3] In some situations (for example, sound recordings), deciding how to adjust the amplifier can be difficult since the amplitude of the sound changes markedly in time. To help, the amplifiers are often equipped with sound meters that indicate the amplitude level with a needle or a row of light-emitting diodes. When the amplifier output is too high, the needle goes from a green zone into a red zone, or the diodes flash red. The objective of the human operator is to adjust the amplifier to be as high as possible while avoiding any excursions into the red zone.

There are two sorts of things that can go wrong:

- The amplification is too low. In this case, the precision of the digital recording is poor. For example, if the amplifier output is only -0.2 V to 0.2 V, rather than the full -5 V to 5 V, then for a 12-bit digitizer the digital output will never go outside of $\pm 0.2 \text{ V} \frac{2047}{5 \text{ V}} \approx \pm 82$. Thus the precision will only be one part in $82 + 82 = 164$ rather than the potential full part in 4096 of a 12-bit digitizer. In general, using only a fraction x of the full range means a loss of $\log_2 x$ in the effective number of bits of resolution; using only a range of ± 0.2 V out of ± 5.0 V means that only $\frac{0.2}{5.0} = \frac{1}{25}$ of the range will be used, resulting in a loss of $\log_2 25 \approx 4.6$ bits. See Figure 13.5 for an example.
- The amplification is too high. In this case, some of the signal is outside of the range of the A/D converter. The sampled signal is said to be clipped; any part of the signal that goes outside of the A/D range is truncated. See Figure 13.6. Clipping is part of the reason for the poor quality of the `fh.wav` file plotted in Figure 13.1.

Whether it is better to set preamplification high in order to gain resolution or to set it low in order to reduce the risk of clipping depends entirely on the specific signal being sampled and the ways that it will be processed after sampling. (Exercises 13.6 and 13.7 explore the consequences of low resolution and clipping for vocal sounds.) If you find that you cannot make the risk of clipping acceptably low and at the same time have acceptable resolution, then it is time to switch to a higher-resolution A/D converter (and perhaps a lower-noise transducer and preamplifier).

---

[3]When sound is being recorded through the microphone on a PC, the preamplifier is usually adjusted via software. The sound-recording software often includes a volume adjustment control.

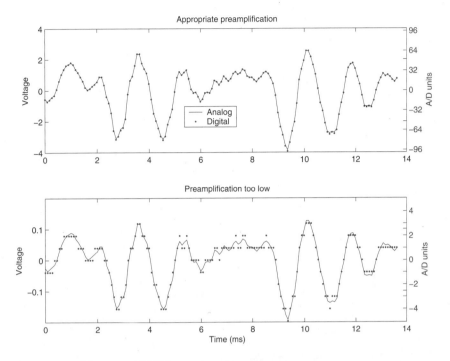

Figure 13.5.   A segment of the sound "OH" transduced to voltage.
Top: The preamplifier has been set appropriately so that the analog voltage signal takes up a large fraction of the A/D voltage range. The digitized signal closely resembles the analog signal even though the A/D conversion is set to 8 bits. Bottom: The preamplifier has been set too low. Consequently, there is effectively only about 3 bits of resolution in the digitized signal; most of the range is unused.

The desire to maintain precision for low-amplitude sound while avoiding clipping for high-amplitude sounds has led to nonuniform amplitude systems for storing measured amplitudes. Speech sounds, for instance, can vary between low amplitude and high amplitude even within a single word, and low-resolution speech is often stored using a floating point scheme called *mu-law* encoding, where 8-bits' storage space is allocated to a single-sign bit on the mantissa, 4 bits of mantissa, and 3 bits of exponent.

In the transduction and preamplification steps, the magnitude of the analog signal is usually made to differ from that of the original signal. Indeed, unless the original signal was in volts, the analog signal will be in entirely different units. For example, we might be measuring an air pressure in Newtons per square meter or a temperature, but in both cases, we transduce the signal to volts and record the voltage in terms of the fraction of the A/D range. For this reason, we often see sampled data presented in terms of *A/D units* or "arbitrary units"; this merely acknowledges that no record

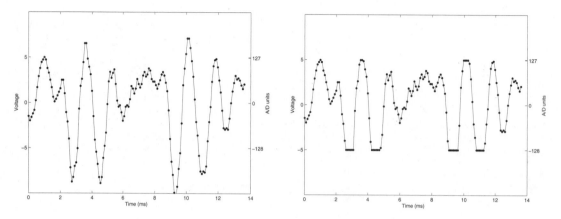

Figure 13.6. Clipping of a signal when the preamplifier has been set too high, so that the signal is outside of the −5 to 5 V range of the A/D converter.

has been kept of the various conversions involved in the transduction and preamplification steps or the range of the A/D converter.

Key Term

In many scientific situations, it is important to know how the A/D units relate to the original signal's physical units. Finding this relationship is a matter of *calibration*. This is often done by imposing a known input or series of inputs on the transducer and writing down in a lab notebook or elsewhere these known values and the corresponding A/D output. With this information, it's possible to find a transformation from A/D units back to the original physical units of interest.

Storing Sounds in Files

Recording sounds on the computer requires an A/D converter and software that drives the converter and stores the samples generated in computer memory. Many desktop and laptop "multimedia" computers include an A/D converter that can sample at frequencies suitable for voice recordings and even high-fidelity music. Such computers generally include built-in microphones or have a place for connecting a microphone. A variety of software is available for driving the A/D converter, which often comes preinstalled on the computer. Using such software, you can record your own signals to computer files. Even if you do not have the required software or hardware, it is easy to find sound files using the Internet.

Variables representing sound can be stored in the usual way (for instance, a .mat file). But for efficiency reasons and to exploit the widely available software for playing sounds, it is helpful to use a conventional format that doesn't depend on MATLAB. There are many such file formats for storing sounds. MATLAB is set up to read one particular format that is common; a format known as PCM or WAV, and whose files typically end in the

.wav suffix. These files include, in addition to the signal samples themselves, information about the sampling frequency and the number of bits of resolution. The operator `wavread` is used to read a `.wav` file into MATLAB variables. For example,

```
>>  [sig,fs,nbits] = wavread('finesthour.wav');
```

reads in the contents of the file `finesthour.wav` and produces an array of numbers `sig` that contains the signal, the sampling frequency `fs`, and the number of bits of resolution `nbits`.

The signal `sig` is an ordinary vector, and can be treated as such. For example,

```
>>  plot(sig)
```

The file operator for writing `.wav` files is `wavwrite`. This takes as arguments a sampled signal (stored as a numerical vector, or a two-column matrix for stereo sounds), a number specifying the sampling frequency at which the signal was created, a number of bits of resolution, and the name of the file to be created. The signal vector, which is stored within MATLAB as a double-precision floating point vector, will be quantized by `wavwrite`. Any values outside the range −1 to 1 will be clipped.

13.5   Aliasing and the Sampling Frequency

Insofar as signals vary in time, it's important to sample them repeatedly in order to capture the variation. But how often should the signal be sampled?

In order to illustrate and explain the situation, let's take a sine wave as a simple model. We want our samples of the signal to contain enough information to represent the original signal accurately. One obvious way to characterize the quality of our signals is the extent to which we can use them to reconstruct the original signal.

Figure 13.2 shows an analog sine wave signal. The signal is characterized by its amplitude A and its frequency f that describes how many cycles of the sine wave occur in one second. The period P is the duration of one cycle, measured perhaps from peak to peak. The period and the frequency are numerical reciprocals of each other, $f = 1/P$, and so contain equivalent information. The period is measured in units of seconds, the frequency in units of Hertz (or cycles per second).

Figure 13.7 shows the effect of sampling from the sine wave at various sampling intervals, together with the reconstruction made by interpolating between the sample points with straight lines. It seems clear that when the sampling interval is very short compared to the period of the sine wave, the reconstruction is much more accurate than when the interval is long.

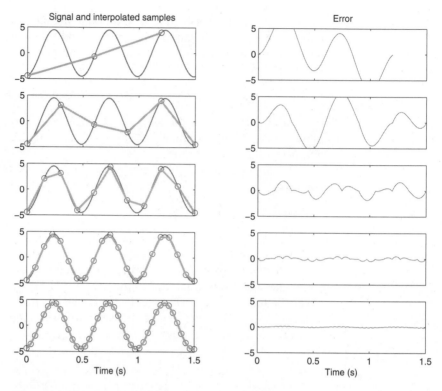

Figure 13.7. The analog sine wave signal from Figure 13.2 sampled at discrete times (circles) for five different sample intervals: .6, .3, .15, .075, and .0375 (from top to bottom). By connecting the samples with straight lines, we can reconstruct the analog signal. On the right is shown the difference between the true analog signal and the reconstruction.

This suggests a simple rule of thumb for setting the sampling interval: For accurate results, make the sampling interval much shorter than the period of the sine wave. Or, in terms of frequencies, the sampling frequency should be much greater than the frequency of the sine wave: at least 5 to 10 times greater.

Key Term There is a famous mathematical theorem, the *Nyquist Sampling Theorem*, that says that the sine wave can be perfectly reconstructed so long as the sampling frequency is higher than twice the sine wave's frequency. This may appear to be contradicted by some of the examples in Figure 13.7 where the reconstruction is not perfect despite satisfying the condition of the theorem. This is because the theorem is not based on interpolating between samples using straight lines but on a more complicated interpolation

Key Term called *sinc interpolation*.

The Nyquist Sampling Theorem, as well as our rule of thumb, applies to all signals, not just sine waves. This raises the question of how to describe the frequency of a signal that is not a single sine wave. The answer, somewhat amazingly, stems from the work of Fourier in the early 1800s in studying the way heat is transmitted through solids.

Fourier discovered that mathematically any continuous function can be broken down into a sum of sine waves of different frequencies. Each of these component sine waves is completely characterized by its amplitude, its phase, and its frequency. An analog signal is a continuous function of time, and so Fourier's theorem applies to signals. For the purposes of setting the sampling frequency for a signal, we describe the frequency of the entire signal by the highest-frequency Fourier component that has a nonnegligible amplitude.

Key Term

Just as the human ear does not respond to very high-frequency sounds, in general, transducers and preamplifiers have a limited range of frequencies to which they respond. When the signal to be transduced has higher-frequency components than these, the components effectively are ignored. The *cut-off frequency* of a transducer or amplifier describes the highest frequency to which the system can respond.[4] In fact, amplifiers are often purposefully designed with a known cut-off frequency so that the user can be certain that the output of the amplifier contains no components faster than the cut-off frequency. The elimination of high-frequency components of a signal is called *low-pass filtering*, and laboratory A/D systems often include adjustable low-pass filters whose cut-off frequency can be set by the user.

Key Term

Key Term

Given a system where the analog transducer and preamplifier have a known cut-off frequency, an appropriate sampling frequency can be set by treating the analog signal as if it were a sine wave at the cut-off frequency. Thus, the sampling frequency is set to perhaps 10 times the cut-off frequency (using the rule of thumb) but certainly at least 2 times the cut-off frequency (based on the Nyquist Sampling Theorem). The process of purposefully applying a low-pass filter is known as *antialias filtering*, a term whose meaning will be explained later.

Key Term

An "alias" is an alternative name or identity. With people, we generally associate aliases with criminals, and use "pen name" or "stage name" for writers or actors. The term also applies well to descriptions of samples from a sine wave. The situation that the term *aliasing* applies to is using discrete-time samples from a sine wave to inferring the frequency of the analog sine wave from which the samples were taken.

Suppose someone gives us some discrete-time samples, taken at a sampling interval ΔT, that lie exactly on a sine wave of frequency $f = \mathcal{F}$, amplitude A, and phase ϕ. The samples have been collected at the discrete times

---

[4]Strictly speaking, there is a more technical definition than this involving the amplitude, but this need not concern us here.

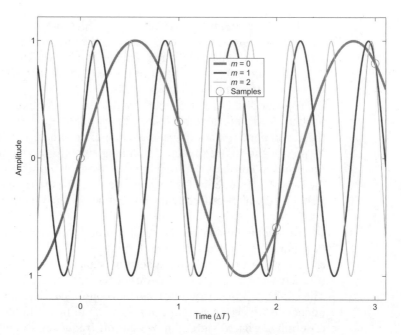

Figure 13.8. Aliasing. A set of samples are marked as a circle. The three sine waves plotted are of different frequencies, but all pass through the same samples. The aliased frequencies are $\mathcal{F} + m/\Delta T$, where m is any integer and ΔT is the sampling interval. The sine waves shown are $m = 0$, $m = 1$, and $m = 2$.

T_0, $T_0 + \Delta T$, $T_0 + 2\Delta T$, and so on, which we can summarize as $T_0 + k\Delta T$ for $k = 0, 1, 2, \ldots$. We can name each sample s_k, where the index indicates what time the sample was taken. Since the samples lie on the sine wave, we know that

$$s_k = \sin 2\pi \mathcal{F} k \, \Delta T + \phi$$

Now someone else comes along who denies that the samples are from a sine wave of frequency $\mathcal{F}$. Instead, this person claims, the samples are masquerading under an alias of $\mathcal{F}$ while they really are from a sine wave of a different frequency, $f = \mathcal{F} + m/\Delta T$ for some integer m. It's easy to test the plausibility of this claim. For any specific value of m, we can simply draw a graph and verify that it also goes through the samples. This is done in Figure 13.8.

We can verify the claim for all m algebraically: Simply check whether it's true that

$$\sin(2\pi f k \, \Delta T + \phi) = \sin(2\pi \mathcal{F} + m/\Delta T) k \, \Delta T + \phi) = \sin(2\pi \mathcal{F} k \, \Delta T + 2\pi m k + \phi)$$

If you remember the trigonometric identity $\sin(a + b) = \sin(a)\cos(b) + \cos(a)\sin(b)$, then you can expand out the right-hand side of the preceding equation to find that it becomes

$$s_k = \sin(2\pi \mathcal{F} k \, \Delta T + \phi)\cos(2\pi mk) + \cos(2\pi \mathcal{F} k \, \Delta T) + \phi)\sin(2\pi mk)$$

Since m and k are both integers, we have $\cos(2\pi mk) = 1$ and $\sin(2\pi mk) = 0$, so the identity is confirmed. That is, a sine wave of frequency $\mathcal{F} + m/\Delta T$ (and appropriate phase) will produce the same samples at time intervals ΔT for any integer value of m. We say that $\mathcal{F}$ is an alias for $\mathcal{F} + m/\Delta T$.

Which is the real sine wave? Just from the samples collected at a sampling interval of ΔT, there is no way to know. But we may have additional information to bring to bear on the issue. For instance, if we have antialias filtered the analog signal, making sure the cut-off frequency of the low-pass filter is less than half the sampling frequency $1/\Delta T$, then we know that any Fourier components of frequency $\mathcal{F}$ must satisfy $\mathcal{F} < \frac{1}{2\Delta T}$. The equivalent frequencies with the same sample values are $\mathcal{F} + m/\Delta T$, and for any $m \geq 1$, these frequencies all are larger than the antialiasing cut-off.

▶

Example: The Sound of Aliasing

Even if you have never heard of the phenomenon of aliasing before, you have probably seen aliasing at work. Perhaps, while watching a Western movie, you have seen a wagon wheel appear to stand still or move backward. The motion picture is, of course, a series of still images, essentially discrete-time samples taken 20 to 30 times a second. Although the wheel is revolving rapidly, the samples of this motion are equivalent to many equivalent revolution speeds. Our eye picks out the slowest of these. The same sort of phenomenon is seen when looking at motion illuminated by a regularly flashing stroboscopic light.

We can also hear the effects of aliasing. To do this, we can generate samples from a high-frequency sound wave, but sampled at a rate that induces aliasing. When the sound is played back, the D/A converter interpolates between the samples, with a result that the sound is perceived at its slowest aliased frequency. The phenomenon is illustrated in Figure 13.9.

The samples of a sine-wave signal of frequency $\mathcal{F}$, at a sampling interval of ΔT, are equivalent to samples of signal of frequency $|\mathcal{F} + m/\Delta T|$ for $m = \ldots, -2, -1, 0, 1, 2, \ldots$. When the samples are interpolated smoothly, the interpolated signal will be the lowest possible frequency consistent with $|\mathcal{F} + m/\Delta T|$. That is, the value of m for the interpolated signal will be the m that makes $|\mathcal{F} + m/\Delta T|$ as small as possible. In the bottom panel of Figure 13.9, for example, $\mathcal{F} = 20$ Hz and $1/\Delta T = 28$ Hz. Setting $m = -1$ minimizes $|\mathcal{F} + m/\Delta T|$ at 8 Hz; thus the interpolated samples resemble a sine wave of 8 Hz.

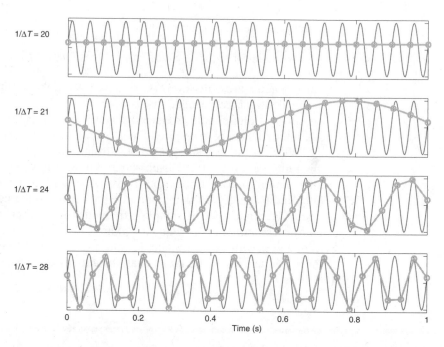

Figure 13.9. An analog sine wave signal of frequency $\mathcal{F} = 20$ Hz (thin line) is shown sampled at four different sampling frequencies $1/\Delta T$: 20 Hz, 21 Hz, 24 Hz, and 28 Hz. The signal created by interpolating between the samples (thick line) is aliased because the sampling frequency is not greater than twice the analog signal's frequency. The aliased signals have frequencies that are $\mathcal{F} - 1/\Delta T$, as can be seen by counting the number of cycles appearing in the 1 second's worth of data shown.

In addition to aliasing, you can see beating; for instance, in the bottom panel, the interpolated samples alternate between relatively large and small sine wave cycles. The beating is most evident when the sampling frequency is close to twice the analog signal's frequency, as shown in Figure 13.10.

The program playalias(analogfreq, sampfreq) will simulate the effect of sampling a pure sine wave tone of frequency analogfreq at a specified sampling frequency sampfreq, playing the interpolated samples through computer loudspeakers or headphones. Since many built-in computer speakers are very cheap, you may prefer to use headphones. If you do use headphones, make sure to adjust the playback volume so that the sound is within a comfortable range.

We explore how the perceived frequency of the interpolated signal depends on the sampling frequency. We use an analog signal frequency of 1000 Hz in this example, since human ears are sensitive to frequencies in this range.

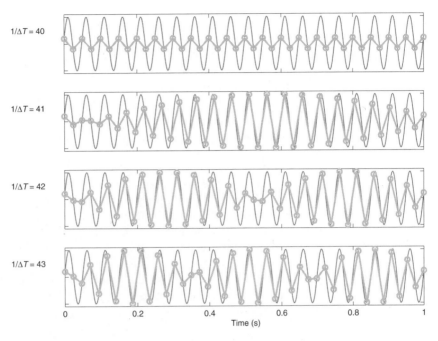

Figure 13.10.   An analog sine wave signal of frequency $\mathcal{F} = 20$ Hz (thin line) is shown sampled at four different sampling frequencies $1/\Delta T$: 40 Hz, 41 Hz, 42 Hz, and 43 Hz. The signal created by interpolating between the samples (thick line) is not aliased because the sampling frequency is greater than twice the analog signal's frequency. However, the amplitude of the interpolated signal varies slowly at a frequency of $\mathcal{F} - 1/\Delta T$. This is beating.

Start by sampling the analog signal at a very high sampling frequency, which should give a faithful reproduction.

```
≫  playalias(1000,15000)
```

The function doesn't return any output; its purpose is to produce a side effect: a sound. Now, gradually scale down the sampling frequency; for instance,

```
≫  playalias(1000,14000)
```

At high sampling frequencies, the sampled signal sounds very much the same regardless of the sampling frequency.

According to the Nyquist Sampling Theorem, the minimum sampling frequency to avoid aliasing is twice the analog signal's frequency. Try this.

```
≫  playalias(1000,2000)
```

The signal you heard was faint; the amplitude depends on the relative phases of the samples and the signal itself. For instance, in the top panel of Figure 13.10, the relative phase is such that the amplitude of the sampled signal is very small, although the frequency is correct.

You can adjust the relative phases of the samples and analog signal by using a third argument to `playalias` that should be in the range 0 to 2π. For instance,

```
≫ playalias(1000,2000,0)
```

produces silence.

Try adjusting the phase argument to find the phase that produces the loudest sound when the sampling frequency is twice the analog signal's frequency. Does the phase make any difference when the sampling frequency is much higher than the analog signal's frequency?

Sampling at frequencies slightly higher than the minimum frequency set by the Nyquist Theorem makes the beating effect audible. Since `playalias` plays one second's worth of signal, you can estimate the beating frequency by counting the number of times the played signal increases from low to high volume. For instance,

```
≫ playalias(1000,2002)
```

should produce two cycles of beating, but the pitch of the sound should be correct.

By sampling at a frequency below the Nyquist frequency, we introduce aliasing. For example, using a sampling frequency of 1300 Hz

```
≫ playalias(1000,1300)
```

creates an aliased sound whose frequency will be $|1000 - 1300|$ or 300 Hz. You can test this by comparing the aliased signal to a high-fidelity sound of frequency 300 Hz:

```
≫ playalias(300,15000)
```

As the sampling frequency approaches the analog signal's frequency, the perceived tone of the interpolated signal will get lower and lower. Beating may also occur when the sampling frequency is close to the aliased frequency.

Try lowering the sampling frequency below the analog signal's frequency, and predict the tone of the played signal by finding the m that minimizes $|\mathcal{F} + m/\Delta T|$. Remember that you won't be able to hear sounds that are below about 50 Hz, even if your loudspeaker or headphones are able to reproduce them.

◀

13.6 Exercises

Exercise 13.1:
The `whistle.wav` waveform is an eight-note ascending scale. Use reversal and concatenation to generate an ascending and descending scale.

Exercise 13.2:
A constant signal, one with no fluctuations, cannot be heard. Find the lowest-frequency signal that you can hear.

Exercise 13.3:
Generate a pure tone sine wave of unit amplitude at 440 Hz. Find the lowest amplitude at which you can hear the played sound. You may want to try using headphones in a quiet room.

Exercise 13.4:
Write a program that plays the seven-note musical scale starting at a given frequency. Note that the frequencies of successive notes in the scale are separated by factors of $2^{n/12}$, where n is either 1 or 2 depending on the position in the scale. (*Hint:* Adjacent notes on a piano—including both black and white keys—are separated by a frequency factor of $2^{1/12}$.)

Exercise 13.5:
Use a sound recording program to record your saying the numbers zero through nine. (The MATLAB function `wavrecord` is suitable for Windows computers.) Extract out the segments of the waveform for each number and store them in 10 appropriately named files.

Write a function that takes a number as input and returns a waveform that, when played, sounds like the number in spoken form. You can do this by concatenating the individual number waveforms together as appropriate.

Exercise 13.6:
Transmitted and recorded signals are often subjected to noise. You can simulate this by adding in random numbers to the signal; for instance,

```
>> noisySig = sig + 0.1.*randn(size(sig));
```

Key Term

The parameter `0.1` in the preceding line is the standard deviation of the noise: a measure of the noise's amplitude. The signal's amplitude can also be represented in this way, with the standard deviation computed by `std(sig)`. The ratio of the signal's standard deviation to the noise's standard deviation is called the "signal-to-noise" ratio. It's conventional to present the signal-to-noise ratio in *decibels* (abbreviated dB), which is 10 times the log base 10 of the signal-to-noise ratio.

Take a clean recording of music and of speech and find the signal-to-noise ratios at which:

1. The added noise is barely perceptible in the speech.
2. The added noise is barely perceptible in the music.

3. The speech becomes barely intelligible.

4. The noisy music becomes unpleasant.

You may have to rescale the signal in order to avoid clipping.

Exercise 13.7:

Repeat Exercise 13.6, but rather than adding noise, amplify the sound to produce clipping. You can synthesize clipping by amplifying the sound and setting too-large values to ±1; for instance,

```
≫  s = 2.0.*sig; % amplification
≫  sigClipped = ones(size(s)); % clipping
≫  sigClipped(abs(s)<=1) = sig(abs(s)<=1);
```

The clipping can be thought of as a sort of nonrandom noise:

```
≫  clippingNoise = sigClipped - s;
```

Try various amounts of amplification, and quantify the clipping by the signal-to-noise ratio of signal s to `clippingNoise`.

Exercise 13.8:

The file `pipe.wav` is a recording of a single, wooden pipe from a pipe organ. The pipe produces a musical note with a frequency of approximately 530 Hz. You can use recordings like this to synthesize musical notes that have a realistically complex sound and use `playmusic` to put together the notes into an entire musical piece.

By default, `playmusic` uses a note generation program named `simpletone`. Any single-note synthesis program to be used with `playmusic` must have the same format as `simpletone`; that is, it must take the same inputs (in the same order) and produce an output of the same size and type.

The `simpletone` function takes five arguments:

1. The desired sampling frequency.

2. The frequency of the note, in Hz. The code for silence (a musical rest) is a frequency of NaN.

3. The duration of the note in seconds.

4. The amplitude of the note. An amplitude of 1 should produce a sound vector in the range −1 to 1.

5. An argument that is reserved for future use.

`Simpletone` returns a vector of samples whose length is the duration times the sampling frequency. For example, here is one-half second of concert A, sampled at 20,000 Hz:

```
≫  r = simpletone(20000, 440, .5, 1, []);
≫  size(r)
  ans: 10000 1
```

Your task in this exercise is to write a program like `simpletone`, but which synthesizes a realistic sound. For this purpose, you can use `pipe.wav` or any other similar recording that you like. (`Pipe.wav` is a recording made by Steve Panizza of a pipe organ. `Mandolin2.wav` is a single plucked note of a mandolin, from Nelson Coates.)

Here are some hints:

- You can read in the sampled sound using the `wavread` operator. This returns both the sampled sound and the sampling frequency.
  In the following hints, we'll use `snd` to denote the sound vector itself and `sf` as the sampling frequency of the sound.
- The recorded sound, if it is from a musical instrument, will have a tone frequency. For `pipe.wav` this is 530 Hz.
- We can translate the tone to another frequency by playing back the signal at a higher or lower playback sampling frequency. For example,

  ```
  ≫   sound(snd,2*sf)
  ```

  plays a note that is one octave higher than the original tone.

  ```
  ≫   sound(snd,(2.^(5/12))*sf)
  ```

  plays a note that is a musical fifth higher than the original, and

  ```
  ≫   sound(snd,(2.^(-5/12))*sf)
  ```

  plays a note that is a musical fourth lower than the original.

- Although we could translate the tone of a single note in any way we like using the preceding technique, `playmusic` needs to play many different notes, all at the same playback frequency. Therefore, rather than playing the sound back differently, we have to simulate sampling the note in a way that the tone will be altered when we play it back at the fixed playback frequency.
  The general idea is that if we resample the sound at a frequency higher than the playback frequency, the tone will be shifted lower. If we resample at a lower frequency and play it back at a higher frequency, the tone will be shifted higher.
- In order to resample a signal that has already been sampled, we can use interpolation. There is an interpolation format that makes this pretty easy:

  ```
  ≫   newsnd = interp1(samptimes, snd, newtimes);
  ```

  where `samptimes` is a vector of times (in seconds) at which each of the original samples was collected, `snd` is the signal, and `newtimes` is a vector of times (in seconds) at which the resamples are to be taken.
- The relative times (in seconds) at which the samples were taken are

```
≫   samptimes = linspace(0, length(snd)./sf, length(snd));
```

- The relative times (in seconds) at which the new samples are to be taken are

```
>> newtimes = linspace(0, alpha*dur, dur*newsf);
```

where dur is the desired duration of the new note (in seconds), newsf is the desired playback frequency of the new note, and alpha is a number that indicates how much to shift the frequency. A value of alpha=2 will shift the tone by one octave. (Which corresponds to a shift to a higher tone, an alpha of 2 or 1/2?) Alpha will be set by the ratio of the desired tone frequency (one of the arguments to the synthesis program) and the nominal tone frequency of the original recorded note.

- It would be inefficient to read in the recorded sound each time the note synthesis program is invoked. Instead, you may want to store the recorded sound and its sampling frequency as persistent variables. You will also need to set the nominal tone frequency in order to compute alpha.

- To test your note synthesis program, you should both listen to the output to make sure that it is reasonable and compare it to the output of simpletone to make sure that the output is in the same format (e.g., size and shape) for a given input.

Once you have your single-note synthesis program working, you can use playmusic to invoke your program.

```
>> playmusic('scale.xls', 'mysynthesisprogram')
```

You may hear clicks at the end of the notes. This can be due to the sharp amplitude transition at the end of the note and can be alleviated by tapering the end of the note.

13.7 Project: The Perception of Beats

Key Terms

In *modulation*, the properties of one signal are set by another signal. For example, in *amplitude modulation* (AM) radio, the amplitude of a pure radio-frequency carrier wave is set by a sound signal. FM radio (*frequency modulation*) is more complicated conceptually: The frequency of the radio carrier signal is varied according to the amplitude of the sound signal.

A simple amplitude-modulated signal is a pure tone of frequency f_b, whose amplitude is varied as a function of time $f(t)$, giving

$$f(t) \cdot \sin(2\pi f_b t)$$

For example, if $f_b = 440$ Hz and $f(t) = 2\cos(2\pi 0.05t)$, this signal will sound like a concert "A" wavering 10 times per second.

The reader may recall the identity from trigonometry:

$$\frac{1}{2}(\sin \alpha + \sin \beta) = \sin(\frac{\alpha + \beta}{2}) \cdot \cos(\frac{\alpha - \beta}{2}) \qquad (13.1)$$

This mathematical identity poses a psychophysical conundrum. The signal $2\cos(2\pi f_d t) \cdot \sin(2\pi f_b t)$—a wavering pure tone— is mathematically equivalent to the sum of two pure tones $\frac{1}{2}(\sin(2\pi f_1 t) + \sin(2\pi f_2 t))$, where $f_b = \frac{f_1 + f_2}{2}$ and $f_d = \frac{f_1 - f_2}{2}$. Seen in this way, the amplitude-modulated signal would be expected to sound like a combination tone: a chord.

So which will we perceive: a wavering pure tone or a chord? The answer depends on f_1 and f_2. When f_1 and f_2 are very close, the ear is unable to distinguish between them, and the sound is perceived as a wavering pure tone of frequency $f_1 + f_2$. When f_1 and f_2 are far apart, you will hear a mixture of pure tones that do not waver.

Write a function that will play 1 second's worth of a mixture of two different frequencies. The function should take as arguments the two frequencies, f_1 and f_2. Make sure to set the amplitude of each of the tones to avoid clipping.

Try the following for several different values of f_1 over a range from 100 to 2000 Hz: For each fixed f_1, vary f_2 and find the smallest $f_2 > f_1$ that you perceive as a mixture of tones. (*Hint:* Usually $|f_2 - f_1|$ has to be somewhere around 100 Hz before you hear a mixture rather than wavering.) Record your results and make a graph of the smallest nonbeating $f_2 - f_1$ versus f_1.

For some pair f_1 and f_2 that produces wavering, compare the pitch of the resulting sound to a single pure tone of frequency f_1. Write a function that takes three arguments, f_1, f_2, and f_3, and plays a single pure tone of frequency f_3 immediately after the sum of the tones f_1 and f_2: one sound followed by the other. Using this function and holding f_1 and f_2 fixed at values that produce wavering, vary f_3 until you can't distinguish between the perceived tones of the first and second sounds. [*Hint:* According to the mathematical identity Eq. (13.1), the tone of the mixed sound should be $f_3 = (f_1 + f_2)/2$.]

If headphones are available, try also playing one of the two frequencies into one ear and the other into the other ear. (*Hint:* Look at the help for the sound function to see how to generate stereo sound.) Do the results differ from what you found when playing both frequencies into each ear?

13.8 Project: Changing the Speed of Sound

Old-timers from the days before compact disks may recall the effects of playing a record at too fast or too slow a speed. We can achieve the same audible effect by using a faster or slower playback frequency. For example, here's Winston Churchill (recorded in the file `fh.wav`):

```
≫  [x,fs,bits] = wavread('fh.wav');
≫  sound(x,fs)
```

and here he is again, sounding like a fast-talking, helium eater:

```
≫  sound(x,fs*1.5)
```

and again in slow motion:

```
≫  sound(x,fs*0.6)
```

Alternatively, we could change the duration of a signal by deleting samples or interpolating samples; for example, taking every nth sample and playing it at the original sampling frequency:

```
≫  y = x(floor(1:1.6:length(x)))
≫  sound(y,fs)
```

It's sometimes desirable to be able to speed up or slow down a recording without changing the quality of the sound. In this project, you are to write a function `speedsound` that takes three arguments: a sound waveform, a sampling frequency, and a fraction that indicates how long the new sound should be relative to the old one. For example,

```
≫  newx = speedsound(x,fs,.8)
```

should produce a waveform that sounds just like x but plays back in only 80% of the time.

Here are some hints:

- To speed up the sound, rather than deleting samples that are evenly spaced throughout the waveform, delete entire segments of perhaps 50 to 100 ms in length. This is an important time scale for human vocal sounds. You can calculate the number of samples in such segments from the sampling frequency `fs`.
- To slow down the sound, insert segments of a similar length. You can construct the inserted segments by copying from the sound just before the insertion point.
- For a better quality of sound, you may want to delete quiet segments of the waveform (e.g., the space between words). Similarly, you can insert quiet segments between words to lengthen the sound. You can identify those segments that are quiet by using the standard deviation (`std`) of the samples in the segment.

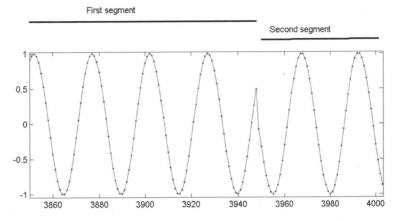

Figure 13.11.   Two segments of a pure-tone sine signal have been concatenated with a mismatch between the endpoints of the segments.

For an even better quality of sound, avoid concatenating segments whose endpoints are very different. For example, Figure 13.11 shows a transition between two segments whose endpoints don't match. Such sharp transitions produce a "pop" in the sound. When such pops occur frequently, the sound picks up a noisy, metallic quality.

Images

A picture is worth a thousand words.

In the previous chapter, we examined signals: quantities that vary continuously with time. In this chapter, we continue with a closely related topic: quantities that vary continuously with two variables. Usually, the two variables are the x- and y-coordinates of spatial position, and the quantity is something that can be measured at each spacial point. Examples include the intensity on a photographic plate of x-rays transmitted through a body; the color of paint at each point on a painting; the temperature at each point on a surface; or the amount of light reflected from each point on the earth. Perhaps the most familiar example is a photograph, where the quantity is the amount of light that falls on each point of a piece of film.

14.1   Black-and-White Images

Consider a black-and-white photographic print as in Figure 14.1: An arrangement of object and shadow, reflection and light, printed on paper or glass. Seen as a whole, we perceive the shapes and figures of the image; we may think that the film has captured the objects being photographed. Yet seen up close, in extreme zoom, the photograph loses its perceived content

Figure 14.1. A photograph of a building in Delhi. (File `sakidelhi.png`. Used by permission, Saki Meir.)

A close-up of a photograph.

Key Term

and becomes an abstract composition in shades of gray: The level of gray varies from place to place in the print.

Before tackling the problem of how to represent an entire photograph, let's concentrate on how to represent the level of gray at a single point in the image. First, recognize that for a printed photograph there are limits: A given point in the image can be devoid of ink, in which case it takes on the white of the background paper, or it can be covered completely with ink, in which case it takes on the blackness of the ink. This suggests that a photograph is a boolean field, either yes or no, blank or ink. A printing technology called *halftoning* exploits the fact that the eye is not perfect and perceives not single points but small regions. A halftoned image covers small regions with even smaller dots of ink as a mixture of black and white blurred by the eye into gray. Computer displays, however, can produce gray directly; they represent gray as a number that ranges from 0 to 1, with 0 standing for black and 1 for white and the various shades of gray in between.

Imagine that we have a penlike device that we can place on a point on the photograph and get a reading of the gray level at that point. We drag the pen in a straight line over the photograph, getting a continuously changing reading of gray level: a signal. The gray level signal from a horizontal track in the photograph is shown in Figure 14.2. The track goes horizontally, and the gray-level read-out fluctuates up and down; the read-out is high where

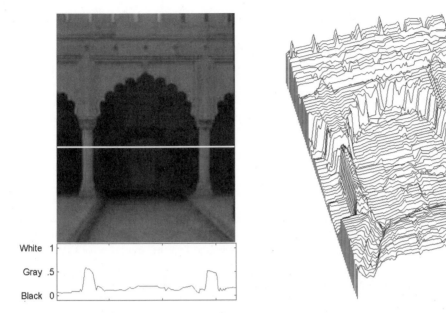

Figure 14.2.   A horizontal track through the photograph of Figure 14.1. The level of gray varies along the track as shown below the photograph. Assembling many such tracks allows the entire image to be sampled.

the track crosses the bright pillars and low where the track crosses the dark regions between the pillars.

We already know how to represent such a signal on the computer: evenly spaced samples. The same rules for setting the sampling frequency that we encountered in Chapter 14 apply here, too. Although now the samples are separated by small increments of space rather than time. To avoid aliasing, we want the samples to be closely spaced.

Capturing an entire image requires more than a single track. Multiple, closely spaced tracks are required, as shown in Figure 14.2. An important issue is how closely these different tracks need to be separated. The answer is seen easily by remembering that our choice of horizontal tracks was arbitrary; we might have decided equally well to make vertical traverses of the image. The spacing between tracks needs to be the same as the spacing of points along a track. The result is a rectangular grid of sample points with a measurement made at each point: a matrix. Each row of the matrix is a horizontal track through the image; each column is a vertical track. Each point in the matrix, at the intersection of a vertical track and a horizontal track, is a single sample of the gray level of the image. Such a sample is called a **Key Term**   *pixel*, short for *picture element*.

It is not easy for our visual system to process image information in the form of the right-hand panel of Figure 14.2. While the figure may successfully convey the idea that the gray-level information in the photograph can be thought of as a function of two variables, the graph of this function as an undulating surface is hard to interpret visually. Our ability to handle the information in a graph of two variables—particularly one as complicated as Figure 14.2—cannot compete with our innate ability to process visual images in the format of Figure 14.1. But the information is the same in the two figures even if the format is different: gray level as a function of two variables.

14.2   Color

In a color photograph, the quantity at each pixel is not a level of gray, but a color. How is the color of a pixel to be represented on the computer? The answer, it turns out, is intimately related to the structure of the human retina.

Underlying our ability to perceive color is the fact that light comes in different wavelengths. The light coming into our eye from any one point in an image is, generally, a mixture of these different wavelengths.

Astronomers measure the mixture of light using specialized equipment called *spectrometers*, which break down the light from stars into several hundred components, each of a different wavelength.

Each individual star is so far away that it looks like a single point in an image. The light from the single-point star is a mixture of different wavelengths; the details of the mixture contain information about the size and type of the star. For example, Figure 14.3 shows the intensity of light of different wavelengths, measured from two different stars by the Very Large Telescope in Paranal, Chile. The stars emit light of all wavelengths, but some of the light is absorbed by atoms in the atmosphere of the cloud. The broad dip near 434 nm is due to the absorption of light by hydrogen. The narrower dips are due to metal atoms, usually iron.

Remember that each of the graphs in Figure 14.3 is a measurement from a single point. This information is rather detailed and informative.[1]

The mixture of different wavelengths of light is not itself color but is part of the psychophysical reality that underlies color. Color is a perception, a complex one that is by no means fully understood.

In the eye, the image is focused by the lens on the retina. Each point of the scene is mapped to a small region of the retina. The retina itself contains

---

[1]The comparatively few absorption dips in AV304 indicate that it is hotter than ICR3287; the hotter stellar atmosphere ionizes the metal atoms so that they do not absorb light in the range of wavelengths shown. The broad hydrogen-absorption dip near 434 nm is shifted to longer wavelengths in AV304 than in ICR3287. This is doppler red shifting and indicates that AV304 is moving away from earth faster than ICR3287.

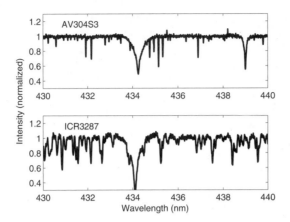

Figure 14.3.   The spectra from two stars in the Magellanic Clouds recorded by a spectrometer attached to the Very Large Telescope in Paranal, Chile. The light emitted by each star is a mixture of light of different wavelengths; the spectrometer measures the intensity of the light at each wavelength. Star ICR3287 is an F-supergiant in the Inter Cloud Region of the Magellanic Clouds. AV304 is a main-sequence B-star in the Small Magellanic Cloud. (*Source:* Data courtesy of Kim Venn, Eline Tolstoy, and Robert Rolleston.)

a field of photoreceptors: transducers from light into the activity of nerve cells. There are two main types of these photoreceptors: *cones* and *rods*.

Only the cones play a role in the perception of color. There are three subtypes of cones: red, green, and blue (so called because each type is most sensitive to a particular range of frequencies of light). The cones in each small region of the retina act somewhat as a spectrometer. But the analysis is rather coarse; rather than the finely detailed analysis of the astronomer that consists of hundreds of measurements, the cones produce a readout that is the equivalent of only three numbers: the amount of red light, of blue light, and of green light.

So far as human perception is concerned, it is these three numbers that matter. Paper-and-ink printers, manufacturers of photographic film, and designers of televisions and computer displays have exploited the coarseness of the human color perception system to store and present the entire spectrum of each image point as a mixture of only three colors. They condense the entire spectral mixture into just three numbers—the intensity of red light, of blue light, and of green light—knowing that this is all the information that the cones will extract from the image and present to the brain.

As a consequence, the storage of color images on the computer involves storing, for each pixel, three separate numbers. There are a number of ways of doing this, intended to save the expense of storing large quantities of data or to interface nicely with various means of transducing the three numbers

Figure 14.4.   The separate color planes of the Russian folk painting in Figure 14.5. One plane gives the intensity of red, one of green, and one of blue.

Figure 14.5.
A Russian folk painting.

into printed ink or displayed light. But, for our purposes, we'll take a simple form of a color image as three separate images aligned pixel by pixel: one image for red, one for blue, one for green. Figure 14.4 shows an example of these three images. Each of these separate images is effectively monochrome: It can be stored as a black-and-white image even though it is intended eventually to be presented in its designated color. Figure 14.5 shows the combined red-green-blue (RGB) image.

14.3   Digital Sampling of Images

The same two questions that we met in Chapter 13 when sampling signals apply to sampling images:

Sampling frequency: How far apart should samples be?

Quantization: How many amplitude levels do we need?

The phenomenon of aliasing arises in images whose samples are too coarsely separated in space, just as it does in signals that are too coarsely sampled in time.

Quantization

Since we have arranged to define black as 0 and white as 1, the quantization level sets the number of different gray levels. With 1 bit, we have only two possible levels: black and white. With 2 bits, there are four levels altogether, providing black, white, and two intermediate levels of gray. A typical level of quantization is 8 bits, giving 256 intensity levels and 256^3 different colors when each plane of a color image is stored with 8 bits.

In sound signals, which fluctuate rapidly, the desired quantization level is tied to the amount of noise in the system. Images raise different issues.

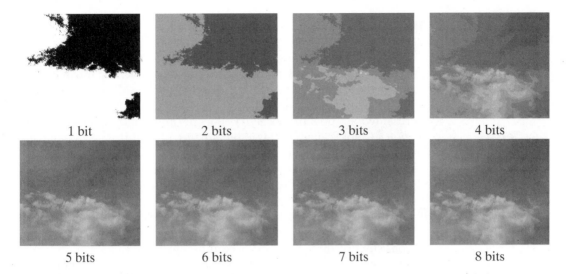

| 1 bit | 2 bits | 3 bits | 4 bits |

| 5 bits | 6 bits | 7 bits | 8 bits |

Figure 14.6. A photo of sky and cloud taken with various quantization levels.

For example, in most images nearby points are correlated strongly in intensity. Too coarse of a resolution turns gradual transitions of shading into sharp edges and often introduces transitions into the image that the eye interprets as edges. Such edges show up in the sky in Figure 14.6. In photographs, one criterion for selecting the quantization level is to avoid the eye's seeing sharp transitions between adjacent levels; 8 bits is conventional, although when it is necessary to distinguish between subtly different shades, as in medical x-rays, a higher quantization, perhaps 12 bits, is needed. For the printed image in Figure 14.6, 5 or 6 bits seems adequate.

When 8-bit quantization is used with color images, there are 8 bits for each color plane, 24 bits altogether. In some cases, this is much more than needed. However, reducing each plane to 2 or 3 bits produces poor results, producing the kinds of sharp transitions seen in Figure 14.6 within each color plane. (See Exercise 14.2.) An effective method involves a *color map*, where a single-number integer is stored for each image pixel. A color map translates this integer into an RGB triple, so that the image can be displayed in full color. By selecting the color map carefully, just a few different colors (perhaps 256, so that the integer is stored in 8 bits) can give a good-quality image.

Key Term

The human visual system is not good at interpreting gray levels in absolute terms; we judge grayness as relative to nearby levels. In those situations where the absolute level of gray is important in reading the image, a good technique is to use color to represent each gray level. This can be done by treating each gray level as an index into a color map. (See Exercise 14.1.)

14.4 Sampling and Storing Images in Files

The equipment for sampling images is widely available. *Scanners* sample an image by successively illuminating strips of the image; each strip is then sampled in the manner indicated in Figure 14.2. The software that drives the scanner generally provides a way to set the sampling frequency, usually presented as "dots per inch" or "dots per mm" and called the image

Key Term *resolution*.[2] Color scanners analyze the mixture of light from each sampled point into three components: red, blue, and green.

Scanner software usually offers a wide number of options for processing the scanned image (for example, as a photograph or as a drawing) and many different types of storage options: jpg, png, tif, gif, and so on. This multiplicity of options obscures a basic simplicity: Black-and-white images are an array of single-number samples; color images are an array of samples each consisting of three numbers. The different file formats and processing options have to do with how these samples are stored and compressed. Commercial issues, such as patent rights, also lie behind the diversity of file types.

The imread file operator in MATLAB is capable of reading many different formats of image files. Imread is able to figure out the format of the file from information stored in the file or from the filename extension. It returns a matrix of the samples in the image.

```
>> a = imread('sakidelhi.png');
```

The returned value is the image matrix:

```
>> size(a)
   ans: 269 176
```

This image is 269 × 176 pixels.

When the file contains an RGB color image, imread returns a matrix with three values for each sample.

```
>> b = imread('monaLisaLouvre.jpg');
>> size(b)
   ans: 864 560 3
```

This is a three-dimensional matrix; it can be thought of as consisting of three planes: one for each RGB color.

Key Term Imread can also read images stored in *index form*, where each sample of the image is a single number used to refer to a colormap: a list of three-number color values. In such cases, imread returns two arguments: the image matrix of index values and the color map. We won't consider such

---

[2]Note that this use of "resolution" as a sampling frequency contrasts with the use of resolution in sampled signals, which refers to the number of bits used to represent each sample.

indexed images further, since they can always be translated into the conventional three-numbers-per-sample image type using the `ind2rgb` operator.

Saving a one-plane or three-plane RGB image to a file is done with the `imwrite` operator. This takes two required arguments: the matrix containing the image (either a single plane for black-and-white or three planes for color) and the name of the file in which to store the image. The filename extension describes the format of the written file. For instance,

```
>> imwrite(a,'figa.jpg')
```

stores the image in a as a JPEG file that can be read by standard software such as Web browsers or photograph editing programs.

```
>> imwrite(a,'figa.png')
```

creates a PNG-format file, also widely used by Web software. There is a variety of other, optional arguments to `imwrite` that allow you to control things such as the level of compression, transparency, and so on. See the documentation for `imread` for information about these.

Sampling Frequency

The Nyquist criterion for avoiding aliasing in images is the same as that for signals: Samples should be spaced closer than half the period of the highest-frequency Fourier component of the images.

For images, the Fourier components are also sine waves but are functions of the x- and y-space coordinates. The general form is $A_{j,k} \sin(2\pi(f_j x + f_k y) + \phi_{j,k})$. If the mathematical form of this function seems daunting to you, the geometry should not: They look like corduroy of various widths and orientations. A few Fourier components are shown in Figure 14.7.

The period of any of these functions can be read off from its graph as the distance between successive crests of the black-to-white sine wave. The x-period is the distance measured horizontally; the y-period is the distance measured vertically. The frequency is the reciprocal of the period. For example, in the upper-left panel of Figure 14.7, the distance between crests is 0.33 inches.

In principle, aliasing can occur in images in the same way as we encountered in signals. A finely striped image, if sampled with too much space between samples, can be severely distorted. Such stripes are seen, for example, in engravings, as in Figure 14.8, or in fingerprints.

To avoid aliasing and beating, the distance between samples should be a small fraction of the spacing between peaks in the Fourier components. In Figure 14.7, the sampling resolution has been set to 0.01 inches. This works well for Fourier components whose stripes are 0.1 inches apart, but some beating can be observed in the images with more finely spaced stripes.

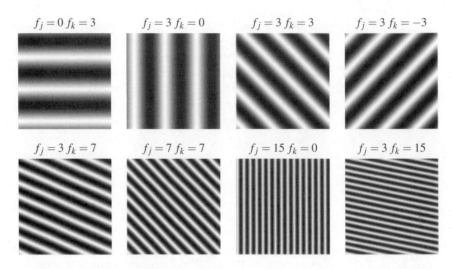

$$f_j = 0 \; f_k = 3 \qquad f_j = 3 \; f_k = 0 \qquad f_j = 3 \; f_k = 3 \qquad f_j = 3 \; f_k = -3$$

$$f_j = 3 \; f_k = 7 \qquad f_j = 7 \; f_k = 7 \qquad f_j = 15 \; f_k = 0 \qquad f_j = 3 \; f_k = 15$$

Figure 14.7. Eight of the Fourier components $\sin(2\pi(f_j x + f_k y))$. x and y both range over the interval 0 to 1, so that f_j gives the number of cycles across the image horizontally and f_k gives the number of cycles vertically. For printing purposes, the images have been sampled with a sampling frequency of 100 samples/inch, corresponding to a sampling resolution—the spacing between samples—of 0.01 in.

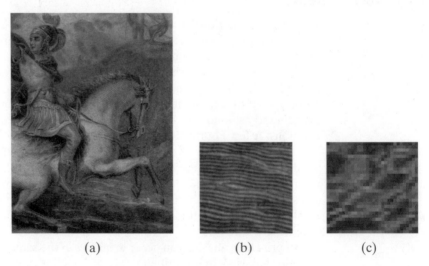

(a)   (b)   (c)

Figure 14.8. (a) A detail of the engraving, *Caesar Passing the Rubicon* by Tucker based on a painting by Smürke. [From [6].] Engravings consist of closely spaced thin lines and small dots. Digitally sampling such an image requires a high-spatial resolution. (b) A close-up of a region about 6 mm × 6 mm of the water near the horse's hoof with a spacing between samples of 0.04 mm (600 dpi). A typical spacing between lines is approximately 0.25 mm. (c) The same close-up region, sampled at 0.24 mm and causing aliasing.

14.5   Manipulating and Synthesizing Images

For the purposes of displaying images or editing them by hand, there is no reason to use MATLAB: Sophisticated and user-friendly software is available. Such photo-editing software makes it easy to crop and rotate images, adjust the colors, correct "red eye," and so on. These same operations can be done within MATLAB, but where MATLAB excels is in allowing you to develop your own algorithms for processing images. We explore some of these later, such as generating synthetic color images or aligning and morphing images, but for now we introduce some simple operations from which the more complicated ones can be built.

Since images are often intended for human consumption, an important operation is displaying an image. This is done with the `imshow` operator. The basic operation of `imshow` is trivial: Just give the image matrix as an argument, whether it be a black-and-white image or a three-plane RGB color image. To illustrate, we use the variables a and b assigned values on page 362.

```
>>  imshow(a)
>>  imshow(b)
```

For indexed images, the color map should be given as a second argument to `imshow`.

There is some additional complexity lurking that has to do with the type of the image matrix. Although we have considered image samples as numbers between 0 and 1, in order to save storage space, images are usually stored in files not as floating point numbers but as 8- or 16-bit integers. When the quantization resolution is 8 bits, the integers range from 0 to 255; when the quantization is 16 bits, the integers range from 0 to 65535. The `imread` operator returns its output as an integer type with either 8 or 16 bits. For instance, the `sakidelhi.png` image assigned to variable a is an 8-bit image. MATLAB refers to the 8-bit type as a `uint8`:

```
>>  class(a)
   ans: uint8
```

Imshow is able to interpret `uint8` or `uint16` images appropriately. But, as we will see, it is often convenient to manipulate images as floating point numbers. You can convert an image from `uint8` or `uint16` to a floating point matrix using the `image2double` operator.

When the image is stored as a floating point matrix, each sample's value is generally a number between 0 and 1. Imshow will also display such images properly. But if the samples are outside of this range, what `imshow` does depends on whether the image is black and white or color. For a color image, `imshow` simply generates an error when any samples fall outside the interval $[0, 1]$. So RGB images, when stored as floating point, need to be scaled so that each sample in each plane is in $[0, 1]$.

For black-and-white images, the situation is different. Imshow will happily display such images but will treat any values greater than 1 as white and values less than 0 as black. A second, optional argument to imshow allows you to set these cut-offs.

What can make the situation confusing is that the uint8 and uint16 image matrices contain values that are interpreted as numbers 0 to 255 or 0 to 65535. If the unit8 or uint16 matrices are converted to floating point using the double operator, then the resulting floating point matrix will not display properly because it will not be in the range 0 to 1. The image2double operator scales things appropriately.

Here are some simple operations on images:

Cropping: Refers to taking out a smaller rectangular section of the image. This can be done using the ordinary indexing operators; for instance,

```
≫ asection = a(50:150, 27:90);
```

For RGB images, it's necessary to indicate that you want all of the three image planes. This is done with a third indexing argument; for example,

```
≫ bsection = b(8:92, 60:140, :);
```

The imcrop program provides a user-friendly interface, allowing you to specify the section of the image to take using the mouse. Imcrop displays the image and activates the computer mouse, which can be click-dragged to define the cropping rectangle. The operator returns the selected part of the image as a matrix. Try the following:

```
≫ bsection = imcrop(b);
```

Extracting a color plane: Applies only to RGB color images. You can use the standard index operators to pull out the plane you want.

```
≫ red = b(:,:,1);
≫ green = b(:,:,2);
≫ blue = b(:,:,3);
```

Each plane of the image, once extracted, is a monochrome image and will display in grayscale. Try

```
≫ imshow(red)
```

Constructing a color image from monochrome planes: With three monochrome images of the same size and type (that is, uint8, uint16 or floating point), the RGB color image can be reconstructed using the `cat` operator, which concatenates the monochrome images, packaging them into the three-dimensional matrix of a RGB image. For example, we can reconstruct the image from its three planes:

```
>> c = cat(3,red,green,blue);
```

The first argument to `cat` will always be 3 in this context; the argument describes the dimension along which to align the images.

The first monochrome image is assigned to the red plane, the second to the green plane, and the third to the blue plane. Changing the order of concatenation changes the colors in the resulting image. Try, for instance,

```
>> c2 = cat(3,blue,red,green);
```

If you would like the monochrome image to appear in a specific tone, you can concatenate it with itself, scaled to create whichever color you like. Putting all zeros in two of the planes gives an image of pure color:

```
>> c3 = cat(3,red, zeros(size(red)), zeros(size(red)));
```

Many operations on images involve arithmetic. There are few numerical operations for `uint8` and `uint16` matrices; applying an arithmetic operator generates an error:

```
>> d = 2*a;
  ans: ??? Error using ==> *
       Function '*' not defined for variables of class 'uint8'.
```

In order to perform such numerical operations, it's necessary to convert `uint8` and `uint16` images to a floating point representation. As described previously, this is done with the `image2double` operator, which performs the appropriate scaling.

```
>> ad = image2double(a);
>> class(ad)
  ans: double
```

Now we can do arithmetic on the image, keeping in mind that any images that we produce should be scaled to be in $[0, 1]$ before displaying them. For example, in a negative image white is represented as black and vice versa. This is accomplished by

```
>> negative = 1 - ad;
```

For those samples where `ad` is 1—that is, white—`negative` will be $1 - 1$. For samples where `ad` is 0, `negative` will be 1.

14.6 Example: The Mona Lisa's Missing Blue

The Mona Lisa, displayed in black and white in Figure 14.9 has a distinct yellowish cast. Even the sky and the river coursing through the background are yellow. Since the painting is 500 years old, it's reasonable to expect that some colors of the paint have faded and the that the varnish has yellowed. Also, the colors may have been altered in the process of photographing and reproducing the image. Can we correct the colors to show what the original painting may have looked like?

Figure 14.9. The Mona Lisa, also called *La Gioconda*, painted by Leonardo da Vinci around 1503–1505. (File: monaLisaLouvre.jpg, from the Louvre, http://www.louvre.fr/img/photos/collec/peint/grande/inv0779.jpg)

One possible clue to the colors of the original image is the eye. We expect the white of the eye to be white, and not the jaundiced yellow of the figure. The color white is an even mixture of red, green and blue, at high intensity. Let's examine the pixels of the eye to see what the actual color mixture is. First, we convert the image to a floating point format:

```
>> mona = image2double(imread('monaLisaLouvre.jpg'));
```

and then grab the pixels near the eye:

```
>> eye = imcrop(mona);
```

Using the mouse, grab the region around Mona's left eye, as shown in the margin: At this scale, it's easier to see the white of the eye, which is covered by only about 4 pixels. Use imcrop to grab just these pixels:

```
>> eyewhite = imcrop(eye)
ans: eyewhite(:,:,1) =
       0.7333  0.8667  0.7843
       0.9647  0.9490  0.8784
       0.9686  0.8902  0.8588
     eyewhite(:,:,2) =
       0.5412  0.6824  0.6078
       0.7686  0.7490  0.6784
       0.7725  0.6784  0.6392
     eyewhite(:,:,3) =
       0.2431  0.3608  0.2902
       0.4196  0.3922  0.3294
       0.3843  0.2863  0.2588
```

A white pixel should have values close to 1 in all three planes, but these pixels have too little green (plane 2) and much too little blue (plane 3). This appears to be a general trend in the image, as a histogram of the three color planes shows in Figure 14.10.

What seems to be required is to increase the amount of green and blue. There are all sorts of ways of doing this: Essentially, what is needed is to transform small values into large ones while keeping the relative relationship between pixels intact. One simple way of doing this is called *histogram equalization*, which transforms the values in such a way that the histogram becomes uniform and flat. The equalize function does this for us, one plane at a time.

```
>> r = mona(:,:,1);
>> g = mona(:,:,2);
>> b = mona(:,:,3);
>> newimage = cat(3,equalize(r),equalize(g),equalize(b));
```

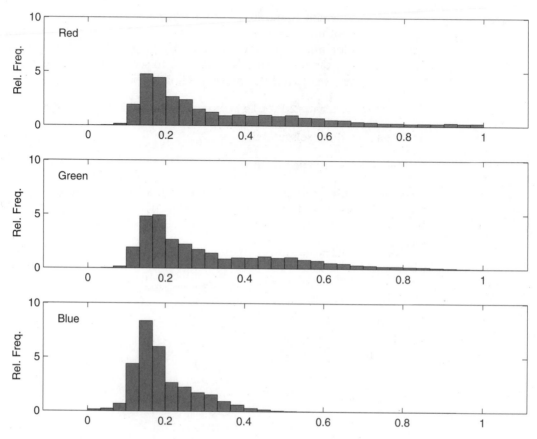

Figure 14.10. Histograms of the three color planes in the Mona Lisa. The histogram can be made by extracting each plane and using the `hist` command:

```
≫   r = mona(:,:,1);
≫   hist(r(:),50)
```

Note the use of the (:) to convert the matrix into a vector: `hist` takes a vector as an argument.

This transformation creates an image that is too blue; it's overcorrected. A compromise is needed; one is shown in Figure 14.11.

The color correction shown in Figure 14.11 isn't perfect. Part of the reason for this might be that the different pigments contained in different parts of the image might fade a different rates, so that some of the blues are being overcorrected. Doing a definitive restoration of the colors would require a study of the pigment types, as well as an art historian's understanding of the styles of the day.

Figure 14.11. A version of the Mona Lisa with the colors adjusted to make the white of the eye appear white. Each plane has been histogram equalized and mixed 50% with the original values of the plane.

14.7 Exercises

Discussion 14.1:
Read in the cloud image from Figure 14.6, stored in the file `cloudBW.jpg`. Display the image using `imshow`. Then set the colormap to `jet` using the `colormap` command. That is,

```
>> a = imread('cloudsBW.jpg');
>> figure(1); imshow(a);
>> figure(2); imshow(a,jet);
```

This technique emphasizes visually small differences in gray levels.

Use a scanner to scan in a printed document, such as a newspaper. Set the scanner to a high-quality quantization, often identified in scanner software as "BW photo quality." The paper background looks like a single color, but it is not in general. Use the color map method to look for variations in the shading of the paper background.

Then try this again, but scanning in the same document as a low-quality quantization, sometimes called "document mode" in scanner software. Do you see the same variations in the background when you use the color map technique?

Discussion 14.2:

Many Internet Web pages contain small, thumbnail-sized pictures. Here is part of one, showing a man in a jacket that appears to be checkerboarded and irregular. The real jacket is neat and regular, however. Explain why the picture appears the way it does. (The picture is in the file `jacket.png` and is 146×182 pixels in size.)

Exercise 14.1:

Write a program `imagequant` that takes two arguments: a single-plane image and a number n of quantization levels. Return a single-plane image where each pixel is restricted to one of the n quantization levels. For ease of display, make the actual levels evenly spaced from 0 to 1, that is, `linspace(0,1,n)`.

Exercise 14.2:

Use `imagequant` from Exercise 14.1 to encode each plane of a three-plane color image to n levels. The total number of colors will then be n^3. For a color photograph, how large does n have to be to give an acceptable image? (Note that the n^3 colors are not independent of one another: for instance, each of the n levels of red shares the same pairs of green and blue values.)

Exercise 14.3:

Use a scanner to scan in the engraving printed in Figure 14.8. Set the spatial resolution of the the scanner to various values and look for aliasing in the scanned image. What regions of the image are most susceptible to aliasing? At what resolutions does aliasing become noticeable? How does this resolution compare to the spacing between ridges in the printed image?

14.8 Project: Landsat Images and False Color

The Landsat 4 Thematic Mapper satellite takes pictures of the earth's surface using seven different wavelengths; it is as if the Landsat's retina has seven different types of receptors in contrast to the human eye's three different color receptors. A Landsat image therefore consists of seven planes

  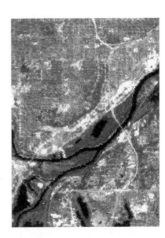

Figure 14.12. Bands 1, 4, and 7 of a Landsat image of the area near the confluence of the Mississippi and Minnesota rivers, near the city of of St. Paul. The Mississippi river cuts across the image from left to right; the Minnesota river comes from the bottom left to join the Mississippi in the middle of the image.

rather than the conventional three planes. This raises the question of how to display the Landsat information to a human viewer. Of course, one can always give seven monochrome images, and this is how the Landsat files are stored. Figure 14.12 shows some Landsat images of the area near the confluence of the Mississippi and Minnesota rivers, near the city of of St. Paul.

Since humans are quite sensitive to differences in color, it is advantageous to combine the information in the seven Landsat planes into a color image. An ordinary three-plane color image constructed from the monochrome Landsat images is called a *false-color image*, since the colors have not been directly observed by the camera.

Key Term

The Landsat images shown in Fig. 14.12, and some related aerial photograph images, are available in file form, as listed in Table 14.1.

The three planes of a standard color image in principal allow one to use an image to encode three independent pieces of information about each pixel. A false-color image is one that uses red, green, and blue to encode other information. For example, let's take the first three planes of a Landsat image:

```
>> one = imread('stp-mac-1.jpg');
>> two = imread('stp-mac-2.jpg');
>> three = imread('stp-mac-3.jpg');
```

Table 14.1   Landsat Thematic Mapper Images of the St. Paul, Minnesota Region Taken on September 4, 1991. Each pixel corresponds to an area of 30 m². The different images record the intensity of different wavelengths of light. (These images were kindly provided by Jean Doyle of the University of Minnesota's Remote Sensing Lab in the Department of Forestry.)

Filename	Description	Size
stp-mac-1.jpg	Band 1, 0.45–0.52 μm	254 × 187
stp-mac-2.jpg	Band 2, 0.52–0.60 μm	254 × 187
stp-mac-3.jpg	Band 3, 0.63–0.69 μm	254 × 187
stp-mac-4.jpg	Band 4, 0.76–0.90 μm	254 × 187
stp-mac-5.jpg	Band 5, 1.55–1.75 μm	254 × 187
stp-mac-7.jpg	Band 7, 2.08–2.35 μm	254 × 187

It's simple enough to combine these into a color image:

```
>> falsecolor = cat(3,one,two,three);
>> imshow( imrescale( falsecolor ) );
```

You will have to look carefully at this image to see much color. It turns out that the first three planes of the Landsat image contain much the same information; when they are combined, they tend to produce gray.

You can play around with the other planes of the image (Landsat plane 6 is not included with the others since it is at a much coarser resolution), combining them in various ways to produce interesting false-color representations.

Exercise 14.4:
Find a false-color representation that shows areas with a lot of vegetation as green, areas covered by water with blue, and paved or built-over areas as red. Some suggestions for how to do this are given, but you can be creative in producing an attractive and informative image.

Let's divide the problem into two parts: (1) deciding whether a given pixel should be colored red, green, or blue; and (2) determining the intensity of pure red, green, or blue of that pixel.

Identifying water in the image is not too difficult for those familiar with the area covered by the image: The Mississippi and Minnesota rivers both appear in the image along with several lakes. In all of the planes, the water appears quite dark. You can create a boolean mask identifying water by thresholding one of the images (e.g., image < threshold) or even a combination of more than one image (e.g., image1 < threshold1 & image2 > threshold 2).

Similarly, vegetation is bright in some of the images, particularly band 4.

Researchers in the field of remote sensing have made studied efforts of how to extract the most useful information from images such as the Landsat images here. One method involves the "normalized differential vegetation index" computed as (band 4 − band 3)/(band 4 + band 3).

Another general approach is based on a statistical technique called *principal components analysis*. In principal components, one makes a linear projection of the entire feature space—in this case six dimensional since we are considering six monochrome images—onto a smaller-dimensional space in such a way as to preserve most of the information in the entire space. Exactly how this is done is beyond the scope of this exercise (but it is closely related to the technique we used in the previous section for constructing ellipses from the covariance matrix). In the case of Landsat images, statistical analysis [23] of many images using principal components leads to the following projections to create three new, synthetic images s_1, s_2, s_3 from the original images θ_1, θ_2, θ_3, θ_4, θ_5, θ_7:

$$s_1 = 0.332\theta_1 + 0.331\theta_2 + 0.552\theta_3 + 0.425\theta_4 + 0.481\theta_5 + 0.252\theta_7$$

$$s_2 = -0.247\theta_1 - 0.163\theta_2 - 0.406\theta_3 + 0.855\theta_4 + 0.055\theta_5 - 0.117\theta_7$$

$$s_3 = 0.139\theta_1 + 0.225\theta_2 + 0.404\theta_3 + 0.252\theta_4 - 0.701\theta_5 - 0.457\theta_7$$

The calculations are done on a pixel-by-pixel basis; simply add together the six original images, each one multiplied by the appropriate coefficients to produce one of the synthetic images. The images s_1 and s_2 are called the "brightness" and "greenness," respectively. These synthetic images are not suitable for direct display; they generally contain negative values. Instead, you will want to use logical thresholding on them to define which pixels should be classified as vegetation, which as built-up, and which as water. Appropriate thresholding of these two images can give a good indication of which pixels cover vegetated areas and which pixels cover built-up areas. (A more complete analysis would also take into account nonvegetated, non-built-up areas; for example, plowed fields.) Pixels that are not identified as built-up or vegetated or water should perhaps be colored gray.

To help check your results, you can reference the aerial photograph of the same region as the Landsat figures using Table 14.2. This photograph has much higher resolution than the Landsat image; the original digitized image has a resolution of 1 meter compared to the Landsat resolution of 30 meters. You can see individual buildings, ships, and even cars in the aerial photograph.

Since the full-resolution image is so large (and may not be usable on computers with less than 64 MB of RAM), I have provided coarser resolution versions. Even so, these images require more pixels for display than available on the computer screen. Consequently, I have provided a program, zoomimage, that allows you to display parts of the aerial photograph by

Table 14.2   Aerial Photographs (from an airplane) from the United States Geological Survey, Taken in 1994, of the Same Area Covered by the Landsat Images in Table 14.1. Each pixel corresponds to an area of 1 m². The high-resolution image may not fit on many computers, so lower-resolution versions are provided also. (From the same source as Table 14.1.)

Filename	Description	Size
stpw-se-doq.jpg	The high-resolution image	7559×5568
aerial2.jpg	A downsampled version of the above	3780×2784
aerial4.jpg	A downsampled version of the above	1890×1392
aerial8.jpg	A downsampled version of the above	945×696

double-clicking on a coarse resolution file such as the Landsat images. For example,

```
>> a2 = imread('aerial2.jpg');
% wait a while, this will be slow
>> zoomimage( one, a2, 15, 300 )
```

will display image one and prompt you to double-click the mouse over the region you want to display. Once you do so, the corresponding region in the high-resolution image will be displayed. The third argument (15 in this case) gives the relative resolution of the fine image compared to the coarse one. The fourth argument gives the size of the region (in pixels) to display in the fine image.

CHAPTER 15

Numbers and Precision

The purpose of computing is insight, not numbers.
— R.W. Hamming

A statement like

```
≫  3 + 7
   ans: 10
```

looks so much like the arithmetic that we learned as children that it is natural to believe that the computer knows how to add numbers. It doesn't. $3+7$ is an engineer's magical incantation. It invokes the computer to move quantum-mechanical pools of electrical charge in a way that merely simulate the behavior of numbers. The engineering is a fantastic bit of technology; the computer numbers behave almost exactly as ideal mathematical numbers behave. "Almost exactly" is the operative phrase here. There are important ways in which the computer-simulated numbers differ in behavior from ideal mathematical numbers. We focus on three:

- Computer numbers take up space in memory.
- Computer numbers have a finite precision that can be surprisingly small.
- Computer numbers don't have units.

15.1 Types of Numbers

In order to use computer memory economically, the designers of computers have adopted several standard ways of representing numbers. Some of these are listed in Table 15.1. These representations can be related to the familiar mathematical typology of numbers:

Positive integers: The counting numbers $0, 1, 2, 3, \ldots$.

Integers: The positive integers along with negative integers $\ldots, -3, -2, -1$.

Rational numbers: Ratios of two integers (e.g., $\frac{1}{3}, \frac{22}{7}$). This includes the integers but extends it to numbers between successive integers.

Real numbers: This extends the rational numbers to include irrational numbers such as π. Real numbers provide a means to represent continuously varying quantities.

Complex numbers: A complex number is a pair of real numbers, one of which is called the "real part" and the other the "imaginary part."

As detailed in Section 3.1, the bits of computer memory provide a natural way to represent positive integers and, with a sign bit or a complement method, negative integers as well. This representation is exact but finite: k bits can represent the numbers 0 to $2^k - 1$, another bit is needed for negatives.

Integers are an appropriate type for representing counts and, as we've seen, the ASCII numbers that represent characters. In addition, some entities (photographs, sounds and signals) can be represented as large sets of integers. (See Chapters 13 and 14.) The storage requirements for such things can be large. A single book might easily have a million characters; a modest-sized photograph might have several million small integers. A convenient

Table 15.1   Some Standard Types of Computer Numbers

MATLAB Name	Size (bytes)	Description	C, C++, JAVA Name
uint8	1	integers 0 to 255	unsigned char
uint16	2	integers 0 to 65,535	unsigned short int
uint32	4	integers 0 to 4,294,967,295	unsigned long int
int8	1	integers −128 to 127	signed char
int16	2	integers −32,768 to 32,767	short int
int32	4	integers −2,147,483,648 to 2,147,483,647	long int
single	4	single-precision floating point	float
double	8	double-precision floating point	double

standard unit of memory for storing the integers encountered in text or images is 1 byte, the `uint8` or `int8` types will do the job. The range of integers encountered in high-fidelity sounds and signals is larger: A 2-byte unit, `uint16` or `int16`, is required.

For large counts, a 4-byte unit, `uint32` or `int32` is employed. Such large integers are needed for many internal purposes in the computer, such as addressing locations in memory. (A modern computer might have a billion bytes of memory, each of which has an integer address.)

For very large counts—As this is being written, the United Nation estimates that world population is $6,138,682,771$ [5]; Avogadro's number is $602,213,670,000,000,000,000,000$—the 4-byte integer does not suffice. Larger-size integers can be defined, but for scientific work, another system is used typically for large counts: the system for representing real numbers.

It might seem natural to represent a real number as a rational number— a ratio of a numerator integer to a denominator integer. This could build on the exact representation of integers. However, rational numbers cannot store any number larger than the numerator and therefore are limited in size by the capacity of the integer type.

Key Term

Instead, a *floating point system* is used that can represent rational numbers, real numbers, and large integers. This system is a compromise between the differing properties of the three different kinds of numbers and the need to keep overall storage space small. The compromise is that the representation is often not exact: For most mathematical numbers floating point is only an approximation. We will return to this important issue in Section 15.2.

Key Terms

A floating point number consists of three integers: the *mantissa*, the *exponent*, and the *base*. The layout of a floating point number is similar to scientific notation. For example, Planck's constant in scientific notation is 6.6260755×10^{34} J $\cdot$ s. Here, the mantissa is 6.6260755, the exponent is 34, and the base is 10. Although the mantissa here is printed with a decimal point rather than as an integer, this is a question of typography. We could represent the mantissa as the integer $66,260,755$ with the convention that the decimal point should be placed after the first digit. In floating point numbers, the mantissa is always arranged so that the leading digit is nonzero (the exponent is adjusted correspondingly).

Key Term

Most modern microprocessors use a standardized scheme for floating point arithmetic called the *IEEE 754 standard* promulgated by the Institute of Electrical and Electronic Engineers. In the IEEE standard, the size of a single-precision float is 4 bytes, with 23 bits reserved for the mantissa, 8 for the exponent, and 1 bit for the sign of the mantissa. The base is 2. The largest magnitude represented by a single-precision float is $\approx 3.4 \times 10^{38}$. The smallest is $\approx 1.2 \times 10^{-38}$.

Although the maximum magnitude represented by a float is impressively large, floats are not terribly precise numbers. They can store only about seven decimal digits. For example, the population of the earth can be represented by a float but only with a precision of plus-or-minus the population of the Republic of San Marino. If you stored a 10-digit telephone number in a single-precision float, the last four digits would be lost.[1]

Much scientific work requires more precision than a float can handle. For this reason, a more capacious collection of 8 bytes—64 bits—is widely used: 52 bits for the *mantissa*, 11 bits for the exponent, and a sign bit for the mantissa. This type of IEEE number is called a *double-precision floating point number* or "double" for short. The 52-bit mantissa provides the equivalent of about 16 significant decimal digits.

Key Term

Key Term

By convention, two bit patterns have been reserved in the IEEE standard to have a special meaning. One of them, which prints out as Inf, stands for *infinity* and is the numerical result from a computation like 1./0. Of course it is not literally infinity—that would take an infinite number of bits! Inf is just a pattern that IEEE-compliant arithmetic operators are set up to return when ∞ would be a sensible result of a computation. Another pattern prints out as NaN, standing for *not a number*. This is the result of a numerically indeterminate computation, such as 0./0.

Key Term

Printing of Numbers

The base-2 representation of integers and floating point numbers is not very easy for humans to read or write. So the interpreter has been programmed to display numbers not in their internal representation but in a format set by the user.

A typical scheme is to use ordinary decimal notation for numbers between 10,000 and 0.001 and scientific notation for larger or smaller numbers. In MATLAB, the format command is used to control how numbers are printed on the screen. You can set the number format to be whatever suits your own purpose. This book usually uses a floating point representation with 5 digits, switching to a 15-digit representation when required. To illustrate some of the formats, we'll print out two numbers in various ways:

$$\sqrt{2000000} \approx 1414.2135623730951356265$$
$$\text{and}$$
$$6138682771$$

First, we assign values to the variables:

```
≫ num = sqrt(2000000);
≫ pop = 6138682771;
```

---

[1] It's debatable whether a telephone "number" is a mathematical number or a series of characters, each of which happens to be a digit. Telephone numbers are not formatted like numbers, and we never apply mathematical operations like + to them.

Here are the short and long formats generally used in this book:

```
≫ format short g
```

which leads to the following display:

```
≫ num
  num: 1414.2
≫ pop
  pop: 6.1387e+009
```

The long format displays more digits:

```
≫ format long g
≫ num
  num: 1414.2135623731
≫ pop
  pop: 6138682771
```

The words long and short indicate how many digits to print: 5 for short and 15 for long. The g indicates that the printing style should switch sensibly between a scientific notation with only one digit to the left of the decimal place (e.g., 6.138e+009) and a decimal style (e.g., 1414.2). In the g style, the decimal point is located in a place that avoids the scientific notation for numbers small enough that the nonfractional part can fit into the 5 or 15 digits allocated for printing the number. Such numbers are not always easy to read, particularly because the large numbers aren't punctuated with commas.

The format command can also produce a strictly fixed-point notation where there is always only one digit to the left of the decimal point and the exponent is used to indicate the scaling of the number.

```
≫ format short
≫ num
  num: 1.4142e+003
≫ pop
  pop: 6.1387e+009
≫ format long
≫ num
  num: 1.41421356237309e+003
≫ pop
  pop: 6.138682771000000e+009
```

Note that the format command controls only how numbers are printed to the screen by the interpreter and not how they are computed or written to files. The long and short format are the most important ones for scientific work, but there are additional formats available (for example, bank, rat, and +).

Numbers and Units

An athlete competes in a triathlon: a 100-km bicycle race, a 10-mile run, and a 2000-meter swim. In computing the total distance traveled, it's easy to fall into a trap:

```
>> 100 + 10 + 2000
   ans: 2110
```

This is wrong. The correct answer is 118.093 kilometers.

In the real world, quantities often have units. Distances are measured in meters, yards, miles or sometimes parsecs. Times are measured in hours, minutes or years. Many quantities have more complicated units: miles per gallon, Joule seconds, or meters per second squared. Some quantities have no units and are called *pure numbers*.

The rules for arithmetic of numbers with units are straightforward. Addition and subtraction must involve numbers in the same units. The output of a sum or difference operation has the same units as the input. The problem in the triathlon computation is that different sorts of numbers (meters, miles, kilometers) were added together without converting them to a common unit.

Multiplication and division can proceed regardless of the units; the output units are the product or quotient of the input units. For example, dividing a distance (in meters) by a time (in seconds) gives a velocity (in meters per second).

Trigonometric operators, logarithms, exponentiation, and other similar operators require inputs that are pure numbers and produce outputs that are pure numbers.

Computer languages could be designed so that the units of each number need to be specified and so that the rules of arithmetic are enforced. But they have not been. It is entirely the responsibility of the programmer to make sure that units are handled sensibly—there is no built-in language support for units, and all numbers are treated as pure numbers.

It's easy to make mistakes. Sometimes these mistakes are costly, as shown in Figure 15.1.

Complex Numbers (Optional)

A complex number generally is written in a rectangular format like $x + iy$ or a polar format like $re^{i\theta}$. In either case, the complex number is set by the two real numbers: x and y in the rectangular case and r and θ in the polar case. To represent complex numbers on the computer, a pair of doubles can be used. The convention in MATLAB is to use the rectangular format, so one of the numbers stands for the real part and the other for the imaginary part.

Complex numbers are so important in many branches of scientific computation that MATLAB allows them to be assigned to variables in the nor-

Missing What Didn't Add Up, NASA Subtracted an Orbiter

By ANDREW POLLACK

LOS ANGELES, Sept. 30 — Simple confusion over whether measurements were metric or not led to the loss of a $125 million spacecraft last week as it approached Mars, the National Aeronautics and Space Administration said today.

An internal review team at NASA's Jet Propulsion Laboratory said in a preliminary conclusion that engineers at the Lockheed Martin Corporation, which had built the spacecraft, specified certain measurements about the spacecraft's thrust in pounds, an English unit, but that NASA scientists thought the information was in the metric measurement of newtons.

The resulting miscalculation, undetected for months as the craft, the Mars Climate Orbiter, was designed, built and launched, meant it was off course by about 60 miles as it approached Mars.

"This is going to be the cautionary tale that is going to be embedded into introductions to the metric system in elementary school and high school and college physics till the end of time," said John Pike, director of space policy at the Federation of American Scientists in Washington.

Lockheed's reaction was equally blunt.

"The reaction is disbelief," said Noel Hinners, vice president for flight systems at Lockheed Martin Astronautics in Denver. "It can't be something that simple that could cause this to happen."

The finding was a major embarrassment for NASA, which said it was investigating how such a basic error could have gone through a mission's checks and balances.

Figure 15.1. From *The New York Times*, October 1, 1999, p. A 1. ©1999 by the *New York Times Co.* Reprinted with permission.

mal way and used in the normal arithmetic expressions, completely hiding the fact that two floating point numbers are involved in the representation.

To indicate to MATLAB that a number is imaginary, you can write the number followed immediately with the lowercase letter i; for example, 3.2i. MATLAB prints out imaginary numbers as complex, with a real part that is zero.

```
≫  3.2i
  ans: 0 + 3.2i
```

To make a complex number, use the + operator with a real and an imaginary number:

```
≫  7 + 3.2i
   ans:  7 + 3.2i
```

The usual arithmetic operations work as do most operators:

```
≫  1/(3 + 6i)
   ans:  0.0667 - 0.1333i
≫  sqrt(-1)
   ans:  0 + 1.0000i
≫  2 + 4.32i - 7 + 2i
   ans:  -5 + 6.32i
```

Although the preceding line looks typographically like the difference of the two complex numbers $2 + 4.32i$ and $7 + 2i$, this is not the case: There are four separate numbers being added together, $2, -7, 4.32i$, and $2i$. If we want to take the difference of two complex numbers, it is necessary either to group the numbers using parentheses:

```
≫  (2 + 4.32i) - (7 + 2i)
   ans:  -5 + 2.32i
```

or assign them to variables and take the difference of the variables:

```
≫  a = 2 + 4.32i;
≫  b = 7 + 2i;
≫  a - b
   ans:  -5 + 2.32i
```

The operator `real` returns the real part of a complex number; `imag` returns the imaginary part.

```
≫  real(a)
   ans:  2
≫  imag(a)
   ans:  4.32
```

The angle and abs operators return θ and r from the polar representation $re^{i\theta}$.

When writing a complex number, the letter i is a part of the numeric token, not a separate variable. Because of the way the parser operates, it is crucial that the i identifier not be separated from the numeral by a space or operator. The number 1i is the imaginary unit, whereas the value of a plain, isolated i is whatever value happened to be assigned to the variable named i. As it happens, by default, the variables i or j are both preassigned to be the imaginary unit, and the printing format is set to be rectangular:

```
≫ i
  ans:  0 + 1.0000i
≫ j
  ans:  0 + 1.0000i
```

This makes it straightforward to write notations that mimic mathematical notation; for example, for $e^{i\pi}$ we can write

```
≫ exp(i*pi)
  ans: -1.0000 + 0.0000i
```

But remember that i and j are just ordinary variables that happened to be preassigned to the value $\sqrt{-1}$. It might happen that you reassign them to be some other value. When you need them to be the imaginary unit, assign them to be the value 1i. Or, write your expressions using 1i to avoid the variable i entirely. To illustrate, suppose the variable i has been redefined to be something other than $\sqrt{-1}$:

```
≫ i=0;
```

Statements using that variable will still be computed properly, but using the variable's value rather than the mathematical i:

```
≫ exp(i*pi)
  ans: 1
```

The preceding expression is completely misleading to the human reader and probably is not what was intended. Instead, use 1i, a single token standing for the mathematical i:

```
≫ exp(1i*pi)
  ans: -1
```

15.2 The Capacity of Floating Point Numbers

In languages such as C, C++, FORTRAN, and Java, there are arithmetic operators for each of the numerical types. For integers, such arithmetic is exact unless the result of the operation exceeds the capacity of the type. It's easy to do this. For example, the numbers 21,220 and 46,783 are representable exactly by an unsigned 16-bit integer, but neither their sum nor their difference falls in the 0 to 65,535 range of that type. Similarly, the results of the multiplication $21,220 \times 46,783$ and the division $21,220/46,783$ cannot be represented adequately in the 16-bit integer type.

MATLAB takes a simple approach: All arithmetic is done using double-precision floating point numbers. So problems like those described previously do not arise:

```
≫   char(21220) + char(46783)
  ans: 68003
≫   char(21220).*char(46783)
  ans: 992735260
≫   char(21220)./char(46783)
  ans: 0.4536
```

In most cases, the use of double-precision floating point arithmetic means that users can get an acceptable performance without worrying about the values involved. In most cases. This section is about how to identify the other cases and deal with them.

Let's consider first the issue of integers. A floating point number can represent quantities as large as 1.7977×10^{308}, but it does not do so exactly. To see the pathology that arises from this fact, we define a big integer number.

```
≫   bigCount = 87392838298342342398398237432343
  bigCount: 8.73928382983423e+032
```

The interpreter has converted the numeral we typed, which was an integer, into a float. Let's add 1 to the count:

```
≫   bigPlus1 = bigCount+1
  bigPlus1: 8.73928382983423e+032
```

This should be a different number, but it isn't: `bigPlus1` and `bigCount` are equal.

```
≫   bigPlus1 == big
  ans: 1
```

The 53 bits of the mantissa of a double-precision floating point number[2] can represent exactly the integers in the sequence 0 to $2^{53} - 1$. Any integers larger than this are not necessarily exact: They overflow the space provided by the mantissa. To represent integers bigger than $2^{53} - 1$, digits are dropped from the mantissa and the exponent is incremented correspondingly. For instance, 2^{53} is too big to fit in the mantissa, but it can be written as 2^{52} with an exponent incremented by 1. To illustrate,

```
≫   2 ^53 == (2 ^53 - 1)
  ans: 0
```

---

[2]The mantissa of IEEE floating point numbers is, for technical reasons, arranged to have a leading 1. Since the leading bit is always 1, there is no need to store it. This gives an effective capacity of 53 bits.

The two large integers are not equal, as expected. Both `2 ^53` and `2 ^53 - 1` are representable exactly as floating point numbers. In contrast, $2^53 + 1$ involves a 1 in the 54th bit of the mantissa. Since there is no such bit, it simply is rounded off. This results in the floating point representation of $2^{53} + 1$ being equal to 2^23:

```
≫ 2 ^53 == 2 ^53 + 1
  ans: 1
```

This is not to say that all floating point numbers bigger than 2^{53} are equal, just that they are not represented necessarily as exact integers. The exact floating point number just greater than 2^{53} is $2^{53} + 2$—integers or real numbers in between these two are rounded off to whichever of the two is closest. This rounding off leads to an error in the number.

15.3 Numerical Error

Rounding-off can introduce a small error into a floating point number. There are other sources of error as well, stemming from approximations used in numerical algorithms and deficiencies in the ways data are collected and stored. In characterizing computations, it's important to be aware of the ways that rounding-off or other sources can introduce or amplify errors.

Suppose the mathematical quantity is x and the floating point representation is $x_\#$. There are two main ways to describe the error in the floating point representation of a number:

- *Absolute error* gives the size of the difference;
  that is, $|x - x_\#|$.
- *Relative error* is the absolute error divided by the size of x;
  that is, $\frac{|x - x_\#|}{|x|}$.

To illustrate, for numbers $2^{53} < x < 2^{53} + 2$, the absolute error is guaranteed to be no larger than 2, regardless of how rounding to a floating point number is performed. The relative error is guaranteed to be no larger than $\frac{2}{x} \approx \frac{2}{2^{53}} \approx 2.22 \times 10^{-16}$.

The relative error is a useful way to describe the error in floating point numbers largely because of the following: Subject to some qualifications, the relative error of any double-precision floating point number is guaranteed to be no larger than $\frac{2}{x} \approx \frac{2}{2^{53}} \approx 2.22 \times 10^{-16}$. This number is so important that MATLAB has a special operator, `eps`, that reminds us of it:

```
≫ eps
  ans: 2.22044604925031e-016
```

The qualification mentioned previously is that the number be within the range of floating point numbers.

The largest double-precision floating point number is given by the MATLAB operator `realmax`:

```
≫  realmax
  ans:  1.79769313486232e+308
```

All numbers larger than this are either rounded down to `realmax` or rounded up to `Inf`. For example,

```
≫  2.*realmax
  ans:  Inf
≫  3.*realmax
  ans:  Inf
≫  (1+0.5.*eps).*realmax
  ans:  1.7977e+308
```

For numbers larger than `realmax`, the relative error is not limited to eps.

The operator `realmin` returns a small double-precision floating point number:

```
≫  realmin
  ans:  2.2250738585072e-308
```

This is not the smallest possible floating point; for example,

```
≫  realmin/10000
  ans:  2.22507385850696e-312
```

But `realmin` gives the smallest floating point number whose relative error is bounded by eps.

The double-precision relative error of 2.22×10^{-16} offers fantastic precision in calculations.[3] It would give, for example, an error of less than one cent in computations involving numbers as large as the value of the total world economy each year. Very high precision measurements in science typically have a relative error substantially worse than 1×10^{-5}. It might seem, therefore, since computations are much more precise than our data, that we do not need to worry about the precision of computation. This is not always the case.

There are four situations that can lead to a dramatic increase in the relative error of calculations:

Overflow: Numbers larger than `realmax` are rounded off to the single value `Inf`. This can result in a large loss of precision. Some computations generate overflow transiently, but this spoils the entire computation. To illustrate, let's define a large number that's still less than `realmax`:

---

[3]By comparison, a single-precision floating point number, which has a 23-bit mantissa, has a relative error of 1.19×10^{-7}.

```
≫  a = 1e200
   ans: 1e+200
```

Here is a computation whose answer is obviously a but where overflow causes a loss of precision:

```
≫  a*a./a
   ans: Inf
```

The difficulty is that a*a has overflowed the allowable range of double-precision floating point numbers.

Sometimes, overflow can be avoided by structuring a computation differently. For example, performing the division before the multiplication eliminates the overflow:

```
≫  a*(a/a)
   ans: 1e+200
```

Underflow: Floating point numbers smaller than realmin are called subnormal and have a relative error larger than eps. The smallest possible nonzero double-precision floating point number is

```
≫  realmin/(2 ^52)
   ans: 4.94065645841247e-324
```

For mathematical values smaller than this, the relative error of the floating point representation can be arbitrarily large. To illustrate, let's define two numbers that are in the range of normal eps precision:

```
≫  big = 1e15;
≫  small = 1e-307;
```

Both big and small are between realmax and realmin:

```
≫  big < realmax & big > realmin
   ans: 1
≫  small < realmax & small > realmin
   ans: 1
```

The quantity small./big, however, is smaller than realmin and its relative precision is poor:

```
≫  small./big
   ans: 9.88131291682493e-323
```

Mathematically, we know that $\frac{1 \times 10^{-307}}{1 \times 10^{15}} = 1 \times 10^{-322}$. The computed result has a relative error of 0.012, much larger than eps.

Things can get even worse; numbers smaller than
`4.94065645841247e-324` are rounded off to zero.

```
≫  2e-324
 ans:  0
```

The relative error here is 1.

Subtraction: Taking a difference between two numbers of similar size causes
the leading digits to cancel and leads to an increase in relative error.
For example, consider the two numbers

```
≫  a = 53242342342237.51
 ans:  53242342342237.5
≫  b = 53242342342237.49
 ans:  53242342342237.5
```

each of which has a relative error less than eps. The mathematical
difference between them is 0.02, but it is computed to be

```
≫  a-b
 ans:  0.015625
```

Poor conditioning: For mathematical reasons, aside from round-off error,
some computations magnify the relative error of the inputs. This mag-
nification is characterized by the condition number of the computation; a large condition number means that the relative error of the
output is much larger than the relative error of the input. We'll see an
example of this in the next section and return to the subject in Chapter 17.

Note that for single- and double-precision numbers the values stored in
the computer generally are much more precise than the values displayed on
the screen. Retyping the displayed numbers will result in a loss of precision.
For example, consider the repeating decimal:

```
≫  1/9
 ans:  .1111
≫  9*.1111
 ans:  .9999
```

Internally, all of the bits of the mantissa are kept, and so the computation is
much more precise:

```
≫  9*(1/9)
 ans:  1
```

15.4   Example: Global Positioning

To illustrate some of the aspects of numerical computing, consider the Global Position System [10]. GPS consists of a constellation of satellites, ground control stations, and receivers that let users determine their position—latitude, longitude, and altitude—with high precision. The GPS can locate a user to within about 6 to 12 meters anywhere on earth.

Each GPS satellite constantly is sending out radio messages containing the satellite's position and the time that the message was sent. The GPS receiver logs the time of receipt of each message so that the receiver can determine the travel time of the message. Since messages travel at the speed of light, the distance to the satellite can be calculated. By measuring the distance to several satellites, the receiver can compute its position.

In Chapter 17, we will consider how to find these intersection points and will treat the entire real-world, three-dimensional problem. To simplify things here, consider the geometry shown in Figure 15.2 involving two satellites and a receiver in two dimensions. The position and distance information for one satellite describes a circle—the receiver might be anywhere on this circle. The other satellite's information describes another circle. Given two such circles, the receiver's position can be localized to one of the two intersection points of the two circles.

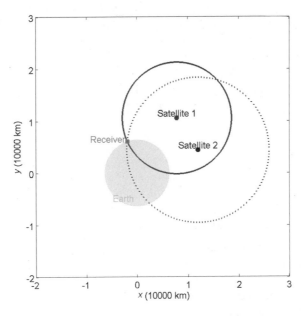

Figure 15.2.   The geometry of a two-dimensional GPS system.

Let us suppose that the satellites report their positions and clock times as double-precision floating point numbers: position (x, y) in kilometers and time in seconds since the GPS system was turned on. We receive messages from two satellites, containing the following information:

```
≫ x1 = 7832.342; % position, km
≫ y1 = 10523.826; % position, km
≫ st1 = 417640801.490407; % time of broadcast, sec
≫ rt1 = 417640801.526187 % time of receipt, sec
≫ x2 = 11932.305; % position, km
≫ y2 = 4387.894; % position, km
≫ st2 = 417640802.322830; % broadcast, sec
≫ rt2 = 417640802.369375; % receipt, sec
```

Each of these quantities is reported to an impressive-looking number of digits: satellite position to the nearest meter (that is, 0.001 km) and the broadcast and receipt times to 15 digits! But, as we shall see, this precision won't be sufficient.

First, we calculate the travel time of the messages:

```
≫ dt1 = rt1-st1
 ans: 0.035780131816864
≫ dt2 = rt2-st2
 ans: 0.046545684337616
```

This looks very precise, since so many digits are being printed out, but only the first six digits after the decimal point are meaningful because, in subtracting two almost equal quantities, we have lost a large amount of precision. The relative error of dt1 and dt2 due to rounding-off is not anywhere near eps; it is roughly

```
≫ .000001/.046545
 ans: 2.15e-005
```

Given the speed of light, the distance to each satellite can be computed:

```
≫ c = 299792458; % speed of light, m/sec
≫ r1 = dt1.*c
 r1: 10726618.8519042
≫ r2 = dt2.*c
 r2: 13954042.8258542
```

There are two things to note about these numbers. First, only the first six digits are meaningful, since dt1 and dt2 are meaningful only to six digits. Second, the units are in meters, not kilometers, since c was given in meters per second and dt1 and dt2 in seconds. The values of r1 and r2 are not precise to a small fraction of a meter, as the printed form of the numbers suggests, but to about 100 meters. This won't be adequate to produce the 6 to 12 meter precision of the working GPS system.

Simply by choosing to represent the broadcast and receipt times in the way we did, we have lost precision. A change in the way we store the information—representing the times as two numbers, seconds since the system was turned on and fractions of a second—would restore the needed precision to the system.

The calculation of the receiver's position from the satellite position and timing information is not elementary, but the program twoCircles will do the work for us. It returns the two possible intersection points.

```
≫  [one,two] = twoCircles(x1,y1,r1,x2,y2,r2)
one =
-2998394770.12448 - 4487347358.56508i
4487366745.66441 - 2998396680.12367i
two =
-2998394770.12448 + 4487347358.56508i
4487366745.66441 + 2998396680.12367i
```

Something is wrong here; the positions are being reported as complex numbers! The problem, however, is not in twoCircles, but in the units used. TwoCircles understandably assumes that all of the distances and positions are in the same units, but whereas x1, x2, y1, and y2 are in kilometers, r1 and r2 are in meters. As specified, the two circles don't actually intersect at any real coordinate, thus the complex-valued result.

We can fix things by translating r1 and r2 from meters to kilometers:

```
≫  [one,two] = twoCircles(x1,y1,r1/1000,x2,y2,r2/1000)
one =  -1921.22587617965       6059.71739034641
two =  15689.0635712837        17826.7211892624
```

It's impossible to tell from the data given which of these two points we are located at, but we have an important auxiliary piece of information: The receiver is near the surface of the earth. Since the earth's radius is approximately 6360 km, point one is our location.

How precise is the reported position? We have a clue in that the numbers r1 and r2 are meaningful only to six digits. But this does not mean necessarily that the position is also meaningful to six digits. In many computations, a small relative change in the input can lead to a larger relative change in the output. The ratio of the output relative change to the input relative change is the *condition number* of the computation. We can examine the condition number by repeating the computation with a slightly changed input. For example, we'll change r2 by one part in a million—a relative factor of 10^{-6}—by multiplying it by 1.000001:

Key Term

```
≫  twoCircles(x1, y1, r1/1000, x2, y2, 1.000001*r2/1000)
  ans: -1921.23699875864 6059.74169199223
```

The output has changed by about 0.015 km in the x-coordinate and 0.024 km in the y-coordinate, so the relative changes are

```
≫  0.015/1921
   ans: 7.808e-006
≫  0.024/6059
   ans: 3.961e-006
```

This is roughly four to eight times the relative change in the input, so the condition number is roughly 4 to 8, a moderate amplification of the input error.

The condition number can vary, depending on the problem inputs. In the case of GPS, when the satellites are arranged approximately along a single line from the receiver, the condition number can become huge, completely destroying the precision of the system. (See Exercise 15.5.)

15.5   Exercises

Discussion 15.1:
Explain the bank, rat, and + formats for printing numbers.

Discussion 15.2:
The isnan operator tests whether a number is NaN. Why can't this be done simply using the numerical equality operator, ==?

Exercise 15.1:
The isfinite operator tests whether a number has an ordinary numerical value, is NaN, Inf, or -Inf. Write the equivalent of

```
≫  isfinite(x)
```

in terms of isnan(x) and isinf(x) and the logical operators.

Exercise 15.2:
By hand, write out the following fractions in binary fixed-point notation as a base-2 number: $\frac{1}{2}, \frac{3}{4}, \frac{3}{8}, \frac{1}{16}$. For example, $\frac{1}{4}$ is written as 0.01.

Discussion 15.3:
Not all "simple" fractions have a simple decimal representation. For example, $\frac{1}{3}$ is 0.3333... and $\frac{1}{7}$ is 0.$\underbrace{142857}\underbrace{142857}$.... Similarly, not all "simple" decimal numbers have a simple binary representation. Write the decimal 0.3 as a binary floating point number.

Exercise 15.3:
How accurate is sqrt? Compute the absolute error $(\sqrt{a})^2 - a$ for a ranging from 1 to 1000 and plot this versus a. Also compute the relative error $\frac{(\sqrt{a})^2 - a}{a}$ and plot this versus a. (*Hint:* When interpreting your graph, look carefully at the axis labels to see the scale.)

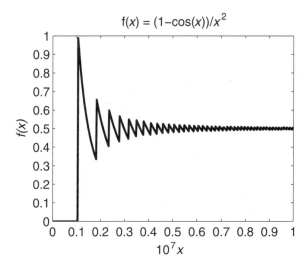

$f(x) = (1-\cos(x))/x^2$

Figure 15.3.

Exercise 15.4:

Figure 15.3 shows a graph of the function

$$\frac{1 - \cos x}{x^2}$$

for x running from 0 to 10^{-7}. The solid line is the computation of the function using floating point arithmetic. The dashed line is the mathematical value of the function. (Calculus students: Apply l'Hôpital's rule twice to see that the limit as $x \to 0$ is $\frac{1}{2}$.) Explain why the numerical result is wrong. (*Hint:* $\cos x$ is approximately equal to $1 - x^2$ for small x.) (I thank Wayne Roberts for drawing this example to my attention.)

Exercise 15.5:

Use the function `twoCircles` to estimate the condition number of finding a GPS position (as in Section 15.4) in the following two situations, both with one satellite at $x_1 = 5$, $y_1 = 8$, $r_1 \sqrt{5^2 + 8^2}$ and the other at

1. $x_2 = 5.1$, $y_2 = 8.1$, $r_2 \sqrt{5.1^2 + 8.1^2}$
2. $x_2 = 5.01$, $y_2 = 8.01$, $r_2 \sqrt{5.01^2 + 8.01^2}$

Assume that the desired answer is the one near $(-3, 3)$.

The cause of the poor condition number is the near alignment of the two satellites with the receiver.

CHAPTER 16

Mathematical Relationships with One Unknown

The nomogram in Figure 16.1 is a graphical calculator for computing gasoline consumption. To use it, we mark the location of known quantities on two of the scales, connect those two points with a straight line, and read off the unknown quantity on the third scale. In this particular nomogram, the three quantities are distance traveled d, gasoline consumed g, and rate of consumption r (e.g., miles per gallon). For example, when $d = 300$ miles, and $g = 7.4$ gallons, we can read off $r = 40.5$ miles per gallon.

A moment's reflection will reveal that the relationship between the three quantities can be expressed algebraically as

$$r = d/g \qquad (16.1)$$

The preceding is only one of many equivalent ways of writing the same relationship: For instance, the following two equations depict the same relationship, but in a form that is unnecessarily verbose:

$$rg = rd/r \qquad \frac{d^2}{r^3} = \frac{g^5 r^2}{d^3}$$

A formula is a particular sort of equation that has one of the variables on one side and all of the other variables on the other side. There are three simple formulas for the gasoline consumption relationship:

$$r = d/g \qquad d = gr \qquad g = d/r$$

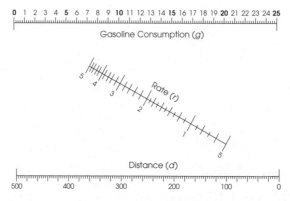

Figure 16.1.   A nomogram for calculating gasoline consumption. Given two of the three variables r, g, and d, the value of the third variable can be found by positioning a straight-edge on the appropriate position on two of the scales and reading off the value on the remaining axis.

These three formulas correspond to the three quantities that we might want to compute. The nomogram cleverly integrates all three formulas into one device, reflecting the fact that all three formulas stem from one relationship.

Relationships can involve inequalities. For example, if the variable r in the gasoline consumption is the automobile manufacturer's estimate of gasoline consumption, then it usually applies under ideal conditions. To take into account the fact that real-world gasoline consumption will typically be worse than the manufacturer's estimate, the relationship should be $d \leq gr$.

This chapter and the following one are concerned with computations involving relationships between variables:

Solving: Given some of the variables, find the values of others consistent with the relationship.

Optimizing: Find values for the variables, consistent with the relationship, that make a desired variable as large as possible (maximization) or as small as possible (minimization).

Where there are data involved, some applications of solving and optimizing are important in their own right:

Interpolation: Given some data relating the variables—and taking the data as depicting precisely the relationship—solve to find a reasonable mathematical form of the relationship exactly consistent with the data.

Fitting: Given some data relating the variables—but data that are perhaps imprecise, conflicting, and contradictory—optimize to find the best mathematical form of a relationship.

In this chapter, we consider the situation where only a single variable is unknown. This is the case, for example, in the gasoline consumption nomogram. In the following chapter, we will deal with the more technically challenging situation where multiple variables are unknown.

16.1   Representing Mathematical Relationships

Students of elementary algebra learn to represent mathematical relationships in terms of equalities and inequalities. They also learn techniques for manipulating equations symbolically to transform them into formulas.

Representing a relationship that is a formula is easy: Use a function. The computer function is exactly analogous to a formula; given the values of the function's arguments the return value can be computed. For example,

```
[1]  function g = gasolineConsumption(d,r)
[2]  g = d./r;
```

This looks very much like the relationship $g = d/r$, but the computer function differs in an important way from the mathematical relationship: Whereas it's clear how to use algebra to translate $g = d/r$ into a different formula, say $d = gr$, it probably is not at all clear how to use gasolineConsumption to compute d given g and r. To be specific, given d and r, the computation for g is straightforward:

```
≫  g = gasolineConsumption(300,40.5)
   ans: 7.4074
```

But, given g=3 and r=32 we can't directly compute d from the function. The statement

```
🐞 ≫  3 = gasolineConsumption(d, 32)
```

makes no sense to the parser and generates an error message.

For gasoline consumption, we can write separate functions to implement all of the particular formulas that we want to use. In addition to the gasolineConsumption function, there are

```
[1]  function r = rateOfConsumption(d,g)
[2]  r = d./g;
```

and

```
[1]  function d = distanceTravelled(g,r)
[2]  d = g.*r;
```

What isn't clear from these three computer functions is that all three formulas stem from the same relationship. Each function/formula depicts the relationship as seen from a particular perspective.

But not every relationship can be presented as a formula. This is not a question of skill with algebraic manipulation, but a mathematical fact established by Cauchy in the nineteenth century. Consider, for example, the complicated-looking relationship

$$\sin(y^2) = \frac{(1-x^r)}{1+x^r} \cos(r\sqrt{x}/y) \tag{16.2}$$

This is a perfectly good relationship among x, y, and r, but there is no general algebraic formula for x in terms of y and r.

How do we represent a relationship that is not a formula? The key is to write the relationship as a formula by adding an additional variable called an *indicator variable*.

Key Term

```
[1]   function indicator = re16point2(r,y,x)
[2]   % Indicator function for Eq. 16.2
[3]   indicator = -sin(y.^2) + ...
[4]       (1-x.^r)./(1 + x.^r).*cos(r.*sqrt(x)./y);
```

When the indicator variable `res` is zero, then line 2 of `re16point2` is equivalent mathematically to the relationship described in Eq. (16.2).

Using this same strategy, the gasoline relationship could be represented this way:

```
[1]   function indicator = gasolineRelationship(r,g,d)
[2]   indicator = g.*r - d;
```

When the indicator variable `indicator` is zero, this function correctly describes the gasoline consumption relationship among g, d, and r. Or, in the case of the inequality $d \leq gr$, the inequality is satisfied whenever the indicator variable is positive.

What may seem strange about `gasolineRelationship` is that it doesn't implement the computation needed to calculate one variable from the other two. But while the function doesn't perform the computation, it does contain all of the necessary information for the computation. We can see this by drawing a graph; plotting the indicator variable while holding the known variables at fixed values and allowing the unknown variable to vary. For example, setting $g = 3$ and $r = 32$, we can use `gasolineRelationship` to create a function of a single variable, d:

```
≫ gasfun = inline('gasolineRelationship(32,3,d)', 'd');
```

and then plot `gasfun` versus d as seen in Figure 16.2. The relationship (with $r = 32$ and $g = 3$) is satisfied at the value of d where the indicator variable is zero.

It is possible to write a mathematical relationship in a different but equivalent form. The solution should be the same even if the gasoline relationship is expressed in a different form, say,

```
[1]   function indicator = gasolineRelationship2(r,g,d)
[2]   indicator = g.^5*r.^2./d.^3 - d.^2./r.^3;
```

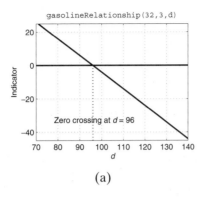
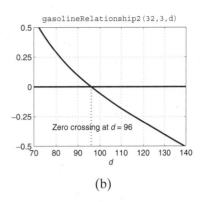

(a)   (b)

Figure 16.2. Graphs of the indicator variable from two different, but equivalent, forms of the gasoline relationship. Variables r and g are held constant at $r = 32$ and $g = 3$ and variable d is allowed to vary. The place where the graph crosses zero indicates the value of d that satisfies the relationship.

Indeed, in Figure 16.2(b), we see that the graph of the indicator variable in `gasolineRelationship2` is different from that of `gasolineRelationship` but leads to the same solution.

For the gasoline relationship, we have a formula for d, namely $d = gr$. We can directly compute d for any given g and r. So there is little point in going through the trouble of graphing the indicator variable versus d and finding the solution from the graph. But for relationships where there is no formula, such as solving Eq. (16.2) for x, the indicator variable approach is valuable indeed. Figure 16.3 shows a graph of the indicator variable from `re16point2` versus x for r and y held at values $r = 3$ and $y = 0.5$. From the graph, it is seen easily that there are several solutions; that is, several values of x for which the indicator variable is zero.

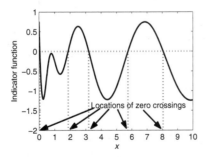

Figure 16.3. A graph of the indicator variable for Eq. (16.2) versus x, holding $r = 3.0$ and $y = 0.5$.

```
>> fplot(inline('re16point2(3.0,.5,x)'),[0 10])
```

Five different solutions are indicated, near $x \approx .048$, $x \approx 1.87$, $x \approx 3.19$, $x \approx 5.75$, and $x \approx 8.05$.

The indicator function approach can be used as well for solving mathematical inequalities. We simply adopt the convention that when the indicator function is positive, the inequality is satisfied.

In those cases where we have a formula for one variable but wish to solve for another variable, an indicator function can be constructed directly from the formula. For example, `gasolineConsumption` computes g when given r and d as arguments. We can nonetheless use `gasolineConsumption` to solve for d given r and g by creating an indicator function that takes d as an argument. Here's an indicator function for finding the value of d when $r = 32$ and $g = 3$:

```
>> indicatorFun = inline('3 - gasolineConsumption(d,32)', 'd');
```

16.2   Zeroing in on Solutions

In the preceding section, we saw one way of solving relationships involving equalities: plot out the relationship's indicator function and look for places where it is zero. For inequalities, look for places where the indicator function is positive. This is a practical and effective strategy for many relationships. Looking at the indicator function quickly provides a lot of information: Is there any solution at all? Are there multiple solutions? Are there "near solutions" where the indicator function comes close to zero but doesn't touch it? Are there discontinuities or sharp transitions?

To make a graph of the indicator function, we need only the indicator function itself (with all variables except one held at known, fixed values), the the endpoints of the domain over which the indicator function is to be plotted. Figures 16.2 and 16.3 are examples.

But typically, a graphical solution is not precise enough. We can, of course, refine the graphical solution by zooming in around the zero crossing. Figure 16.4 shows this being done near one of the zeros in `re16point2`. The technique is to select iteratively an area near the zero crossing and redo the plot around this area. (MATLAB provides a mouse-driven tool for narrowing the range of a plot, but this doesn't increase the level of detail in the plot.) With only two iterations, the zero crossing is located to better than one part in 1000 of the original 0-to-10 interval.

Bracketing Algorithms

When we use a graphical method to find a solution, as in Figure 16.4, we can employ common sense to make sure that we are focusing on an interval where a solution actually exists. When we seek to develop algorithms that find solutions automatically, we need some way of implementing this common sense. Fortunately, there is a simple mathematical condition that guarantees that any continuous indicator function will have a zero crossing in an interval so long as the the following condition holds: The indicator

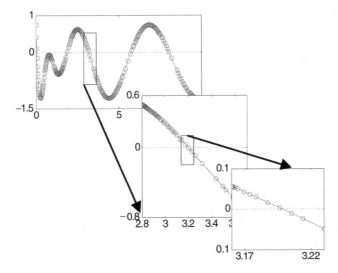

Figure 16.4.   Finding a solution to Eq. (16.2) using a series of plots of the indicator function
re16point2 holding $r = 3.0$ and $y = 0.5$. The first plot,
   ≫   fplot(f,[0 10],'o-') shows several zero crossings. The second,
   ≫   fplot(f,[2.8 3.7],'o-') zooms in on an area near the zero crossing.
The third plot zooms in even more: ≫   fplot(f,[3.16 3.23],'o-')

Key Term   function must have one sign at one end of the interval and the opposite sign
at the other end of the interval. A *bracket* of a solution is a pair of values
x_1 and x_2 such that $f(x_1)$ and $f(x_2)$ have different signs. Figure 16.5 shows
several cases.

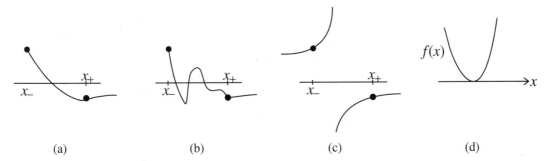

(a)   (b)   (c)   (d)

Figure 16.5.   Four situations encountered with brackets. (a) The "ideal" situation where there
is one zero crossing within the bracket. (b) More than one solution within the
bracket. (c) A singularity. The indicator function doesn't cross zero, but
the discontinuity looks, numerically, like a zero crossing. This situation is easily
identified by evaluating f at the point where the crossing was identified.
(d) A solution where the indicator function touches zero but doesn't cross it.
There is no bracket in this case.

A bracketing algorithm is an algorithm that takes as arguments an indicator function and an initial bracket and homes in on a solution by producing iteratively narrower brackets. We denote the initial bracket by its endpoints x_+ and x_-. As the notation suggests, x_+ is the endpoint where $f(x_+) > 0$ and x_- is the other endpoint. One of the endpoints is designated as our guess of the location of the root. We call this guess x_{old}.

Figure 16.6 shows two common ways to narrow a bracket, both of which produce a new endpoint x_{new}.

Bisection: The endpoint x_{new} is defined as the midpoint of the old bracket:

$$x_{new} = \frac{1}{2}(x_+ + x_-)$$

The new bracket will be half the size of the original bracket.

False position: The endpoint x_{new} is defined as the zero crossing of the line that connects $(x_+, f(x_+))$ and $(x_-, f(x_-))$. The new bracket will always be smaller than the original bracket, but it may not be much smaller. For example, in Figure 16.6, the new false-position bracket is almost the same size as the original.

To produce the new bracket, the new endpoint replaces whichever one of the old endpoints had a function value of the same sign as $f(x_{new})$. Similarly, the old guess x_{old} of the position of the zero crossing is replaced by x_{new}.

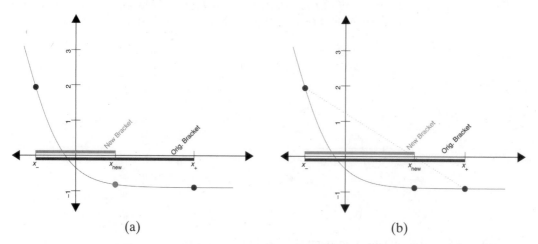

(a) (b)

Figure 16.6. Two ways to narrow a bracket around the zero of a function $f(x)$. The original bracket ranges from x_- to x_+. (a) In bisection, a new point $x_{new} = \frac{1}{2}(x_- + x_+)$ is computed. The sign of the function at x_{new} determines which side of x_{new} the new bracket will be on. (b) In the false-position method, x_{new} is taken as the point where a line drawn between $f(x_-)$ and $f(x_+)$ crosses zero.

The process of narrowing the bracket and updating the guess x_{new} continues iteratively. We need to have some criterion for stopping the iteration. The obvious criterion, stop when x_{new} equals the true solution $x_\star$, is impractical because we don't know $x_\star$, and in general, $x_\star$ may not be exactly a floating point number. There are three practical stopping criteria:

1. Stop when $|x_{new} - x_{old}|$ becomes small. A reasonable definition of small is $|x_{new} - x_{old}| < \varepsilon x_{new}$, where ε is an acceptable relative precision (which should be substantially bigger than the floating point relative precision eps).
2. Stop when $f(x_{new})$ is close to zero.
3. Stop after a fixed number of iterations. This is a simple criterion and avoids many problems with numerical pathologies.

There are additional variations on bracketing algorithms described in numerical methods texts [17, 19]. MATLAB provides an effective hybrid method, fzero. The details of the method are in some sense unimportant to the user, since bracketing methods come with a guarantee to converge to some small bracket that contains a zero crossing (or a discontinuity). Here is an example, using re16point2:

```
≫  f = inline('re16point2(3.0,.5,x)');
≫  fzero(f,[3 3.5])
   ans: 3.19779669247292
```

Bracketing methods provide no guarantee that the found solution is the only one within the bracket. If there is more than one solution, the methods will zero in on one of them and provide no indication that there are others. For example,

```
≫  fzero(f,[0 10])
   ans: 0.04845415155794
```

This is one reason why it is often worthwhile to examine a graph of the indicator function before using automated methods.

Nonbracketing Methods

Since bracketing algorithms provide a guarantee that a zero crossing of some type will be found (if one exists within the starting bracket), you might wonder why there is interest in any other sort of algorithm for finding zero crossings. There are three main reasons:

- Nonbracketing algorithms can be faster.
- When searching for a solution involving two or more simultaneous relationships (a subject discussed in the next chapter), bracketing is not a possibility. While a one-dimensional interval can be bracketed with only two points, a region in two or more dimensions cannot be bracketed with a finite number of points.

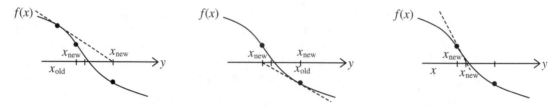

Figure 16.7.   Three iterations of Newton's method.

- Sometimes, a guarantee that the algorithm will converge to a zero crossing is provided by some knowledge of the function itself: The bracket is irrelevant.

Key Term

The most famous of the nonbracketing algorithms is *Newton's method*, which works from only a single guess, x_1, of the location of the zero crossing. Newton's method uses the derivative $f'(x_1)$ to extrapolate a linear approximation to $f(x)$ as the line that passes through the single point and has the specified slope. That line crosses zero at

$$x_{new} = x_1 - \frac{f(x_1)}{f'(x_1)} \tag{16.3}$$

The updating rule at each step is simply

$$x_1 \leftarrow x_{new}$$

as illustrated in Figure 16.7. Termination of the algorithm can be based on the criteria on page 405.

Key Term

Another important nonbracketing algorithm is the *method of secants*. This is described in Exercise 16.1.

▶

Example: Computing Square Roots

The square root is an extremely common computation: It needs to be done quickly, reliably, and rapidly. In computing $\sqrt{y}$, we are looking for a value x such that

$$x = \sqrt{y}$$

This relationship is equivalent to $y^2 = x$, corresponding to the indicator function

$$f(x) = x^2 - y$$

which we've written as a function of a single variable x, since y is known. Note that the indicator function involves only arithmetic computations: multiplication and subtraction. A bracket for the zero crossing of this indicator function is always easily found: 1 and y will do.

Table 16.1 A Comparison of the Bisection and Newton's Methods for Finding Zeros of the Indicator Function $x^2 - 4$. The number of significant digits doubles at each iteration for Newton's method, resulting in extremely fast convergence.

Iteration	Bisection Method		Newton's Method	
	x_+	Significant Digits	x_1	Significant Digits
initial	4.000000000000	0	1.0000000000000000	0
1	2.500000000000	1	2.5000000000000000	1
2	2.500000000000	1	2.0499999999999998	2
3	2.125000000000	1	2.0006097560975609	4
4	2.125000000000	1	2.0000000929222947	8
5	2.031250000000	2	2.0000000000000022	15
6	2.031250000000	2	2.0000000000000000	16+
7	2.007812500000	3		
8	2.007812500000	3		
9	2.001953125000	3		
10	2.001953125000	3		
11	2.000488281250	4		

$$\vdots$$

To use Newton's method, we need to compute $f'(x_1)$. From calculus methods, this is found to be $f'(x_1) = 2x_1$. This gives the updating rule

$$x_{\text{new}} = x_1 - \frac{x_1^2 - y}{2x_1} \quad \text{or, more simply} \quad x_{\text{new}} = \frac{1}{2}(x_1 + \frac{y}{x_1})$$

Table 16.1 compares the bisection and Newton's methods for the computation of $\sqrt{4}$. Each line shows the left end of the bracket, x_+, from the bisection method and x_1 from Newton's method. Since $\sqrt{4}$ is known to be exactly 2, we easily can see how close each method is to the correct answer. For example, the entry 2.00060975609843 has four significant digits (2.000). Similarly, 1.999997471100 has six significant digits (1.99999).

For this computation, Newton's method converges very rapidly; the number of significant digits roughly doubles at each iteration. It takes only six iterations to reach the precision of a floating point number. This doubling of significant digits at each iteration is termed *quadratic convergence*. Bisection is substantially slower. It takes roughly three steps to achieve each additional single significant digit. This is because bisection halves the length of the bracket at each iteration; it takes $\log_2 10 \approx 3.3$ iterations to add a significant digit. This linear growth in the number of significant digits is called *linear convergence*.

Key Term

Key Term

The superiority of Newton's method does not hold for all computations. Indeed, in some circumstances, Newton's method may not work at all. (See Exercise 16.1.)

◀

16.3   Derivatives

Key Term

Key Term

Key Term

The derivative of a function $f(x)$ is another function, often denoted $f'(x)$, that gives the slope of $f(x)$ at each point x. In calculus, the derivative is sometimes defined by a *forward formula*

$$\lim_{h \to 0} \frac{f(x+h) - f(x)}{h}$$

and sometimes by a *backward formula*

$$\lim_{h \to 0} \frac{f(x) - f(x-h)}{h}$$

and sometimes by a *centered formula*

$$\lim_{h \to 0} \frac{f(x+h) - f(x-h)}{2h}$$

Figure 16.8 shows how the slope is calculated using the forward and centered formulas as the slope of a line connecting two points on the graph of the function. The true derivative is the slope of the tangent to the graph.

The limit in the preceding equations makes all of the formulas equivalent (at least so long as the derivative is continuous). Calculus students learn a symbolic technology for computing derivatives that doesn't directly involve any of the formulas. For example, the function $f(x) = x^2$ has derivative $f'(x) = 2x$. Or, to take a more complicated-looking example, function $f(x) = \frac{1}{2}\exp\left(-(x-1)^2\right)$, graphed in Figure 16.8, has a derivative that can be computed using the chain rule as $f'(x) = -(x-1)\exp\left(-(x-1)^2\right)$. We can find the numerical value of this derivative at any value of x simply by evaluating $f'(x)$. For example, to find the derivative at $x = 1.9$ we can define

```
≫  x = 1.9;
```

```
≫  f = inline('exp(-(x-1).^2)./2');
```

and define the derivative of f as found using the chain rule:

```
≫  fprime = inline('-(x-1).*exp(-(x-1).^2)');
```

Now, finding $f'(x)$ is simply a function evaluation:

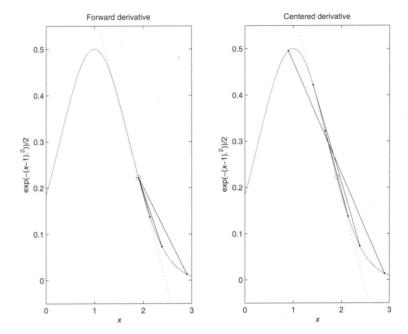

Figure 16.8. The function

$$f(x) = \frac{1}{2}\exp\left(-(x-1)^2\right)$$

and the computation of the slope using the forward formula (left) and the centered formula (right) at $x = 1.9$. The true tangent to the graph at $x = 1.9$ is shown as a dotted line.

```
≫  byCalculus = fprime(x)
   byCalculus:  -0.40037225960065
```

Derivatives can also be computed numerically, using the aforementioned finite-difference formulas directly rather than the symbolic rules of differential calculus. This can be important if the function does not have a simple expression; for instance, if there are conditionals or tables of data.

Although the formulas are simple to implement as a computer program, care has to be taken in giving a numerical meaning to $\lim_{h\to 0}$. The obvious interpretation, making h as small as possible, doesn't work. Remember from Section 15.2 that the next floating point number after x is $(1+\varepsilon)x$, where ε is the floating point relative precision. So the smallest possible h is $(1+\varepsilon)x - x = \varepsilon x$. At $x = 1.9$, this is impressively small:

```
≫  h = eps*x
   ans:  4.218847493575595e-016
```

With a numerically small value of h, we can directly implement the forward-difference formula

```
≫ byFiniteDiff = (f(x+h) - f(x))./h
  byFiniteDiff: -0.39473684210526
```

Although there are many digits of precision, comparing the value of byFiniteDiff to that from the calculus result, we see that the leading digit is wrong. The relative error is

```
≫ (byCalculus - byFiniteDiff)./byCalculus
  ans: 0.01407544444014
```

The finite-difference formula gives a result that is in error by 1.4% even though h is as small as it can be numerically.

The central difficulty stems from the finite precision of floating point numbers. Recall from Section 15.2 that the computation of $f(x)$ as a floating point number typically involves round-off and truncation error. Round-off error is set by the floating point relative accuracy, eps, and is about the size of $eps \cdot f(x)$. When h is very small, then $f(x+h) \approx f(x)$.

```
≫ f(x)
  ans: 0.22242903311147
≫ f(x+h)
  ans: 0.22242903311147
```

But when two almost equal numbers are subtracted, there can be a dramatic increase in the relative error. If $f(x+h)$ and $f(x)$ differ only by round-off error, or if they are rounded off to exactly the same floating point value, then the value of $(f(x+h)-f(x))/h$ will be entirely erroneous.

Ironically, by making h large, we can improve the situation. For example,

```
≫ h = sqrt(eps)*x
  ans: 2.831220626831055e-008
```

This h is bigger than eps*x by a factor of 10^8. Now the forward-difference formula will be correct out to 8 digits:

```
≫ (f(x+h) - f(x))./h
  ans: -0.40037225637781
```

The reason for this is that the increased h makes $f(x+h)$ substantially different from $f(x)$:

```
≫ f(x+h)
  ans: 0.22242902177605
```

Note that the last seven digits of f(x+h) differ from f(x).

But this can easily be taken too far. If h is too large, then the computed derivative can be quite different from the tangent to the line, as seen in Fig-

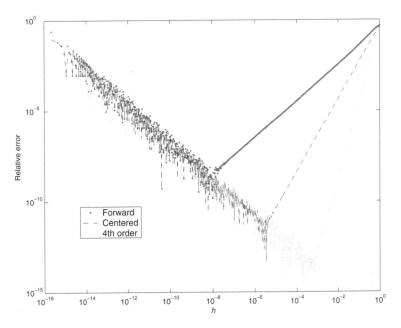

Figure 16.9. The relative error of the numerically computed derivative of $f(x) = \frac{1}{2}\exp\left(-(x-1)^2\right)$ at $x = 1.9$ for the forward-difference, centered-difference, and fourth-order centered-difference methods, graphed as a function of h. The lowpoint in the graph indicates the best value of h. The jagged fluctuations on the left side of the graph reflect the sensitivity of the result when h is too small.

ure 16.8. Figure 16.9 shows the results of an experiment, where the numerically computed derivative is compared to the known value for a range of h. For the forward-difference formula, the h that leads to the most accurate answer is near `sqrt(eps)*x`, while for the centered-difference formula the best value of h is near `eps.^(1/3).*x`. A theoretical analysis of the situation [17] (page 187) shows that these expressions can serve as general guides for the selection of h (unless x is very near to 0).

The function `derivfun` is a simple implementation of a numerical derivative function, using the forward formula. This is not a particularly accurate way to compute a derivative; more accuracy can be gained at the expense of more computation. (See, e.g., [17].) The trade-off between accuracy and expense will vary from problem to problem. Despite the apparent simplicity of the finite-difference approach, there is no single best approach to computing a numerical derivative. Perhaps this is why there is not a standardized, built-in function for computing numerical derivatives. The point of presenting `derivfun` is more to show that a derivative function can be packaged with a simple interface. In Chapter 17, we will make extensive use of such functions.

Make the step small

h ordinarily

h when *x*0 is close to zero

```
[1]   function res = derivfun(fun,x0,varargin)
[2]   % derivfun(fun,x0,params)
[3]   % num derivative of a function at position x0
[4]   seps = sqrt(eps);
[5]   h = x0.*seps;
[6]   h(abs(x0)<seps) = seps;
[7]   x1 = x0 + h;
[8]   f0 = feval(fun,x0,varargin{:});
[9]   f1 = feval(fun,x1,varargin{:});
[10]  res = (f1-f0)./(x1-x0);
```

16.4 Going to the Extremes: Optimization

Key Term

When you look at a submerged object from above the water's surface, the object is distorted. Figure 16.10 shows a photograph of a straight stick inserted in water: The stick looks bent at the waterline and appears narrower under water. The reason for the distortion is that light rays are bent when crossing the interface between air and water, as shown in Figure 16.10. This bending, or *refraction*, has a simple physical origin: Light travels the shortest path between two points; that is, the path that it takes the least time to traverse. The time it takes to traverse a path is, of course, the distance along

Figure 16.10.   A straight rod has been placed in a glass of water. The rod appears bent at the water's surface, due to refraction. Additional refractions appear at the facets of the water glass.

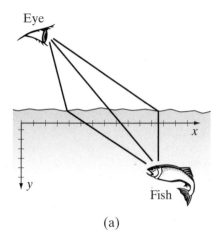

Eye

x

y

Fish

(a)

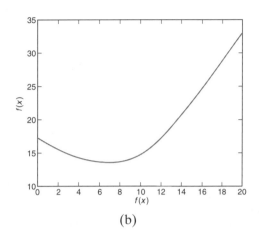

(b)

Figure 16.11. Refraction of light by water. (a) Paths connecting the two marked points. The path of refracted light is the one that takes the shortest time to traverse; this may not be the straight-line path. (b) The traversal time of paths parameterized by the position where each crosses the surface of the water.

the path divided by the velocity; the slower the velocity, the longer it takes to traverse the path. A line is the shortest distance between two points, but since the speed of light in water is slower than the speed of light in air, the overall path is shorter when the light travels somewhat farther in the faster medium in order to save time in the slower medium. The result is a bent line.

Figure 16.11 shows several such bent-line paths connecting the two marked points. We can describe each path in terms of the value x where the light crosses the surface of the water. That is, x *parameterizes* the various paths.

Key Term

The path actually followed by light is the one with the shortest travel time. The physicist Richard Feynman [27] gives a simple explanation of how a photon can determine the shortest path. We humans can use the techniques of geometry to do the calculation of the travel time $T(x)$ for any value of x:

$$T(x) = \frac{\text{dist. from eye to } x}{\text{velocity in air}} + \frac{\text{dist. from } x \text{ to fish}}{\text{velocity in water}}$$

or, in terms of the variables,

$$T(x) = \frac{\sqrt{(x_{\text{eye}} - x)^2 + (y_{\text{eye}} - y)^2}}{v_{\text{air}}} + \frac{\sqrt{(x - x_{\text{fish}})^2 + (0 - y_{\text{fish}})^2}}{v_{\text{water}}} \qquad (16.4)$$

Note that the style of this notation involves leaving parameters such as x_{eye} in symbolic form on the right-hand side and not identifying them explicitly as inputs to the mathematical function. This allows us to highlight our interest in the single variable x.

In the computer notation, we list all the parameters as inputs to the function:

```
[1]  function time = eyeToFish(x,y,...
[2]       xeye,yeye,xfish,yfish,vair,vwater)
[3]  %  Travel time on the bent line between the eye
[4]  %  and the fish.
[5]  time = sqrt((x - xeye).^2 + (y-yeye).^2)./vair + ...
[6]       sqrt((x-xfish).^2 + (y-yfish).^2)./vwater;
```

Since we know the values of all of the variables except x (for instance, $y = 0$ is the surface of the water), we can define a function of a single variable by giving specific numerical values for the other variables:

```
>> f = inline('eyeToFish(x,0,0,5,10,3,1,.85)');
```

Figure 16.11 shows the graph of $f(x)$, made with `fplot(f, [0,20])`, from which it's easy to see that the shortest path goes through $x_\star \approx 7$.

Key Term

This example of finding the value $x_\star$ of the shortest path is an example of *minimization*. The function $f(x)$ is called the *objective function* and $x_\star$ is the argument for which $f(x)$ takes on the smallest value. $x_\star$ is called the *argmin* of $f(x)$. The value of $f(x_\star)$ is called the *minimum* of $f(x)$.

Key Term

Minimization is an operation that takes a function $f(x)$ as an argument and returns the argmin $x_\star$. Given $x_\star$, the minimum is easily computed by applying f to $x_\star$; that is, $f(x_\star)$. *Maximization* is basically the same operation as minimization; the *argmax* of $f(x)$ is exactly the same as the argmin of $-f(x)$.

The MATLAB operator for minimization, `fminbnd`, needs additional information beyond the function $f(x)$; it needs an interval in which to search for the minimum.

```
>> fminbnd(f, 0, 20)
   ans: 7.1004
```

This is similar to the use of `fplot` to find the minimum. Fminbnd returns the argmin of its first argument within the interval between the two specified bounds.

Key Term

The minimum of a function is the low point in the graph of the function. It's common for functions to have more than one low point, each of which is called a *local minimum*. The lowest of all the local minimum is called the *global minimum*; possibly two or more local minima will tie for the global minimum.

As an example, consider the mass hanging from a pair of bunjee cords as illustrated in Figure 16.12. A bungee cord is a sort of spring; as it is stretched, it exerts more force. But there is a limit. If the cord is stretched too much, then it will weaken permanently and exert less force.

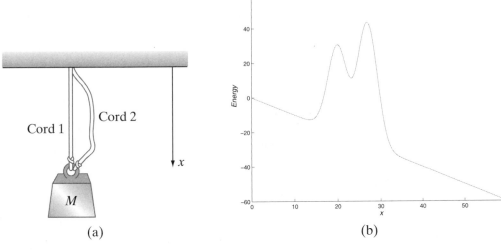

(a)   (b)

Figure 16.12.   A mass hanging from two ropes. Rope 1 starts to weaken and break at length 20, rope 2 at length 27. (a) A schematic of the two-cord system. (b) The potential energy of the system, including gravity.

Such a system can be studied in terms of its potential energy. In this case, there are three potential energy terms: Two due to the two cords, and one due to gravity. The resting position of the system is a point of minimum potential energy. From the graph, we can see that there are three local minima: one argmin at about $x \approx 15$, one at about $x \approx 25$, and one at $x \approx 60$. The first argmin, at 15, corresponds physically to both cords being intact. At the second argmin, one cord has stretched beyond its useful length and only the second cord is providing an upward force to resist gravity. The third point, at 60, isn't really a local minimum at all—it's just the end of the range in the plot. At this point, both cords are stretched beyond use and there is virtually no resistance to gravity. (In this third case, the real-world equilibrium point will be where the mass hits the ground, something not shown in Figure 16.12.)

The operator fminbnd will find the two local minima to high precision. We need, however, to specify an interval that contains the particular minimum of interest.

```
≫  fminbnd('tworopes', 0, 20)
   ans: 13.5443
≫  fminbnd('tworopes', 20,30)
   ans: 23.1939
≫  fminbnd('tworopes', 30,60)
   ans: 60
```

We can see that this last isn't really a local minimum, because it's at one of the ends of the interval, and if we extend the interval by a bit the answer changes:

```
>> fminbnd('tworopes', 30,61)
  ans: 61
```

If an interval is specified that includes more than one local minimum, fminbnd will find one of them, but not necessarily the best one. For instance, in the interval $0 \le x \le 35$, the lowest point of the function is at $x = 35$, where the function has a value

```
>> tworopes(35)
  ans: -34.8837
```

but fminbnd finds another local minimum:

```
>> fminbnd('tworopes', 0,35)
  ans: 13.5443
>> tworopes(13.5443)
  ans: -12.7698
```

Key Term

In Chapter 17, we will examine optimization methods for functions of several variables; these can be complicated. But for functions of a single variable, finding the argmin is generally straightforward. The main complication comes from the possibility of *local minima*. Finding the global minimum (the best of all the local minima) can be hard since it generally involves finding all of the local minima (or knowing enough about the shape of the function to see which local minima can be disregarded).

Algorithms for Minimization

The MATLAB fminbnd operator provides a convenient way of finding the argmin of a function.[1] To understand how fminbnd and other optimization operators work and how they are limited, we explore some of the algorithms for minimizing a function of one variable. There are strong analogies to the algorithms used for solving equations and the same trade-offs: the guarantees provided by slow-but-steady bracketing methods versus the potential unreliability of fast nonbracketing methods.

A bracket is a set of x values and the corresponding $f(x)$ values that constitute a guarantee that a (local) minimum falls within an interval. The bracket for a minimum involves three x-values. The first two of these, which we'll denote x_- and x_+, define an interval. However, $f(x_-)$ and $f(x_+)$ do not tell us whether there is a minimum in the interval: We can always draw a

---

[1]Similar operators are provided in many other systems, even ones that are not oriented toward scientific work. For example, the Excel spreadsheet includes a maximization operator.

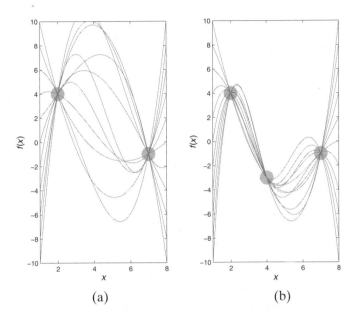

Figure 16.13.   A bracket of three points is needed to trap a minimum. (a) A few of the infinite number of continuous functions that can be drawn through two points. Some of these have minima in the interval between the two points; some do not. (b) The three-point bracket includes a middle point lower than the endpoints. All continuous functions drawn through these three points will have at least one local minimum in the interval between the two endpoints.

graph through the two points $(x_-, f(x_-))$ and $(x_+, f(x_+))$ that doesn't contain a local minimum.

To complete the bracket, a third value x_{middle} is needed. In order guarantee that there is a minimum in the interval x_- to x_+, we require that $f(x_{\text{middle}}) < f(x_-)$ and $f(x_{\text{middle}}) < f(x_+)$. The situation is shown in Figure 16.13.

The basic idea of searching for a minimum using brackets is iteratively to construct smaller and smaller brackets that trap the local minimum. When the bracket is small enough, then we stop.

It's easy to take a bracket x_-, x_{middle}, x_+ with the associated $f(x_-)$, $f(x_{\text{middle}})$, (x_+) and use it to construct a smaller bracket. Pick any point x_{new} that is between x_- and x_{middle} and compute $f(x_{\text{new}})$. If $f(x_{\text{new}}) > f(x_{\text{middle}})$, then the new bracket is x_{new}, x_{middle}, x_+. Otherwise, if $f(x_{\text{new}}) < f(x_{\text{middle}})$, then the new bracket is x_-, x_{new}, x_{middle}. [In the unlikely event that $f(x_{\text{new}})$ exactly equals $f(x_{\text{middle}})$, then pick a new x_{new} and start over.]

Of course, we could also pick x_{new} between x_{middle} and x_+, updating the bracket in the analogous way: If $f(x_{\text{new}}) > f(x_{\text{middle}})$, then the new bracket is $x_{\text{middle}}, x_{\text{new}}, x_+$. Otherwise, if $f(x_{\text{new}}) < f(x_{\text{middle}})$, then the new bracket is $x_{\text{middle}}, x_{\text{new}}, x_+$.

In some sense, it doesn't matter exactly how x_{new} is selected: The new triplet of points is guaranteed both to be a bracket and to be smaller than the old one. Iterating the bracket refinement enough times will make the bracket small, locating the minimum to whatever precision is needed.

Key Term

A simple way of choosing x_{new} is to select the midpoint of whichever of the two subintervals x_- to x_{middle} or x_{middle} to x_+ is bigger. It turns out that a slight modification of this, called a *golden section search*, is superior. Instead of the midpoint exactly dividing the longer subinterval, it should be divided unevenly, respecting the golden ratio $\frac{\sqrt{5}-1}{2} \approx 0.61803$. If the longer subinterval is on the left, then

$$x_{\text{new}} = x_- + 0.61803(x_+ - x_-) \qquad (16.5)$$

If the longer subinterval is on the right, then reverse the subscripts in the preceding formula. Using a golden section search, each new bracket will be smaller than the previous one by a factor of 0.61803. After k iterations, the bracket will be 0.61803^k times as big as the original bracket. For instance, for $k = 20$, the final bracket will be only 0.000066 as long as the original.

Golden section search is not based on any model of the shape of the function $f(x)$. It uses only a boolean comparison $f(x_{\text{new}}) > f(x_{\text{middle}})$ to determine how to refine the bracket. The 1-bit result of this comparison sets just one feature of x_{new}: whether it is in the left subinterval or the right. The detailed value of x_{new} is set by Eq. (16.5), which does not use any information from $f(x)$.

The situation here is analogous to that of the bisection search for a minimum in Section 16.2. There, we selected either the left or right subinterval depending on the sign of $f(x_{\text{new}})$; that is, the boolean comparison $f(x_{\text{new}} > 0)$. There is information in the value of $f(x_{\text{new}})$ compared to $f(x_-)$ and $f(x_+)$. In the method of false position, we exploited this information by using a straight-line model of the function $f(x)$ and setting x_{new} to be the argument at which the line crossed zero.

For a local minimum, a line is not an appropriate model. A linear function will always have a local minimum at one of the endpoints of the interval. Indeed, a linear function can never have a bracket because there can be no points $x_-, x_{\text{middle}}, x_+$ at which $f(x_{\text{middle}}) < f(x_-)$ and $x_{\text{middle}} < f(x_+)$.

Key Term

Instead, a quadratic function provides an adequate model of a local minimum. The method of *quadratic interpolation* finds the unique parabola that passes through the three bracketing points and picks x_{new} to be the value at which this model parabola is minimized. This can be found using calculus techniques and is

$$x_{\text{new}} = \frac{f(x_-)(x_{\text{middle}}^2 - x_+^2) + f(x_{\text{middle}})(x_+^2 - x_-^2) + f(x_+)(x_-^2 - x_{\text{middle}}^2)}{2f(x_-)(x_{\text{middle}} - x_+) + 2f(x_{\text{middle}})(x_+ - x_-) + 2f(x_+)(x_- - x_{\text{middle}})}$$

$$(16.6)$$

Quadratic interpolation is a bracketing method. Equation (16.6) will give x_{new} between x_- and x_+ and so a new, narrower bracket can be constructed using x_{new} and $f(x_{new})$. If the function $f(x)$ is well modeled by a parabola, quadratic interpolation will narrow brackets faster than a golden section search.

Newton's method is a nonbracketing optimization search method. Like quadratic interpolation, it models the function $f(x)$ as a quadratic. But instead of using three bracketing points to define the model, it uses a single point x_1 and the values of the first and second derivative of $f(x)$ at x_1.

$$x_{new} = x_1 - \frac{f'(x_1)}{f''(x_1)} \tag{16.7}$$

The updating rule is $x_1 \leftarrow x_{new}$.

Newton's method for optimization is closely related to Newton's method for finding zero crossings. To find a minimum, we can look for a zero crossing of $f'(x)$. Equation (16.7) is equivalent to the zero-crossing method of Eq. (16.3), but with $f'(x)$ replacing $f(x)$—in the optimization method we find the zero crossing of $f'(x)$ rather than $f(x)$.

Since it is not a bracketing method, Newton's method is not guaranteed to work. Indeed, if $f''(x_1) > 0$, then Eq. (16.7) will be looking for a maximum rather than a minimum. But when the function $f(x)$ is closely approximated by a quadratic with a minimum, then Newton's method can be very fast.

Operators like `fminbnd` use a combination of methods, bracketing and nonbracketing, switching from one to another to improve speed without losing the guarantees provided by brackets.

Optimization with Constraints

The optimization methods introduced previously all deal with a scalar-valued objective function; that is, an objective function that yields a single number. This is not a realistic situation in many ways; real-world decisions usually involve several different factors and not just a single factor as modeled by the scalar objective function.

These multiple factors are perhaps most evident in economic decision making, even in familiar decisions. For instance, deciding where to live often involves a trade-off between price and location. Picking a car involves a balance between performance and price or between capacity and gas economy.

One approach to dealing with multiple factors is to combine them into a single, scalar-valued objective. The location versus price issue, for instance, can be resolved by accounting for the price of transportation or the wage value of time spent commuting. When this can be done, the multiple factors can be combined into a single one by first converting them to common units (e.g., money in the housing decision) and adding them into the objective function.

Another, more flexible approach is to model the various factors as constraints. For example, an engineer designing a bridge strives to make the bridge as economical as possible to construct and maintain but can't lose sight of the main objective—the bridge must be strong enough to carry its load safely. In terms of an optimization problem, the bridge should be as inexpensive as possible subject to the constraint that it carry a specified load. Additional constraints can model other factors. For example, it's common that bridges over water be required to leave a specified span and clearance for the passage of ships; this is a constraint.

The mathematical representation of constrained optimization is this: Set a decision variable x in order to minimize the value of the objective function $f(x)$ but subject to the conditions that $g(x) = 0$ and $h(x) \leq 0$. The functions $g(x)$ are called *equality constraints*; $h(x)$ are *inequality constraints*.

Key Term

Constrained optimization is most important in optimization problems involving several variables, which will be treated in Chapter 17. But the one-variable problem helps to illustrate some of the concepts.

Key Term

The values of x that satisfy the constraints are called *feasible solutions*. For one-variable constrained optimization, an effective solution strategy is (1) to locate the feasible solutions and (2) optimize $f(x)$ restricting the search to feasible solutions.

▶

Example: Constrained Optimization

Minimize $f(x) = \sin(x)$ subject to the equality constraint $x + 3\cos(x) = 1.3$.

The first step is to set up and solve the constraint function $g(x) = x + 3\cos(x) - 1.3 = 0$:

```
≫ g = inline('x + 3*cos(x) - 1.3');
```

A plot of $g(x)$ (Figure 16.14) quickly shows that there are three solutions near $x \approx -1$, $x \approx 2$, and $x \approx 4$. These solutions can be found precisely using `fzero`. We store them in an array `feasible`:

```
≫ feasible = [];
≫ feasible(1) = fzero(g,[3,5])
   ans: 3.7543
≫ feasible(2) = fzero(g,[1,3])
   ans: 3.7543 1.7068
≫ feasible(3) = fzero(g,[-2,0])
   ans: 3.7543 1.7068 -0.7969
```

Now, we evaluate the objective function at each of these feasible solutions:

```
≫ sin(feasible)
   ans: -0.5751 0.9908 -0.7152
```

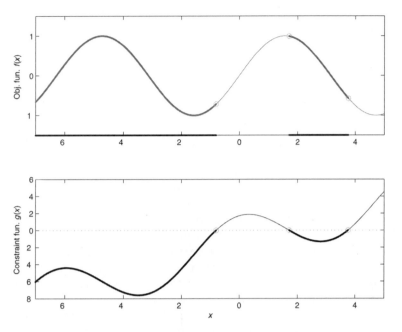

Figure 16.14.   The objective function and constraints for the constrained optimization example.
The constraint function sets the feasible regions (shown with a heavier
line)—optimization is constrained to the feasible regions.

It's evident that the last solution produces the minimum value of the objective function, so the argmax is `feasible(3)` or -0.7969.

Inequality constraints are handled in an analogous way. To illustrate, we repeat the optimization where the constraint is $x + 3\cos(x) \leq 1.3$. Now the feasible solutions are regions where the inequality holds, as in Figure 16.14. A plot quickly establishes that there are two feasible regions, $x \in (-\infty, -.7969)$ and $x \in (1.7068, 3.7543)$. A plot, such as that in Figure 16.14, indicates that the left region has multiple local minima. We can use `fminbnd` to refine a search for the argmin inside the feasible regions

```
≫  fminbnd('sin(x)', -3, -0.7969)
   ans: -1.5708
```

which is $-\pi/2$, as might be expected for the argmin of a sine function. The value of the objective function at this argmin is

```
≫  sin(-1.5708)
   ans: -1.000
```

Key Term

In this case, the constraint is an *inactive constraint*: The location of the local minimum is determined by the objective function and not by the endpoints of the feasible region.

For the other region,

```
≫  fminbnd('sin(x)', 1.7068, 3.7543)
   ans: 3.7542
```

and the value is

```
≫  sin(3.7542)
   ans: -0.5750
```

Key Term

Key Term

Two things should be observed here. First, this local minimum is not as good as the one in the other feasible region. Second, the local minimum is at the edge of the feasible region.[2] This is an *active constraint*, because the argmin would change if the boundary of the feasible region were moved.

An important concept when constraints are involved is the *shadow cost* of a constraint. This describes how much the objective function could be improved if the constraint were relaxed; that is, if the feasible region were made bigger. Inactive constraints always have a shadow cost of zero since the objective function, not the constraint, sets the location of the local minimum. Only active constraints have a nonzero shadow cost.

The active constraint in Figure 16.14 has a nonzero shadow cost. Suppose the constraint were eased by a small amount, say, from 1.3 to 1.31. Then the boundary of the feasible region would be moved

```
≫  fzero('x + 3*cos(x) -1.31', [3,5])
   ans: 3.7580
```

At this new boundary, the value of the objective function is

```
≫  sin(3.7580)
   ans: -0.5781
```

The shadow cost is the ratio of the improvement in the objective function to the change in the constraint value, in this case

```
≫  (-0.5750 - -0.5781)./(1.31-1.3)
   ans: 0.31
```

◀

---

[2] Although `fminbnd` returns an argmin that is a small distance inside the feasible region; that is, due to the numerical issues involved in minimization. From the shape of the objective function, we can see that the local minimum is on the edge of the feasible region.

16.5 Fitting

The simple equation

$$d = rg \qquad (16.8)$$

describes the relationship between the distance d traveled by a car, gasoline g consumed, and the rate r of gasoline consumption. Of the quantities in this relationship, d and g can be directly measured: distance from the car's odometer and consumption by the volume of gasoline used.

In this section, we are concerned with how data can be used to describe such relationships. Such data might look like Table 16.2, which shows part of an actual record of gasoline use and mileage. Each line records one stop at a gasoline station, giving the date, odometer mileage, and the amount of gasoline purchased. The amount purchased is a measurement of the gasoline consumed since the last filling (particularly if the tank is completely filled at each purchase). The distance traveled is the difference in odometer readings from one filling to the next.

From this record, it's easy to compute d and g. How do we describe the relationship between them? Since there are just two variables, we can display the relationship using a scatter plot of g versus d (or vice versa) as in Figure 16.15. It's clear from the plot that as d increases, g also increases.

Equation (16.8) provides a model with which to interpret the data in Figure 16.15. According to Eq. (16.8), the data points should lie on a line $g = \frac{1}{r}d$ that passes through $(0,0)$ and has slope $\frac{1}{r}$. Three such lines are shown

Table 16.2 Part of a Record of Gasoline Consumption from the Author's Family Car. The complete record is in the file `gasoline.dat`. **(In that file, the date is stored as serial integers, with 1 corresponding to Jan. 1, 1900.)**

Date	Odometer Reading (miles)	Gasoline Purchased (gals)
8/7/2000	25,907	14.5
8/21/2000	26,151	12.5
8/21/2000	26,439	12.8
9/10/2000	26,691	13.6
9/30/2000	26,910	13.81
10/25/2000	27,127	13.58
11/15/2000	27,307	12.5
11/29/2000	27,507	13.7

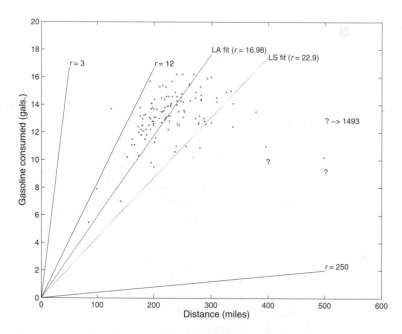

Figure 16.15.   The gasoline consumption g between successive fillings versus the distance traveled d for the data in the file gasoline.dat. The lines show the relationship $g = d/r$ for several different values of the parameter r. Three outliers are marked with "?" (One of the outliers is not plotted, since it would fall far to the right of the other data.)

in Figure 16.15. Evidently, the data points do not lie on any single line, but some lines clearly are better than others. The problem of fitting in this case is to find the the best line; that is, the line that the data points are closest to. This line will be described by the parameter r. Although we say that we "fit r to the data," the model of Eq. (16.8) plays a crucial role: It provides the meaning of r.

The general framework of fitting as follows:

Key Terms

Model: describes the relationship between one or more *explanatory variables*, and a *response variable*. In the gasoline example, we have selected d as the explanatory variable and g as the response variable. In some cases (this gasoline economy example is one), it's not clear which should be the response and which the explanatory variables. We'll return to this later.

Data: consist of several or many simultaneous measurements of both the response and explanatory variables. Each such measurement is called a *case*. In the gasoline example, each filling is one case. That is, each line of Table 16.2 is one case: a simultaneous measurement of the two variables d and g. It's helpful to introduce a subscript to denote the case, so that d_i and g_i are the measurements for case number i. Note that in this example the raw data in the table don't directly give d_i, which has to be calculated as the difference between odometer readings on successive fill-ups.

Parameters: are quantities in the model that are not measured. It is these parameters that we seek to estimate. In the gasoline example, there is one parameter r, but in many models there are multiple parameters. In Chapter 17, we'll study models with more than one parameter.

Given a model and data, the fitting problem can be seen as an optimization task where the objective function that describes, for any value of the parameters, how close the measurements are to the model.

Key Term

One way to measure the closeness of the data to the model is the size of the *residuals*. The residual is the difference between the response variable and the value that would be predicted from applying the model to the explanatory variables. In the gasoline example, the model dictates that $g = d/r$, so for the ith case, the predicted value is $\hat{g}_i = d_i/r$. Since the actual value of g for case i is g_i, the residual for that case is $R_i = g_i - \hat{g}_i$ or, plugging in the model for $\hat{g}_i$, $R_i = g_i - d_i/r$. The better the model, the smaller the residual.

The usual way to measure the size of the residual is to square it. At a simple level, the motivation behind squaring is that a small residual could be either positive or negative; we want either situation to count the same. A more subtle consideration is the probability distribution of the residuals.

There is a residual for each of the cases; we want to consider all of the cases when evaluating how well a model matches the data, so sum over them. The objective function then becomes

$$\mathcal{E}(r) = \sum_{i=1}^{N} R_i^2 = \sum_{i=1}^{N} (g_i - d_i/r)^2 \tag{16.9}$$

Note that the objective function takes the parameter that we see to optimize, r, as an argument.

For the purposes of setting up the objective function on the computer, let's generalize Eq. (16.9) a bit. Our model of gasoline consumption returns g_i for any d_i and r. We implement this as an m-file, gasModel. Here it is:

```
[1]   function g = gasModel(r,d)
[2]   %  A model of gasoline consumption
[3]   g = d./r;
```

This is such a simple model that it hardly seems worth its own m-file, but remember that other models might be much more complicated. What's essential here is that the model returns the response variable as a function of the explanatory variable (or variables) and any parameters.

The residual is simply the difference between g and gasmodel(d,r). Squaring and summing give us the objective function. We could put this objective function into an m-file, but it's so simple that an in-line function will do nicely:

```
≫ objfun = inline('sum((g-gasModel(r,d)).^2)', 'r', 'g', 'd');
```

It's important to note the arguments to objfun closely. Recall that inline takes an expression as its first argument. In this case, the expression is 'sum((g-gasmodel(r,d)).^2)'. The remaining arguments to inline are the names of the variables in the expression that will be the arguments to the function created by inline. In the preceding invocation of inline, we've declared r to be the first argument of objfun, followed by g and d.

Ultimately, the purpose of the objective function is to find the optimal value of the parameter r; the values of g and d are data themselves and therefore fixed. But in using objfun, we have to pass the data values as arguments, so objfun includes both the data and the parameters as arguments.

We can, by trial and error, find the value of r that minimizes the objective function:

```
≫ objfun(10,g,d)
  ans: 3336.2
≫ objfun(20,g,d)
  ans: 492.13
≫ objfun(30,g,d)
  ans: 553.58
```

Note that the middle value, for $r = 20$, is lower than the other two values. This means that the three points $r = 10$, $r = 20$, $r = 30$ define a bracket, so we know that there is a local minimum between $r = 10$ and $r = 30$.

We can use fminbnd to do the work for us. It was in anticipation of using fminbnd that we arranged for objfun to take as a first argument the variable, r, over which the optimization will be performed. But since objfun takes d and g as arguments, we need to arrange to have fminbnd automatically pass them as arguments to objfun. This is accomplished with the following syntax:

```
≫ fminbnd(objfun, 10, 30, [], g, d)
  ans: 22.8855
```

The first three arguments are the objective function[3] and the bounds in which we want to search. The next argument is a placeholder for the various

---

[3]Since objfun is an inline function, it doesn't need to be placed in quotes as would be the case if it were an m-file.

settings of how `fminbnd` works internally—we'll use the defaults by specifying the empty vector. Any additional arguments to `fminbnd` are passed as arguments to the objective function.

Key Term

This type of fitting, where the objective function is based on the square of the difference between the actual value of the response variable and the value predicted by the model, is called *least squares fitting*. The term "least" refers to the minimization, and "square" to the form of the objective function.

But there are possibilities other than least squares. We might, for example, decide to minimize the absolute deviation:

$$\mathcal{E}(r) = \sum_{i=1}^{N} |g_i - d_i/r| \qquad (16.10)$$

This is easily accomplished by specifying another objective function

```
≫ objfun2 = inline('sum(abs(g-gasmodel(r,d)))', 'r', 'g', 'd');
```

Again, the optimum can be found with `fminbnd`:

```
≫ fminbnd(objfun2, 10, 30, [], d, g)
   ans: 16.9767
```

Why do we get such different answers, $r = 17.0$ versus $r = 22.9$ miles per gallon, from the same data? The answer in this case has to do with the outliers marked with ? in Figure 16.15. These outliers presumably are due to failing to record gasoline purchases so that the measured amount of gasoline used is much less than the real amount. Outliers, as the word suggests, are very far from the rest of the data. Since the least squares fitting penalizes heavily large deviations from the data (the square amplifies large deviations) it is sensitive to outliers. The least squares fitted model, shown the line marked LS in Figure 16.15, tries to go close to the outliers. Using the absolute value reduces sensitivity to outliers; the line marked LA in Figure 16.15 shows the least absolute value fit, which goes closer to the main body of data points than the LS fit.

Some of the gasoline outliers can be identified using common sense. For example, since the manufacturer describes the range of the car as 350 miles, cases where $d > 350$ are suspect. We can mark these as outliers and delete them from the fitting.

```
≫ outlier = d > 350;
≫ fminbnd(objfun, 0, 30, [], g(~outlier), d(~outlier))
   ans: 17.6170
≫ fminbnd(objfun, 0, 30, [], g(~outlier), d(~outlier))
   ans: 16.7961
```

Note that with some of the outliers deleted, the least squares and least absolute value fits are much closer together.

As we shall see in Chapter 17, the least squares fitting can be performed rapidly by matrix manipulations rather than an optimization search. This enables fitting multiple parameters easily. For example, the model

$$y = a + bx$$

describes variable y as a linear[4] function of x, using the two parameters a and b. (In the gasoline model, a was not included since the appropriate relationship between g and d is one of proportionality.)

Why did we choose g as the response variable rather than d? The answer lies in our understanding of the errors in the measurement of each variable. The quantity that we minimized to conduct the fit is based on the deviation of the response variable from the prediction. The assumption here is that the explanatory variable is known precisely but that the response variable may differ from the model value due to errors in measurement or the influence of other factors. In the gasoline example, it's reasonable to think that the mileage measurement is precise and that the gasoline measurement is not (perhaps we didn't fill the tank all the way) or is influenced by factors such as the weather, the type of driving, and so on. (There is an advanced approach, called *total least squares*, that can handle situations where the explanatory variables are also imprecise.)

Key Term

16.6   Interpolation

We have often represented mathematical functions using arithmetic combinations of variables. The function `gasmodel` is a simple example of this; we compute the output by an arithmetic operation on the input.

Sometimes a function is represented by a table of data and not arithmetic operations. For example, a sound wave is a function of time but is represented as a list of numbers. This list is implicitly a table: The time column could be explicitly constructed using the sampling frequency information. Similarly, a monochrome image—light intensity as a function of x, y position—is represented as a matrix of numbers. Written explicitly as a table, this would have three columns: for each pixel x, y, and the intensity. In either case, a continuous function (of time or position) has been sampled at discrete locations.

Key Term

Interpolation is the process by which a continuous function implicit in the discrete location values of a table can be evaluated at other locations. As an example, consider the table of logarithms in Figure 6.1 on page 115. This table gives the base-10 logarithm of the integers 1 to 50; that is, the table is $\log_{10}(x)$ evaluated at the discrete values $x = 1, 2, \ldots, 50$. With this table, we can easily find the logarithm of any value x in the table—find the x value in

---

[4]More properly, we should say "affine."

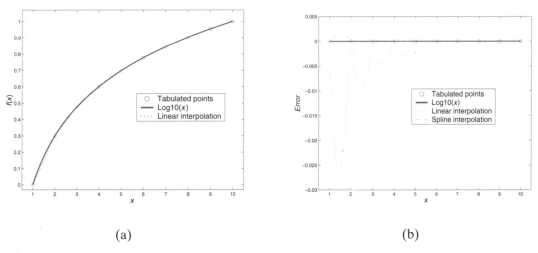

(a)   (b)

Figure 16.16.   (a) The function $\log_{10}(x)$ and the piecewise linear interpolating function of the discrete tabulated points $x_{\text{table}} = 1, 2, \ldots,$. (b) The difference between the interpolating function and the true function, using linear interpolation and cubic spline interpolation [with derivatives matched to those of $\log_{10}(x)$].

the left column (labeled "Num.") and look up its value in the middle column (labeled "Logarithmi.").

But how to find the function at an intermediate value of x, say, $x = 3.1415926$?

Henry Briggs, the seventeenth-century author of the table, had a solution. The function $\log_{10}(x)$ can be approximated, as in Figure 16.16 as a piecewise linear function. To approximate $\log_{10}(3.1415926)$, we find the adjacent values in the table—these are $x = 3$ and $x = 4$. Then we find the value of the linear function that connects these two points in the graph. (See Goldstine [3] for details of Briggs's calculation.)

In general, to evaluate at x a piecewise linear approximation to a function tabluated at x_{left} and x_{right}, we compute

$$f_{\text{linear}}(x) = f(x_{\text{left}}) + \frac{x - x_{\text{left}}}{x_{\text{right}} - x_{\text{left}}}(f(x_{\text{right}}) - f(x_{\text{left}})) \qquad (16.11)$$

Note that $f(x_{\text{right}})$ and $f(x_{\text{left}})$ are already in the table; we only need to perform additions, subtractions, multiplications, and divisions.

Briggs's table makes things even more convenient: The quantity $x_{\text{right}} - x_{\text{left}}$ is always 1, and $(f(x_{\text{right}}) - f(x_{\text{left}}))$ is included in the table as the entries written in small type between successive logarithm values. For example, to compute $\log_{10}(3.1415926)$ we find the surrounding points $x_{\text{left}} = 3$ and $x_{\text{right}} = 4$. We then look up in the table that $f(3) = 0.4771212547$ and $f(4) - f(3) = 0.1249387366$. Since $3.1415926 - x_{\text{left}} = 0.1415926$, we get

$$f_{\text{linear}}(3.1) = 0.4771212547 + 0.1415926 \times 0.1249387366 = 0.4948116552$$

The preceding gives an algorithm for linear interpolation. Thinking about interpolation as an operator, we see that it requires three inputs:

1. The location x at which we would like to evaluate the interpolated function. We will call this the *target point*.
2. The set of discrete locations x_{table} included in the table. These are the *tabulated points*.
3. The values $f(x_{table})$ in the table. These are the *tabulated values*.

To illustrate, we can reconstruct Briggs's table in MATLAB with just two commands:

```
>> xtable = 1:50;
>> ftable = log10(xtable);
```

The reconstructed table is

```
>> [xtable; ftable]'
ans: 1. 0
     2. 0.30102999566398
     3. 0.47712125471966
     4. 0.60205999132796
     5. 0.69897000433602
     and so on ...
```

We don't need the "Differ." column since the interpolation operator, `interp1` will take care of this for us:

```
>> interp1(xtable,ftable, 3.1415926)
ans: 0.49481165527675
```

Key Term

We can use `interp1` to evaluate the linear interpolation of the table $x_{table}, f(x_{table})$ for any value of x. But when x is outside the interval bounded by the smallest and largest values in x_{table}, we are doing *extrapolation* rather than interpolation. To signal this, `interp1` will return NaN:

```
>> interp1(xtable,ftable, -1)
ans: NaN
```

The linear interpolation of the log table at $x = 3.1415926$ is impressively precise, but it isn't accurate. Compare our interpolated value of 0.48961512838049 to the true value:

```
>> log10(3.14152926)
ans: 0.49714986528587
```

Only the first two digits are right!

Briggs was interested in high-precision computing and recognized that linear interpolation with a small table did not provide adequate accuracy. Briggs's approach to improving accuracy might be regarded as brute force: He created a large table with 100,000 entries: x_{table} was the integers 1 to

100,000. This means that in Briggs's table, $\log_{10}(x)$ of any x with five digits is represented to full precision without any need to interpolate. This extremely large table—the size of a telephone book for a large city—has the advantage that, for large x, the piecewise linear segments are very short and the linear approximation accurate. For example, to find $\log_{10}(3.1415926)$ we can multiply by a power of 10 to move to the high-accuracy part of the table. Multiplying by 10,000 gives $x = 31415.926$, which can be interpolated between $x_{\text{left}} = 31,415$ and $x_{\text{right}} = 31,416$. First, make the table:

```
≫  xtable = 1:100000;
≫  ftable = log10(xtable);
≫  interp1(xtable, ftable, 31415.926)
   ans: 4.49714986527079
```

Now the answer is accurate to 10 digits! [Since we multiplied x by 10,000 and $\log_{10}(10,000) = 4$, we would subtract 4 to get the final answer.]

Briggs would have marvelled at our ability to construct a 100,000-entry log table in a second's time. His table was years in construction and his computers were people laboring over the extraction of roots and multidigit multiplication by hand. Certainly Briggs would have known that given a fast `log10` operator, it's really absurd to make a table and do linear interpolation—it's much better just to compute the logarithm directly. Briggs used linear interpolation as a tool to leverage increased precision from the hard-earned entries in his massive table. The attractiveness of linear interpolation is that the computation, for Briggs's table, involved just one addition and one multiplication operation. [In general, a division and subtraction are also needed, per Eq. (16.11).]

With today's powerful computers, we do not need to select linear interpolation as a tool because it is fast to compute. Instead, we choose an interpolation method so that the mathematical function that results from the interpolation has the properties we desire.

Key Term

Two types of properties are of general importance: smoothness and monotonicity. By the word *smooth*, we have in mind a function that varies slowly; it doesn't indulge in wild oscillations or sudden breaks between tabulated points. A linear interpolating function is extremely smooth in this regard.

But another aspect of smoothness has to do with derivatives of the function. The linear interpolating function has a constant first derivative, or slope, between tabulated points, but at the tabulated points, the slope generally changes sharply. This means that the first derivative of the linear interpolating function is discontinuous and that the second derivative is spikelike (see Figure 16.17).

Key Term

Monotonicity has to do with whether the function changes slope from positive to negative. A monotonically increasing function $f(x)$ is a function that increases in value as x gets bigger; a monotonically decreasing function gets smaller in value as x gets bigger. There are also monotonically

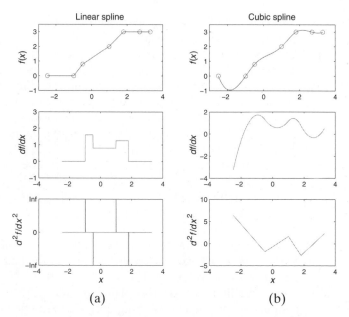

Figure 16.17. Splines through some tabulated points. The tabulated points are scatter plotted as circles. (a) The linear interpolating function, and its first and second derivatives. The first derivative is piecewise constant; the second derivative is zero except at the tabulated points. (b) The cubic spline. The first and second derivatives are both continuous. Note that although the tabulated points are monotonically nondecreasing, the derivative of the cubic spline curve is negative in some places.

nondecreasing and nonincreasing functions. For example, in classical economic theory, the supply of a product is a monotonically nondecreasing function of price; an increase in the price will not lead to less being produced. Similarly, demand for a product is a monotonically nonincreasing function of price—increases in price do not (by themselves) lead to increases in demand.

The tabulated values of a monotonic function will themselves be monotonic. A linear interpolating function respects monotonicity. One consequence of this is that all of the extrema of the interpolating function are tabulated points.

Monotonicity may or may not be a desirable property of an interpolating function, depending on context. While an economic supply or demand curve arguably should be monotonic, a sound signal should not.

Through any finite set of tabulated points,[5] an infinite variety of interpolating functions can be drawn; all that is required to be an interpolating function is that the function pass through the tabulated points. There are,

---

[5] Assuming that there are no repeats in the set x_{table}.

however, just a few classes of functions that are in general use. These have the important virtue of being fast to compute while offering a range of possibilities as regards the properties of smoothness and monotonicity.

Key Term

A important class is the *splines*. The word "spline" historically referred to a thin, stiff piece of wood or other material. The spline is flexible enough that it can bend smoothly to follow a curve; it is a smooth interpolant of points on the curve.

In the most common sort of mathematical spline, a polynomial function is created in the interval between tabulated points. The parameters of the polynomial are set so that the polynomial intersects the two tabulated points at the ends of the interval. We have already encountered a linear spline, where the polynomial has a linear form: $a + bx$. A formula for the values of the parameters a and b can be derived from Eq. (16.11).

In a cubic spline, the local polynomial over each interval has the form $a + bx + cx^2 + dx^3$. The four parameters a, b, c, and d offer more than enough flexibility for the polynomial to go through the two tabulated points at the end of the interval; these parameters are set so that the first and second derivatives of the polynomial at each tabulated point match the derivatives at that point of the polynomial from the adjacent interval.

The continuous first and second derivatives give a cubic spline a shape pleasing to the eye. Beyond this, a cubic spline can be the appropriate interpolating function to use if it is believed that the mathematical function underlying the tabulated points has continuous first and second derivatives. For example, if the tabulated points set the position of a robot arm as a function of time, a continuous second derivative means that the force applied to the arm to accelerate it can be varied steadily, with no sharp transitions.

The MATLAB cubic spline operator `spline` takes the same arguments as the linear interpolation operator `interp1`. Whereas there is a unique linear interpolating function, the cubic spline is not defined uniquely by the tabulated points; there is an infinite number of different cubic splines that pass exactly through the tabulated points, all of which have continuous first and second derivatives.

Two additional parameters are needed to specify a unique cubic spline. For convenience, these parameters have default values that set the second derivative of the interpolating function to be zero at the tabulated endpoints—this eliminates any curvature of the default cubic spline function at the endpoints. But this is not appropriate for all functions. (See Exercise 16.4.) In using `spline`, you can override the default parameters and explicitly set the first derivative of the spline function at the endpoints. Figure 16.18 shows several interpolating functions through a common set of tabulated points. These functions differ in their endpoint derivatives.

Another important class of interpolating functions is based on Hermite polynomials: locally cubic functions that are knit together in a way that respects the monotonicity of the tabulated points [18]. The cost of this, however, is that the second derivative is not necessarily continuous. (See Exercise 16.8.)

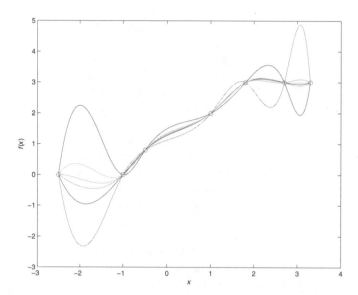

Figure 16.18. Six of the infinite number of cubic splines that pass through the tabulated points. All of them have continuous first and second derivatives. The different splines are parameterized by their slopes at the leftmost and rightmost of the tabulated points.

Interpolation can also be performed on tabulated values of functions of more than one variable. When the tabulated values form a regular array, as in the case of intensity values for an image, then interpolation of multiple independent variables can be performed one variable at a time using any of the methods described previously. The sequence of steps is shown in Figure 16.19. The function `interp2` performs interpolation of functions of two variables such as images. The same issues of smoothness and monotonicity apply as with one-variable interpolation.

When the tabulated values of a function are known at irregularly scattered tabulated points, not on a grid, interpolation can be accomplished by triangulation: finding the nearest tabulated points that form a triangle around the target point. Although a different method is used for nongridded tabulated points, the functional interface is the same: arguments specifying the locations of the tabulated points, the tabulated values, and the locations of the target points. The function `griddata` implements interpolation on nongridded tabulated points.

It's important to remember that interpolation cannot recover information that is not in the tabulated points. If the tabulated points do not provide a good sample of the function, as in Figure 16.20, the interpolated function will not resemble closely the function from which the samples were taken, although the interpolant will match exactly at each of the samples.

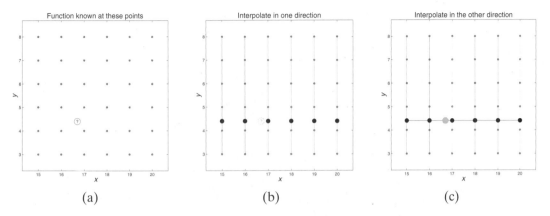

Figure 16.19. Interpolation in two variables on a regular grid. (a) The function $f(x,y)$ is known at a gridded set of points (x,y) (small dots), and it is desired to interpolate a value at the target point marked "?". (b) Interpolation in the y variable is done separately for each of the x-values, and each of these interpolants is evaluated (larger dots) at the y-coordinate of the target point. (c) The interpolation of the points evaluated in the previous step is evaluated at the x-coordinate of the target point.

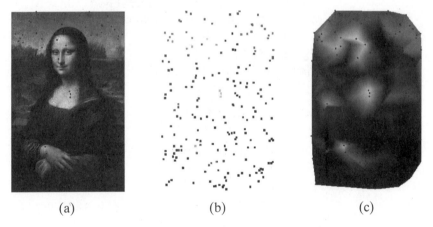

Figure 16.20. Interpolation when the tabulated points are not on a grid. (a) The 200 tabulated points are marked as dots. The tabulated values are the image intensities at each of the tabulated points. (b) An image consisting of only the tabulated points. (c) Interpolation via triangulation has been performed using only the information at the tabulated points. A linear interpolation method has been used resulting in ridges in intensity connecting the tabulated points. For a color image, separate interpolation is done in each color plane.

16.7 Exercises

Discussion 16.1:

Consider the function $f(x) = \frac{\tanh(x)}{|x|+1}$. This has only one zero, which is at $x = 0$. Explain why Newton's method will not work for finding this zero when the initial point, x_1 is outside of the interval $[-1, 1]$. (*Hint:* Make a graph of the function.)

Exercise 16.1:

The method of secants is a nonbracketing algorithm very similar to the method of false position. The algorithm is initialized with two guesses about the location of the zero crossing, x_1 and x_2, which don't need to provide a bracket. [That is, the signs of $f(x_1)$ and $f(x_2)$ can be the same.] A new guess, x_{new}, is generated in the same way as in the false-position method.

What's different in the method of secants from the false-position method is the updating rule. In the method of secants, the updating ignores the sign of $f(x_{new})$ and simply involves, at each iteration,

$$x_1 \leftarrow x_2$$
$$x_2 \leftarrow x_{new}$$

as shown in Figure 16.21. Termination of the algorithm can be based on the same criterion described on page 405.

Write a program, `secantmethod(fun,x0)`, that implements the method of secants. Test it by computing $\sqrt{10}$ using the indicator function $x^2 - 10$.

Exercise 16.2:

Write a program, `deriv4(fun,x)`, that computes a derivative numerically by the "fourth-order centered" formula

$$\lim_{h \to 0} \frac{-f(x+2h) + 8f(x+h) - 8f(x-h) + f(x-2h)}{12h}$$

Check whether your function gives the same relative accuracy depicted in Figure 16.9.

Exercise 16.3:

It's tempting to think about solutions to equations in terms of boolean op-

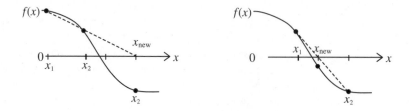

Figure 16.21. Two iterations of the method of secants.

erations: A value x either is or is not a zero of the function $f(x)$. Either–or is the boolean dichotomy. Writing boolean functions of x is straightforward. For instance, consider a version of the gasoline usage relationship that takes into account that the actual mileage will generally be less than the manufacturer's estimate, but not very much less. The relationship among the three variables is something like $.75rg \leq d \leq rg$. This translates directly into the boolean condition implemented in the following function, `booleanGas`:

```
[1]  function res = booleanGas(r,g,d)
[2]  res = .75.*r.*g <= d & d <= r.*g;
```

In contrast, we might choose to write a nonboolean indicator function as in `smoothgas`:

```
[1]  function res = smoothGas(r,g,d)
[2]  res = (d - .75.*g.*r ).*(g.*r - d);
```

Both of these functions correctly represent the mathematical relationship. `BooleanGas` will have a value larger than 0 when the three variables satisfy the relationship, and the same is true of `smoothGas`. But the very different shapes of the functions have important consequences for finding solutions.

1. Graph both `booleanGas` and `smoothGas` as a function of r for $g = 12$ and $d = 250$.
2. Use the bisection method to solve for r for both functions.
3. Explain why Newton's method or the secant method (Exercise 16.1) won't work with the `booleanGas` function.

Exercise 16.4:
Here is a small table of logarithms:

```
>> x = 1:10;
>> y = log10(x);
```

We can use spline interpolation to compute logarithms of values not in the table (although in practice, since there is a `log10` function, we would not do this); for instance,

```
>> spline(x,y,1.5)
   ans: 0.1715
```

This is not very accurate, as can be seen by comparing to

```
>> log10(1.5)
   ans: 0.1761
```

The spline technique can make use of information about the slope of the function at the endpoints of the tabulated interval. In the case of $\log_1 0x$, we know the slope at any value x; it is $\frac{1}{x \ln 10}$, where $\ln 10$ is the natural log of 10—that is, `log(10)`. A spline with the correct slopes at the endpoints can be constructed with

```
≫ spline(x,[1/(x(1)*log(10)),y,1/(x(end)*log(10))],1.5)
  ans: 0.1782
```

Show that over the interval $x = 1$ to $x = 10$, the spline with the appropriate endpoint slopes, is more accurate than the spline without the endpoints specified. Do this by computing the relative error at many points in the interval. (Remember that `spline` can take a vector of values at which to evaluate the spline.)

Exercise 16.5:

Write a function, `secondDeriv(fun,x)`, to compute the second derivative. A simple formula for the second derivative of $f(x)$ is

$$\frac{f(x+h) - 2f(x) + f(x-h)}{h^2}$$

Exercise 16.6:

Consider collecting data on the water level of a river in a flood state. By measuring the water level at different points, we can construct a profile that allows us to determine where the crest of the flood is and how fast it is moving downstream. Such water-level measurements can be taken by recording the height of the water on a stationary object such as a bridge abutment. Here are some measurements as might be taken by an observer:

Time	Water Level (ft) at Ford Bridge Mile 23	Water Level (ft) at Long Bridge Mile 32
5:50 AM		25
6:10 AM	26	
6:30 AM		26
8:10 AM	29	
8:35 AM		27
10:15 AM	32	
10:30 AM		25.5
1:10 PM	32	
2:50 PM	31	
3:10 PM		24
5:05 PM	28	
5:20 PM		23
7:10 PM	27	
7:50 PM		23

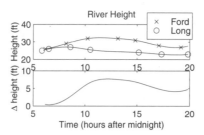

Figure 16.22. The water level measured at each of the two bridges and the difference in height (Δ height) between the bridges.

From these data, we want to compute how fast the crest of the flood is moving downstream and what the water-level difference is between the two points as a function of time.

1. Use `interp1`, for linear interpolation, to produce a graph (as shown in Figure 16.22) showing the water levels at the two bridges as well as the difference in water levels. Repeat this using `spline`.

2. Find the time of the crest at each of the bridges, first using linear interpolation and then spline interpolation. Make an estimate of how precise your estimate of the crest time is.

Exercise 16.7:
Suppose we want to move a robotic arm along a piece of paper. The pen should move from $(0,-1)$ to $(6,0)$ and thence move along a straight line to $(0,4)$. However, we want to avoid sharp accelerations of the arm, so we want the movement to be smooth. One way to do this involves splines. Make a sequence of time points t, and the corresponding x-coordinates and y-coordinates where the robot arm should be at each time point. Then use a spline to interpolate the motion finely for x and y independently. A plot of one such path is shown in Figure 16.23.

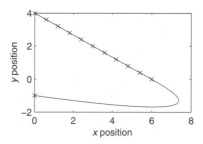

Figure 16.23. A path for the robotic arm in Exercise 16.7.

Since the spline necessarily will not give exactly a straight line, it is help-ful to place some additional points in the t, x, y sequences that fall along the straight line. Such points are shown in Figure 16.23.

Construct a set of t, x, y points and graph out the smooth trajectory that results from a spline.

From your trajectory, compute the acceleration (the second derivative). Plot the magnitude of the acceleration versus time.

Exercise 16.8:

You work for a petroleum refining plant that produces a special synthetic lubricant. Your state-of-the-art equipment runs continuously at maximum capacity, but your competitors have equipment of various degrees of obsce-lescence that can also produce the product but at a higher cost.

Based on your past experience with the equipment, you have tabulated the cost of production for the most expensive batch of product versus total amount produced. This function of total amount produced versus cost is termed a "supply curve." The marketing department of your company has collected some data on the demand for the product as a function of cost: This is the "demand curve." The available data are given in Table 16.3.

Simple economic theory suggests that the market will settle at the price where supply exactly equals demand.

Find the price at which supply equals demand. You can use a spline to interpolate the supply and demand curves. The pchip operator will give a monotonic spline.

In solving this problem, you should write an m-file, supplyvsdemand, that takes price as an argument and returns the difference between supply and demand; that is, how much greater is supply than demand. Then use fzero to find the price at which this difference is zero.

Table 16.3

Supply		Demand	
Marginal Cost	**Total Amount**	**Market Price**	**Total Amount**
10	3	12	22
13.4	6	14.7	19
17.9	12	18.3	14
18.2	13	21.4	11
22.6	18	24.3	8
24.3	21	25.2	7.3
28.2	29	28.7	6.1
		30.3	5.8

Exercise 16.9:

An environmental lobby is pushing for a tax on your product as described in Exercise 16.8, claiming that it pollutes. The tax would be imposed on the purchase of the product and would effectively raise the price to consumers but will leave the production cost unchanged.

You want to explore what effect the tax will have on the price of the product. The environmental lobby claims that the tax will just force the inefficient, highly polluting producers out of the market without penalizing the efficient producers. Your company, however, realizes that the tax will reduce profits for the company. They want to know how much the loss will be so that they can decide whether it's worthwhile to oppose the tax.

Use your previous data and the software you already wrote. Modify supplyvsdemand so that it can take a second numerical argument that is the amount of the tax. This amount should be added to the price used in the demand calculations but not to that used in the supply calculations.

- Make a graph of the price consumers will see as a function of the amount of the tax.
- Find the amount of the tax that will reduce the equilibrium cost (to the producer) to $15.

CHAPTER 17

Mathematical Relationships with Two or More Unknowns

In this chapter, we revisit the topics studied in Chapter 16: finding solutions to equations and inequalities, finding maxima and minima, interpolation, and fitting. But where Chapter 16 was concerned with situations in which only a single variable was unknown, here we deal with a richer and more realistic setting where multiple variables are to be determined.

This higher-dimensional setting imposes significant difficulties. Functions of a single variable can be drawn on a piece of paper and the important algorithms of optimization and solution can be graphed and grasped intuitively. Functions of multiple variables cannot so readily be graphed on paper; even the graphic artist's tricks of perspective and shadow ultimately rely on the processing capabilities of our paperlike retina and our brain, capabilities that evolved to deal with a three-dimensional physical world and not the higher-dimensional mathematical spaces in which dwell many of the relationships we study here. Our intuition about the geometry of functions is much less sharp in three or more dimensions, and we will need to rely on the technical apparatus provided by mathematics.

Another loss, as we move to higher dimensions, is the concept of a bracket. In Chapter 16, we used brackets to guarantee success. A bracket of two points traps a solution; a bracket of three points traps a local minimum or maximum. The analogous encircling capability in two dimensions is provided by a closed curve and in three dimensions by a balloonlike closed surface. In either case, we are dealing not with two or three points but an infinite number: an impracticality on a finite-state automaton such as a computer. Without brackets, we will have to endure extra labor and accept that some searches may be fruitless; our quarry is not cornered.

Since our intuition will fail in high dimensions, we need to rely on more formal techniques, and these may have an abstract flavor. But the situations to which the techniques will be applied can be concrete, as the following examples illustrate.

▶

Example: Gasoline Mileage

In Chapter 16, we used a simple model relating a car's gasoline use g and distance d traveled, $g = d/r$, to estimate the single parameter r that summarizes the car's gasoline consumption performance. This model is unrealistically simple; the performance of a car depends not just on the design of the car but how it is used. For instance, it is well known that cars are more efficient in highway than in stop-and-go city driving. Less well known perhaps is that gasoline economy depends on weather conditions (for example, summer versus winter conditions).

A more comprehensive model of gasoline use would take these factors into account. For example, if we let c stand for the fraction of driving done in the city and t stand for the day number (defining 1 to be Jan. 1, 1900), then the rate of gasoline use might be modeled as follows:

$$r_{total} = r_{car} + \alpha c + \beta \cos(2\pi t/365.25) + \gamma \sin(2\pi t/365.25) \qquad (17.1)$$

That is, we imagine that the effective fuel use is a combination of the car's ideal highway performance r_{car}, the loss of efficiency due to city driving, and a seasonal effect that is itself modeled by a sinusoidal function of the day of the year. The quantities c and t are things we can measure. Along with g and d, they constitute the data that we want to use to find the parameters r_{car}, α, β, and γ.

Finding the parameters that best fit the data is an optimization problem similar in many ways to those encountered in Chapter 16 but now involving not just a single parameter but four parameters to be found simultaneously from the data.

Example: Stress in a Truss Bridge

The small, planar truss bridge seen in Figure 17.1 supports a load of mass M at node 3. This load imposes stresses on each member of the truss. Computing the magnitude of these stresses and whether they are compression or tension is essential to designing the bridge.

If the bridge is to maintain its shape, the forces at each of the nodes must add to zero. Each member exerts a force on the node; that force is directed along the member. Both the horizontal and vertical components of the force vector must add to zero at each node. For example, at node 3 the vertical forces are

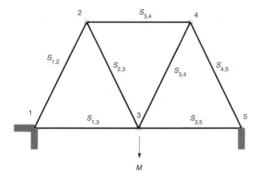

Figure 17.1. A planar truss bridge.

$$-Mg + S_{2,3} \sin \theta_{231} + S_{3,4} \sin \theta_{635} = 0$$

and the horizontal forces are

$$-S_{1,3} + S_{3,5} + S_{3,4} \cos \theta_{635} - S_{2,3} \cos \theta_{231} = 0$$

where $S_{i,j}$ denotes the tensile force in the member connecting node i to node j. The sin and cos terms reflect the geometry of the truss and the direction of the force; θ_{ijk} is the angle formed by the nodes i, j, and k. Each member might be in either compression or tension; a negative $S_{i,j}$ denotes compression. A member in compression exerts a force on each node away from the other node; tension pulls the nodes toward each other. In either case, the force is positive if it is upward or to the right. For instance, member $(2, 3)$ would, if in tension, pull node 3 upward and to the right—a positive vertical force and a negative horizontal force.

In nodes 2 and 4, the vertical and horizontal forces are

$$
\begin{array}{ll}
-S_{1,2} \sin \theta_{213} - S_{2,3} \sin \theta_{231} = 0 & \text{node 2, vertical} \\
-S_{1,2} \cos \theta_{213} + S_{2,4} + S_{2,3} \cos \theta_{231} = 0 & \text{node 2, horizontal} \\
-S_{3,4} \sin \theta_{635} - S_{4,5} \sin \theta_{453} = 0 & \text{node 4, vertical} \\
-S_{3,4} \cos \theta_{635} - S_{2,4} + S_{4,5} \cos \theta_{453} = 0 & \text{node 4, horizontal}
\end{array}
\tag{17.2}
$$

Node 5 is held vertically by the bridge support, so only the horizontal forces need to be considered:

$$S_{3,5} + S_{4,5} \cos \theta_{453} = 0$$

Node 1 is completely fixed by the bridge support, so the forces exerted by the members do not need to add to any particular value.

How do we solve the preceding seven equations for the several unknown variables $S_{1,2}$, $S_{1,3}$, $S_{2,3}$, $S_{2,4}$, $S_{3,4}$, $S_{3,5}$, and $S_{4,5}$?

◀

17.1 Visualizing Functions of Two Variables

Key Term

In drawing a function of a single variable x, we use one axis to represent the x variable and the other to represent the function $f(x)$. The graph itself is a one-dimensional curve (see Figure 17.2). x is called the *independent variable* and the quantity $f(x)$ is the *dependent variable*.

A function $f(x,y)$ of two variables, x and y, requires three axes: one axis for x, one axis for y, and a third for the function's value $f(x,y)$. The independent variables x and y make up a plane—the (x,y) plane. Any pair of values, say, $(x = 7, y = 3)$, corresponds to a point in that plane. The value of the function $f(7,3)$ at that point is a number and is plotted by placing a point at the height $f(7,3)$ over the position $(7,3)$. The graph of the function as a whole is a surface: the set of points of height $f(x,y)$ above position (x,y) in the plane of independent variables. You can imagine laying out a circus tent on the (x,y) ground, placing vertical poles through the tent. At each pole the tent is elevated to a height corresponding to the value of f evaluated at the point where the pole is seated. In a real circus tent, there are only a few poles; the heights of these poles and the drape of the tent material determine the height of the other points on the tent's surface. In mathematical functions, the height is defined at each possible (x,y) position.

To bring things down to earth, take the independent variables x and y to be latitude and longitude: They describe geographic position on the two-dimensional surface of the earth. Here are several examples of functions of these independent variables:

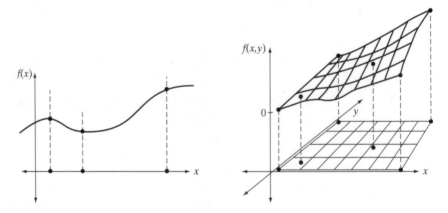

Figure 17.2.   (a) The graph of a function $f(x)$ of one variable, x, is a one-dimensional curve.
(b) The graph of a function $f(x,y)$ of two variables, x and y, is a surface.

Elevation: The height (compared to sea level) of the ground.

Atmospheric pressure: Although this varies from hour to hour, at a fixed time atmospheric pressure at ground level is a function of latitude and longitude.

Total rainfall over the past month: Again, this varies in time, but at any given time, rainfall is a function of latitude and longitude.

Wind velocity: Velocity has both a magnitude and a direction (e.g., 15 miles per hour from NNE). This is an example of a *vector-valued function.* An important vector-valued function is the gradient function, which we introduce later in this section.

Key Term

A convenient way of displaying functions of two variables is as *contour plots* as in Figure 17.3. This style has the advantage of avoiding the complexities of perspective that arise in plots such as Figure 17.2. Each contour is associated with a number, β: Every (x,y) point on the contour is a position where $f(x,y) = \beta$.

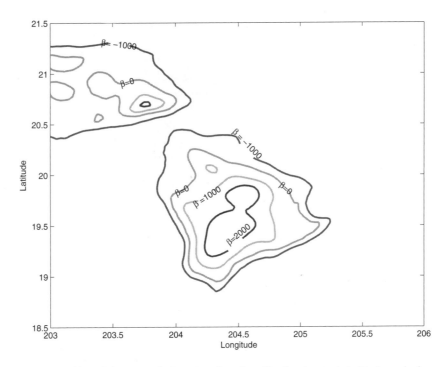

Figure 17.3. A contour plot of the ground elevation function $f(x,y)$, where x is latitude and y is longitude. Each contour is labeled with a number β and is the set of points (x,y) where $f(x,y) = \beta$.

Figure 17.4. A topographic map of the Red Blanket valley to the southwest of Crater Lake, Oregon. A river runs through the valley floor and is fed by streams coming in from the sides. Note that the streams cross the contour lines at right angles.

Hikers are familiar with reading contour plots: topographic maps. In a topographic map of the sort illustrated in Figure 17.4, the contours give useful information about the landscape. Where contours are far apart, the terrain is comparatively level. Where contours are close together, the terrain is steep. A trail that follows a contour is easy going: level. Lake shores always parallel contours since lakes are level. Trails that cross contours are not level; the steepest path through a given point is at right angles to the contour through that point. Rivers and streams, running their downhill course, are at right angles to contours.

Graphs of Functions of Two Variables

To illustrate computer techniques for graphing functions $f(x,y)$, consider the function

$$f(x,y) = x^2 + 3xy - y^2$$

To make a graph, we need to compute the value of $f(x,y)$ at each of many (x,y) points. It's often convenient to use a rectangular array of points in the (x,y) plane.

Step 1 Decide on the limits of the x- and y-variables. For this example, we take $-4 \leq x \leq 4$ and $-3 \leq y \leq 3$

```
>> xmin=-4; xmax=4; ymin=-3; ymax=3;
```

Step 2 Decide how many points to include in the grid. For perspective plots, a reasonable choice is in the small tens. We use 50 for this example, but a larger number might be appropriate in graphing a function that has considerable detail.

```
≫  xpts = linspace(xmin,xmax,50)
   xpts: -4.0000 -3.8367 -3.6735 ...
≫  ypts = linspace(ymin,ymax,50)
   ypts: -3.0000 -2.8776 -2.7551 ...
```

Step 3 Create matrices of x-values and of y-values for each point in the grid. This can be done easily with the meshgrid operator, which takes as arguments vectors giving the x-coordinates and y-coordinates. We generate these vectors using linspace in order to get 50 evenly spaced points within the chosen limits:

```
≫  [x,y] = meshgrid(xpts, ypts);
```

Note that x and y are both matrices:

```
≫  size(x)
   ans: 50 50
≫  size(y)
   ans: 50 50
```

Step 4 Evaluate the function at each of the points on the grid. Depending on how the function is implemented, this might have to be done using a loop. Our simple function can easily be written to exploit parallel arithmetic on vectors, so we can simply compute

```
≫  z = x.^2 + 3*x.*y - y.^2;
```

Alternatively, we might implement $f(x,y)$ as an m-file or an inline function:

```
≫  f = inline('x.^2 + 3*x.*y - y.^2', 'x', 'y');
```

In this case, we would have computed z by applying f to the x and y coordinate matrices:

```
≫  z = f(x,y);
```

Step 5 Use the MATLAB graphing operators to make the graph.

There are three main models for drawing graphs of functions of two variables $f(x,y)$.

Image model: This displays the value $f(x,y)$ as a color or shade in just the same way that an image is displayed.

```
>> pcolor(x,y,z); shading('flat')
```

The `shading` operator controls how the individual grid-point pixels are related to one another. `shading('interp')` gives a smooth picture.

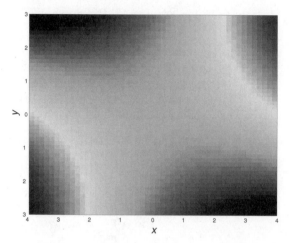

Elevation model: This displays $f(x,y)$ as a surface—the circus tent picture.

```
>> surf(x,y,z);
```

The left graph below is made with the default shading, which draws in grid lines. These lines help to give a three-dimensional quality to the surface, but even with 50 points there are so many of them that they clutter the picture. The center graph has been made with flat shading, which removes the grid lines, but the grid implied by the pixel boundaries still suggests the shape of the surface. The right graph was made with interpolated shading, which gives the smoothest effect but makes it hard to perceive the shape of the surface.

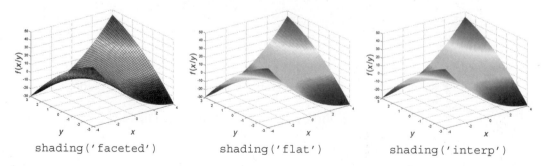

shading('faceted')   shading('flat')   shading('interp')

Contour model: This displays the contours of $f(x, y)$.

```
>>  [c,h] = contour(x,y,z);
```

Interpreting this requires the most skill, but acquiring the skill pays off as many of the techniques in this chapter can be visualized easily in terms of movement along or across contours. In the event that labels for the individual contours are needed, the `clabel` operator will do this, as shown on the right:

```
>>  clabel(c,h);
```

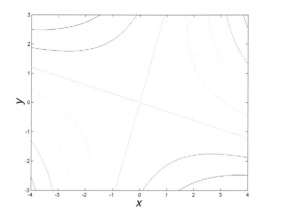
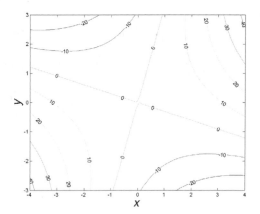

The three techniques can be combined in various ways. By default, the image model technique is used to shade the elevation and contour plots. But we can make more explicit combinations. Here are some examples:

Combine an elevation plot with contours: By making the surface translucent (with the `alpha` operator), we avoid the problem of some parts of the surface obscuring other parts. The `contour3` operator draws contours at the right elevation to match with the elevation plot. The contours provide enough shape information that interpolated shading is effective.

```
>>  surf(x,y,z);
>>  shading('interp');
>>  alpha(.7); % make surface translucent
>>  hold on;
>>  contour3(x,y,z); % add the contours
>>  hold off;
```

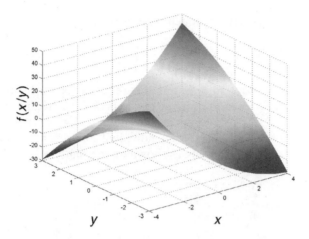

Enhancing contour plots with continuous color information:

The `colorbar` command draws a scale giving the z-value corresponding to each contour. This is often an effective alternative to labeling the contour with numbers.

Combining the image and contour models gives a result much like a topographic map.

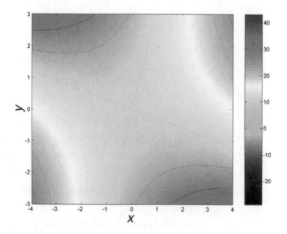

The preceding graph was made with these commands:

```
>> pcolor(x,y,z);
>> shading('interp');
>> alpha(.4);
>> hold on;
>> contour(x,y,z);
>> hold off;
>> colorbar
```

17.2   Notation for Functions of $\geq$ 2 Variables

Up to now, we have emphasized functions of two variables mainly because they effectively can be graphed. But in realistic applications, we may encounter functions of multiple variables, much more than merely two. In order to accommodate such functions, we will change notation a bit. Rather than denoting the independent variables as x, y, and so on, we will call them $x_1, x_2, \ldots$. For N variables we simply use $x_1, x_2, x_3, \cdots, x_{N-1}$, and x_N. Note that the N variables constitute a vector of length N.

It's convenient to give the collection $x_1, \cdots, x_N$ a simple name. In traditional mathematical notation, boldface letters are used (for instance, **x** to refer to $x_1, \cdots, x_N$). On the computer, we can use an ordinary variable name to refer to the collection. The individual elements can be accessed using the familiar indexing operations [for example, x(3) is the equivalent in computer notation of the mathematical x_3].

As an example, consider the spring-mass system shown in Figure 17.5. Two weights are suspended from springs. In the equilibrium configuration, the springs are stretched until the upward force they exert balances the downward pull of gravity. This configuration can be determined from the potential energy of the system; at equilibrium, the potential energy is at a minimum.

The potential energy is a function of the two variables, x_1 and x_2, that describe the position of the weights. There are also several parameters involved: the rest length of the springs, the stiffness of the springs, the mass

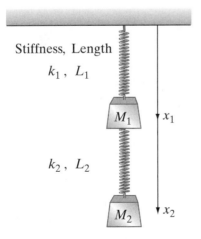

Figure 17.5.   A simple system of weights hanging from springs. The positions of the weights are given by variables x_1 and x_2. The mass of the weights is given by parameters m_1 and m_2, the rest lengths of the springs by L_1 and L_2, and the stiffness of the springs by k_1 and k_2.

of the weights, and the acceleration due to gravity. The form of the function will be familiar to physicists, but for our purposes here, it suffices to say that the potential energy is a given function of the two variables and the several parameters. Here it is, implemented as an m-file, with a vector argument for **x** variables and and argument for each of the parameters.

See Figure 17.5

```
[1]  function E = twosprings(x,m1,m2,L1,L2,k1,k2,g)
[2]  % Potential energy of the spring-mass system
[3]  E = .5*(k1*(x(1)-L1).^2 + k2*(x(2)-x(1)-L2).^2) ...
[4]      - g*(m1*x(1) + m2*x(2));
```

Figure 17.6 shows the potential energy graphed versus the two variables x_1 and x_2. The perspective plot shows a smoothly curved sheet. The contour plot is more informative: The function is bowl shaped with a minimum near $x_1 \approx 25$ and $x_2 \approx 50$.

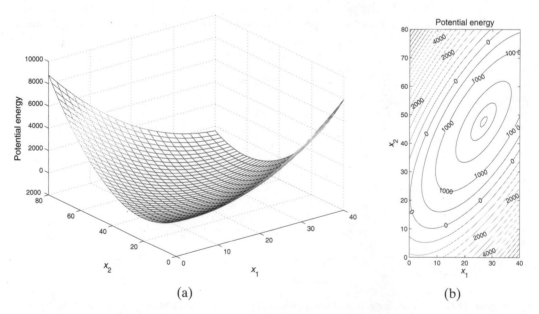

(a)   (b)

Figure 17.6. The potential energy of the two-spring system in Figure 17.5 as a function of x_1 and x_2. (a) A 3D perspective plot of the function. (b) The corresponding contour plot. The parameters of the system are $m_1 = m_2 = 3$, $L_1 = L_2 = 15$, $k_1 = k_2 = 5$.

17.3 Geometry of Functions: The Gradient

Key Term
The *gradient* of a function $f(x_1, x_2)$ is a vector that points in the steepest direction uphill and whose length is proportional to the slope in that direction. Since the slope varies from place the place, the gradient also varies: The gradient is a vector-valued function of x_1 and x_2. In terms of a picture, the gradient function is a field of arrows. At each point, the arrow is perpendicular to the contour through that point. This makes sense: The direction in which the contour runs is the direction of a level path, whereas the gradient points in the direction of steepest change.

Figure 17.7 shows the gradient—depicted using arrows—of the potential-energy function from the two-spring system. When the function is steep, the gradient is large, and the arrow is long. Note that at places where the function is steep, the contour lines are closely spaced.

Near the minimum of the function, the function is level, and the gradient is small. At the exact argmin, the gradient has zero length.

For functions of N variables, the gradient is an N-dimensional vector. This vector tells the direction in which the value of the function increases most.

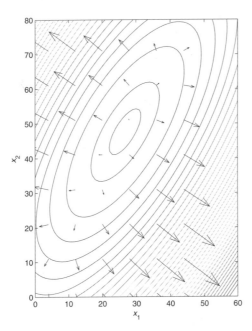

Figure 17.7. The gradient (arrows) and contour lines of the two-spring system in Figure 17.5. The gradient is a function of x_1 and x_2; at each point, it gives the direction in which the function increases most steeply. It has been plotted as an arrow whose length indicates the steepness. Note that the gradient is perpendicular to the contours.

Computing the gradient is straightforward. One technique is to use the tools of calculus to find the gradient function symbolically. For a function of N variables, $f(x_1, x_2, \ldots, x_N)$, the gradient function, denoted $\nabla f(x_1, x_2, \ldots, x_N)$,

Key Term

is a length-N vector of *partial derivatives*

$$\nabla f(x_1, x_2, \ldots, x_N) = (\frac{\partial f}{\partial x_1}, \frac{\partial f}{\partial x_2}, \ldots, \frac{\partial f}{\partial x_N}) \tag{17.3}$$

This function is evaluated at a given point to find the gradient at that point.

In symbolic calculus, we compute the partial derivative with respect to some variable x_k in the same way as the ordinary derivative, but treating all variables other than x_k as constant. This same logic applies to the *numerical partial derivative*. To compute the numerical partial derivative, we can take the finite difference

$$\frac{\partial f}{\partial x_k} = \frac{f(x + h_k) - f(x)}{||h_k||}$$

where h_k is a vector that is zero in every component except component k, and $||h_k||$ is the length of h_k. The same important issues of the size of the step h and the type of finite-difference approximation that we encountered when studying ordinary derivatives apply here. A simple implementation of a numerical gradient is given by gradfun, which is quite similar to the ordinary derivative program derivfun given on page 412:

```
[1]    function res = gradfun(fun,x0,varargin)
[2]    % gradfun(fun,x,params)
[3]    % gradient of a function at position x0
[4]    seps = sqrt(eps);
[5]    h = x0.*seps;
[6]    h(abs(x0)<seps) = seps;
[7]    f0 = feval(fun,x0,varargin{:});
[8]    res = zeros(size(x0));
[9]    for k=1:length(x0)
[10]        x1 = x0;
[11]        x1(k) = x0(k)+h(k);
[12]        f1 = feval(fun,x1,varargin{:});
[13]        res(k) = (f1 - f0)./h(k);
[14]    end
```

Marginal notes aligned to code lines:
- Make the step small in each direction — [4]
- But not too small! — [6]
- Find $f(\mathbf{x}_0)$ — [7]
- Consider each direction — [9]
- Displaced in that direction — [11]
- The partial derivative — [13]

▶

Example: Computing The Potential Energy Gradient

The twosprings function on page 454 computes the potential energy of the two-spring system at any point (x_1, x_2). This function takes as arguments the position vector $\mathbf{x}$ and several parameters. The first step in computing the gradient numerically is to assign specific numerical values to the parameters. We follow the values listed in Figure 17.6.

We can repackage `twosprings` and the parameter values using the `inline` operator:

```
≫  f = inline('twosprings(x, 3,3,15,15,5,5,9.8)');
```

In Figure 17.7 we can see that the minimum of f is somewhere around $\mathbf{x} = (30,40)$. Let's try a few different points:

```
≫  xtry = [30,40];
≫  f(xtry)
   ans: -1433
≫  xtry = [25,45];
≫  f(xtry)
   ans: -1745.5
```

It looks like $\mathbf{x} = (25,45)$ is a better guess.

Rather than searching at random, it would be better to find the way downhill and search in that direction. Finding the gradient of f at any given point is merely a matter of invoking `gradfun` on f and the point. For example, at $\mathbf{x} = (25,45)$ the gradient is

```
≫  g1 = gradfun(f,xtry
   g1: -4.4  -4.4
```

Since the gradient vector points uphill, we know that if we move a little bit in the direction of g1, the value of f should increase.

```
≫  littlebit = 0.1;
≫  nexttry = xtry + littlebit*g1;
≫  f(nexttry)
   ans: -1741.1
```

This confirms that the function value is larger in the uphill direction.

To move downhill, we want to go in the direction opposite to the gradient: This is just the negative of the gradient vector.

```
≫  nexttry = xtry - littlebit*g1;
≫  f(nexttry)
   ans: -1748.9
```

In Section 17.4, we will see how to automate the search for a minimum using the gradient. Care needs to be taken in defining `littlebit` and in choosing search directions.

An alternative to using numerical finite differences to compute the gradient is to use calculus techniques. The potential energy function $f(x_1,x_2)$ implemented in `twosprings` is, when written in traditional mathematical notation,

$$f(x_1,x_2) = \frac{1}{2}k_1(x_1 - L_1)^2 + \frac{1}{2}k_2(x_2 - x_1 - L_2)^2 - g(m_1x_1 + m_2x_2) \qquad (17.4)$$

Using calculus, the gradient is

$$\nabla f(x_1, x_2) = \left(\frac{\partial f}{\partial x_1}, \frac{\partial f}{\partial x_2}\right)$$
$$= \left(k_1(x_1 - L_1) - k_2(x_2 - x_1 - L_2) - gm_1, \ k_2(x_2 - x_1 - L_2) - gm_2\right)$$

$$(17.5)$$

This mathematical expression can be translated into a special-purpose m-file that we call twospringsgradient:

```
[1]  function df = twospringsgradient(x,m1,m2,L1,L2,k1,k2,g)
[2]  % The gradient of potential energy of the
[3]  % spring-mass system in Fig. XXX
[4]  df = [k1*(x(1)-L1)  - k2*(x(2)-x(1)-L2)  - g*m1,  ...
[5]        k2*(x(2)-x(1)-L2)  - g*m2 ];
```

We use inline to repackage the arguments to this function and assign the specific numerical values to parameters:

```
≫ df = inline('twospringsgradient(x,3,3,15,15,5,5,9.8)');
```

The gradient computed in this way is the same as that computed using numerical methods:

```
≫ df([25,45])
  ans: -4.4 -4.4
≫ df([30,40])
  ans: 70.6000 -54.4000
```

In an advanced technique, this process of using calculus to find the gradient and writing the corresponding computer program can be automated. The technique is called "automatic differentiation".

◀

A Matter of Style:

Many of the operations we consider in this chapter—optimization, finding zeros, and the gradient—involve varying a single vector argument of a mathematical function. It's convenient to write programs implementing these operations, assuming that the mathematical function will be represented as a computer function of a single-vector argument.

In the previous example, we used inline to reformat a function of many arguments into a function of a single-vector argument; for example,

```
≫ f = inline('twosprings(x, 3,3,15,15,5,5,9.8)');
≫ x0 = [30, 40]; % a vector
≫ f(x)
  ans: -1443
```

This style is practical when each of the fixed parameters is a number; we simply type that number as part of the `inline` statement.

In many situations, however, the parameters may themselves be arrays of data and it is impractical to pass them as numerals to `inline`. In such situations, it's convenient to allow the computer representation of the mathematical function to take additional arguments representing these parameters. These additional arguments then need to be handed off to operators such as `gradfun`.

For example, suppose we wish to retain one of the parameters in `twosprings` as an argument, rather than fixing its value numerically. This involves creating a function of both a vector argument and another argument

```
>> f2 = inline('twosprings(x, 3,3,15,15,5,5,g)', 'x','g');
>> f2(x, 9.8)
   ans: -1443
```

To use a function like `f2` as an argument to an operator like `gradfun`, we need to pass the value of the second argument to the operator. A helpful style for doing this involves the `varargin` syntax for handling variable numbers of arguments. The `gradfun` program, for instance, has two mandatory arguments, `fun` and `x0`, but accepts additional arguments. These additional arguments are simply passed on line 12 to `feval`, which in turn uses them when invoking `fun`.

```
>> gradfun(f2, x, 9.8)
   ans: 70.6000 -54.4000
```

When using this style, it is imperative that that the order of the additional arguments to `fun` match the order of the values provided to `gradfun`.

17.4 Optimization Using the Gradient

The gradient provides an important tool for optimization. A search for a local minimum can proceed in a very intuitive way from a given initial point: From the initial point, walk downhill. When we reach a point where there is no direction that points downhill, we are at a local minimum. The gradient tells us, at each point, which direction points most steeply uphill. To walk downhill, we simply go in the opposite direction: the direction of the negative gradient.

To put things in symbolic form, let us denote an initial guess about the location of the minimum as x_0. From this initial guess x_0, we will take a step in the direction of the negative gradient, $-\nabla f(x_0)$. This will bring us to a new, improved guess,

$$x_{new} = x_0 - \alpha \nabla f(x_0) \tag{17.6}$$

Key Term

The quantity α is a *scalar*—a single number—that controls how far we go in the direction of the negative gradient.

To complete this algorithm, we need only say how far to step downhill; that is, how big α should be. There are many possibilities here, but we adopt a very simple one that exploits the methods developed in Chapter 16. We will choose α to make $f(\mathbf{x}_{new})$ as small as possible. That is, α will be the scalar variable of a one-dimensional minimization problem, and we can solve this problem using the one-dimensional optimization tools from Chapter 16.

Key Term

This process of reducing a multivariable optimization problem to a one-dimensional problem is called a *line search*, refering to the fact that we are finding the minimum of the function along a single line as defined by Eq. (17.6); we walk along a linear path until we get to a locally low point on that path. Of course, there is no guarantee that the minimum of the function lies on the path. In general, it will not. But once we have reached the minimum along the path, we can choose a new direction and search in another path in that direction. Insofar as each line search brings us downhill, iterating this process of line search and selection of a new direction eventually will bring us to a local minimum.

Note that the line search algorithm is not the one followed by a hiker searching for the locally low point in a landscape. A hiker would not pick a downhill direction and walk steadily in a straight line, persisting until starting to head uphill. The gradient varies from place to place, and the hiker takes single steps, choosing the direction at each step to go in the steepest downhill direction. For a hiker, this is straightforward because the gradient can be sensed by the feet.

Mathematically, though, we need to compute the gradient. In n dimensions this requires n function evaluations. It may be cheaper to use these function evaluations to perform a line search rather than reconsidering the direction at each step. On the other hand, it seems wasteful to continue to walk along a line that is not pointing straight downhill. The trade-off between using function evaluations to choose search directions and using them to perform a search along a line is a fundamental issue in numerical optimization. For our purposes here, we will choose to perform a complete line search before selecting a new direction, but more advanced methods than we will consider here consider the trade-off in more sophisticated ways.

Key Term

Steepest descent refers to the algorithm of using the negative gradient as a search direction and performing a complete line search in that direction. The operator `steepest` implements the following algorithm:

1. From an initial position $\mathbf{x}_0$, perform a line search in the direction $-\nabla f(\mathbf{x}_0)$ to find a new position $\mathbf{x}_{new}$.
2. If $\mathbf{x}_{new}$ and $\mathbf{x}_0$ are very close together, return $\mathbf{x}_{new}$ as the argmin. Otherwise, update $\mathbf{x}_0$ by setting it to $\mathbf{x}_{new}$, and repeat the algorithm.

Since the gradient can be computed using the `gradfun` operator, the only required arguments to `steepest` are the function to be minimized and an initial position. For example, for the two-springs problem, starting with an initial guess of $x_1 = 10, x_2 = 10$,

```
≫  xmin = steepest(f,[10,10])
   ans: 26.7596 47.6391
```

The value of the objective function at this argmin is

```
≫  f(xmin)
   ans: -1755.2
```

The algorithm of steepest descent has advantages and disadvantages. One important advantage is that it is guaranteed to work, since every step is downhill. Unfortunately, it can be quite slow. The reason for this is that taking steps in the steepest downhill direction, which seems like simple common sense, can be very inefficient. Figure 17.8 shows why. The line search at each iteration takes the hiker to a point where the line parallels the contours. At this point, the steepest direction is perpendicular to the previous direction. But this perpendicular direction doesn't point toward the function's

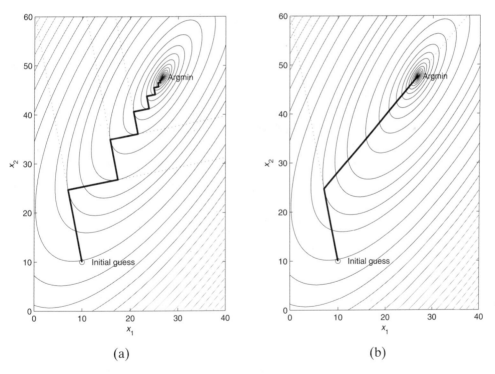

(a)   (b)

Figure 17.8. Line search minimization of the two-spring system in Figure 17.5. (a) Steepest descent. (b) Conjugate gradient. In each case, the dashed line shows the direction of the line searches; the heavy line marks the path to the local minimum along each linear path.

Key Term

minimum, and so the steepest descent path switches between perpendicular directions. The overall path will slowly and inefficiently twist downhill. Steepest descent is an example of a *greedy algorithm*, which tries to pick, at each iteration, the locally best direction without taking into account nonlocal information.

Key Term

A better algorithm for line search minimization takes into account the disadvantage of successive perpendicular search directions and chooses a search direction that is not necessarily in the steepest downhill direction. Looking at Figure 17.8, we can imagine how this might be done (for example, by choosing a search direction that is intermediate to the two previous perpendicular directions). The *conjugate gradient* algorithm involves a more careful selection of search directions, exploiting the information collected by previous line searches; it involves a simple, but sophisticated, calculation of a search direction. Although the derivation of this calculation is beyond the scope of this book, we can employ the method, implemented in conjugategradient, which is packaged in the same way as steepest:

```
≫ xmin = conjugategradient(f,[10,10])
  ans: 26.7600 47.6400
```

The efficient search paths generated by the conjugate gradient method are shown in Figure 17.8.

It may seem remarkable that the conjugate gradient method is able to pick a direction that heads directly toward the argmin of the objective function. This is because the spring-mass system's potential energy function involves only linear and quadratic terms in x_1 and x_2 and, as a result, the contours are concentric ellipses. For such functions, involving n variables, the conjugate gradient method will find the minimum in n line searches. So, for the two-variable spring problem, the conjugate gradient takes only two line searches. For other sorts of functions, with nonelliptical contours, the conjugate gradient generally requires more line searches.

▶

Example: Coffee Cooling

Newton's law of cooling states that the rate of change of the temperature of some object is proportional to the difference between that object and the ambient environment. (See Chapter 5.) When the ambient temperature is constant, the law implies that the object's temperature changes according to the following function of time:

$$T(t) = T_{\text{env}} + (T(0) - T_{\text{env}}) \exp(-kt) \qquad (17.7)$$

The parameter k governs how fast the cooling occurs and depends on the properties of the object.

Newton's law is never enforced; it is merely a model. The real world can be considerably more complicated and there can be multiple, distinct mechanisms of cooling whose properties change as the temperature changes. Recall the cooling cup of coffee described on page 47. In addition to cooling by conduction through the cup wall, there is a different mode of cooling through the liquid-air interface and the internal mixing of temperature due to convection.

In collecting the coffee data, we measured the initial tempterature $T(0)$ to be 93.5 °C and the ambient temperature T_{env} to be 24.5 °C. The only unknown parameter in Eq. (17.7) is k.

The function `coffeemodel` implements our Newton's law model of temperature:

```
[1]   function temp = coffeemodel(k,Tzero,Tenv,time)
[2]   %   coffeemodel(k,T0,Tenv,time)
[3]   %   Simple Newton's Law model of coffee cooling
[4]   temp = Tenv + (Tzero-Tenv).*exp(-k.*time/1000);
```

Coffeemodel takes k, T_{env}, and $T(0)$ as arguments as well as the time values. It returns the model prediction of temperature at each of the time values. To fit k, we need merely to try various values of k and find the difference between the model prediction and the measured temperature at each time point; these are the residuals of the model. We try to find values of k that make the residuals as small as possible. There are, of course, many ways to define "as small as possible." For this example, we use the sum of the absolute value of the residuals.

To start, we read in the recorded data:

```
≫   data = csvread('coffeecooling.csv');
```

and create separate explanatory and response variables:

```
≫   time = data(:,2)/60; % in minutes
≫   temp = data(:,1);
```

The objective function has to be packaged so that the quantities involved in the minimization search are packaged together into a single argument and the fixed parameters (e.g, T_{env}) are given specific numerical values.

```
f = inline('sum(abs(temp - coffeemodel(k,93.5,24.5,time)))',...
    'k', 'time', 'temp');
```

The function `f` takes as its first argument the quantity we will vary to find the minimum: `k`. Since the `time` and `temp` data are not single numbers, it's convenient to give them as parameter arguments to the objective function (as discussed on page 458).

Now we are in a position to find the optimal k. Since this is one-dimensional optimization, we can even try to find a value "by hand."

```
≫   f(10,time,temp)
  ans: 575.48
≫   f(50,time,temp)
  ans: 215.32
```

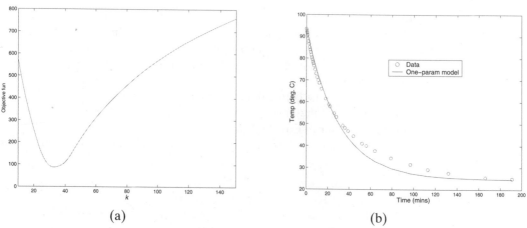

(a)   (b)

Figure 17.9.   Fitting k in the model of Eq. (17.7) given $T(0) = 93.5\,°C$ and $T_{env} = 24.5\,°C$. (a) The value of the fitting objective function versus k. (b) The theoretical temperature curve compared to the measured data.

```
≫  f(150,time,temp)
    ans: 760.85
```

Since the middle value is lower than the endpoints, we have a bracket. Success is now guaranteed! We can find the optimum using the built-in function fminbnd, passing the time and temperature data as parameter arguments.[1]

```
≫  fminbnd(f,10,150,[],time,temp)
    ans: 32.98
```

Figure 17.9(b) compares the coffee-cooling data to the model at the optimal value of k. The fit is pretty bad: The early data are systematically lower in temperature than the model; the late data are systematically higher.

The poor fit is not due to a bad choice of the parameter k—the fit is the best possible value of k. The problem is in the model. Newton's Law is unequal to the task of giving a detailed account of a cooling cup of coffee. There are too many mechanisms involved in cooling for the simple model of Eq. (17.7).

Newton's law can do a better job if we simplify the situation somewhat. We can envision, for instance, that very hot coffee cools by a fast "steaming" mechanism, but that after a while, it cools mainly by conduction. To study the conduction mechanism, we can discard the hottest data, when steaming is most active. This means, though, that we will have to estimate an initial temperature $T(0)$ to use in the conduction-only model.

---

[1]See the documentation for fminbnd for an explanation of the empty vector passed as a fourth argument.

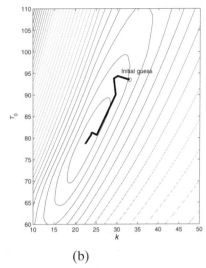

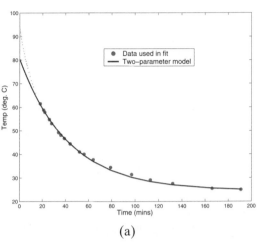

(a)   (b)

Figure 17.10.   (a) Contours of the the objective function for the two-parameter coffee-cooling fitting problem and the search path used by steepest descent starting from $k = 33$, and $T(0) = 93.5$. (b) The data and the model with parameters set at their optimal values.

We start by identifying those data we want to participate in the fitting process. We define very hot coffee to be $> 65\,°C$.

```
≫  keep = temp < 65;
```

We redefine the objective function to include the $T(0)$ parameter:

```
f2 = inline('sum(abs(temp - coffeemodel(x(1), x(2), 24.5,time)))',...
    'x', 'time', 'temp');
```

Since this is a function of two parameters, it can be graphed, as in the right panel of Figure 17.10.

We can get a pretty good reading of the optimal values of k and $T(0)$ from the contour graph in Figure 17.10, but if we didn't have this graph, a good initial guess would be the values found in the one-parameter fit.

```
≫  initialx = [33, 93.5];
≫  conjugategradient(f,initialx, [], time(keep),temp(keep))
  ans: 23.467 80.405
```

These fitted parameters give a much nicer fit—at least to the part of the data used for fitting—as seen in Figure 17.10(a).

A more comprehensive model tries to fit all of the data by expanding the number of mechanisms. Exercise 17.2 shows that three exponential mechanisms follow the data closely.   ◀

17.5 Finding Solutions

In Chapter 16, we found solutions to relationships by constructing indicator functions and finding their zero crossings. The same approach applies in higher dimensions. The zero crossing of a function of two variables is the set of contours where $f(x,y) = 0$.

Typically, the zero crossings of a function of two variables are one-dimensional closed curves. Consider the zero crossing of the function giving ground elevation above sea level as a function of latitude and longitude. The coastline is the set of locations where elevation goes from above sea level to below. There is, of course, the theoretical possibility of other configurations: an isolated point (an under-sea mountain whose top barely touches the surface of the sea); a nonclosed curve (the ridge of a sea-mountain range that runs exactly horizontal at exactly sea level); a two-dimensional area (an under-sea mesa exactly at sea level). These possibilities are exceptional, not generic, and susceptible to annihilation by even the slightest change in sea level or the shape of the functions.

For functions of three variables (somewhat harder to visualize), the generic zero-crossing contours are not closed curves but closed surfaces: balloons. For functions of four and more variables, our intuition fails us.

Finding a zero crossing of a function of two variables is straightforward. If you can find one point above sea level and one point below sea level, all you need do is walk on any continuous path that connects the two points, as in Figure 17.3. Eventually, you will come to the shore. This is, essentially, the one-variable situation of Chapter 16: The distance along the path is a single variable, the above-sea and below-sea endpoints constitute a bracket, and success is practically inevitable. (Of course, walking on an arbitrary path across a real-world landscape is difficult or impossible. But "straightforward" is not the same as "easy".)

Things become difficult when there are two relationships to be satisfied simultaneously. For example, think of trying to find a point whose elevation is at sea level and where the annual rainfall is exactly 30 inches per year. We have two contour maps involved here, one giving elevation and the other rainfall. (See Figure 17.11.) The ground-elevation and rainfall contours are both generically curves. Simultaneous solutions are (x, y) points where the two functions' zero contours intersect; this may or may not be the case, depending on the functions themselves. Generically, such intersections will be single, isolated points. Finding such points is not straightforward; it's extremely unlikely that an arbitrary path of the sort in Figure 17.3 will encounter such an isolated path. We need another way to find solutions.

If, rather than the two relationships shown in Figure 17.11, there are three relationships to be satisfied simultaneously (say, a point at zero ground elevation with 30 inches of rainfall in the last year and with a barometric pressure of 700 torr) there usually will not be any solution at all. Although

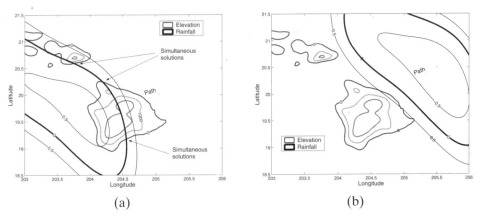

(a)   (b)

Figure 17.11.   Contour maps of the ground-elevation function and the rainfall function
superimposed. (a) A situation where the two sets of zero contours intersect.
(b) A situation where the zero contours don't intersect. The points ($\times$) mark places
where both functions are simultaneously above zero or simultaneously below zero.
An arbitrary path between two such points crosses the zero contours of each
function but is unlikely to cross at a simultaneous solution.

an intersection of two contours at a single point is commonplace, the intersection of contours from three functions at a single point is exceptional.

For functions of three variables, two balloonlike contours typically intersect at a one-dimensional curve and three contours typically intersect at a
point. In general, for N variables, we typically expect simultaneous solutions
to N functions, but there is no guarantee of this.

17.6   Solutions to Systems of Linear Equations

Despite the difficulty of finding simultaneous solutions to general functions,
as illustrated in Figure 17.11, there is one situation where finding solutions
is often easy: when the functions are linear. Consider the function of two
variables

$$f(x,y) = 3.2x - 7.9y + 4$$

In a perspective plot, $f(x,y)$ is an inclined plane. The contours of $f(x,y)$ are
straight lines given by $f(x,y) = \beta$. In particular, the zero contour of $f(x,y)$
can be found by setting $f(x,y) = 0$, or, in the familiar slope-intercept form:

$$y = \frac{3.2}{7.9}x + \frac{4}{7.9}$$

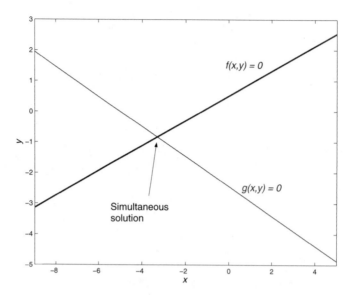

Figure 17.12.   The zero contours of linear functions are straight lines. The simultaneous solution occurs where the lines intersect. The zero contours here are for the functions $f(x,y) = 3.2x - 7.9y + 4$ and $g(x,y) = 2x + 4.1y + 10$. The contours intersect at the point $(x,y) \approx (-3.2988, -0.8299)$.

Finding a solution in this case is trivial: Plug in any value for x and compute the resulting y.

What's significant here is not that we can easily find x and y that satisfy $f(x,y) = 0$. As we saw in Section 17.5, it's easy to find solutions to a single function. The difficult thing is to find simultaneous solutions to multiple functions.

But with linear functions, finding simultaneous solutions is fairly easy. For example, consider solving simultaneously $f(x,y) = 0$ and $g(x,y) = 0$, where

$$g(x,y) = 2x + 4.1y + 10$$

The function $g(x,y)$ has zero contours that are lines: $y = -\frac{2}{4.1}x - \frac{10}{4.1}$.

Geometrically, the simultaneous solution lies at the intersection of the straight-line zero contours of $f(x,y)$ and of $g(x,y)$, as shown in Figure 17.12. From a basic understanding of lines, we can see that, in general, any two linear functions of two variables will intersect at a single point. The exceptions occur when the lines are parallel—same slope but different intercepts—or when the lines overlay each other—same slope and intercept. But such nonpoint solutions are exceptional; they require that the slopes be exactly equal. We can easily compute, by examining slopes and intercepts, whether a unique solution exists.

Linear functions of three variables have zero contours that are planes. More generally, linear functions of N variables have zero contours that are

$N-1$ dimensional hyperplanes. The contours of such N functions will intersect at a single point, unless the hyperplanes are parallel. For most people, though, our intuitive powers of visualization are limited when dealing with three variables and fail completely when more than three variables are involved. But the rules of algebra provide a way to cope with high-dimensional situations.

Computing Solutions

Consider the indicator function of the single-variable relationship $ax = b$

$$f(x) = ax - b$$

We could use the iterative methods of Chapter 17 to find the solution $x_\star$. But there is a better way in this case: Use algebra to find

$$x_\star = \frac{b}{a}$$

Such a solution is guaranteed to exist so long as $a \neq 0$.

We can easily implement this solution on the computer. Since the relationship is fully specified by the numerical values of a and b, we can dispense with the indicator function $f(x)$ and instead write the solution as an operator that takes the parameters a and b as arguments.

```
[1]   function xstar = linearSolve1(a,b)
[2]   if a == 0
[3]       xstar = Inf*b;
[4]   else
[5]       xstar = b./a;
[6]   end
```

No solution exists

It may seem far more difficult to find simultaneous solutions to multiple linear functions; for example, two equations in two variables:

$$
\begin{aligned}
ax + by &= c & (17.8) \\
dx + ey &= f & (17.9)
\end{aligned}
$$

Rather than the simple arithmetic approach that suffices to solve $ax - b = 0$, the pair of linear functions in Eq. (17.8) seems to call for labor and ingenuity: a fresh approach for each particular set of numbers a, b, c, d, e, and f. In fact, this isn't true; there is a formula:

$$x_\star = \frac{ec - bf}{ae - db} \tag{17.10}$$

and either

$$y_\star = \frac{f}{e} - \frac{d}{e}x_\star \quad \text{or} \quad y_\star = \frac{c}{b} - \frac{a}{b}x_\star \tag{17.11}$$

The two forms are needed because a solution might exist even if either e or b is zero. If both $e = 0$ and $b = 0$, or if $ae - db = 0$, there is no solution.

We can implement the solution on the computer as a formula that takes the six coefficients (the numbers $a, b, c, d, e,$ and f) as arguments and returns the value of $x_\star$ and $y_\star$:

```
[1]   function [xstar,ystar] = linearSolve2(a,b,c,d,e,f)
[2]   if a*e - b*d == 0
[3]       xstar = Inf; ystar = Inf;
[4]   else
[5]       xstar = (e*c - b*f) ./ (a*e - b*d);
[6]       if e == 0
[7]           ystar = c./b - xstar.*a./b;
[8]       else
[9]           ystar = f./e - xstar.*d./e;
[10]      end
[11]  end
```

Such formulas could, in principle, be written for N linear equations in N variables for any N. But the complexity of checking for cases where a solution exists, even though some coefficients are zero, would become overwhelming. There is also an interface problem: It is difficult to manage long sequences of arguments as in `linearSolve2`. In general, for N equations in N variables there will be $N^2 + N$ coefficients. For $N = 2$, this gives six coefficients (a manageable number) but even for N as small as 6 there are 42 coefficients required. We need a consistent scheme to name the coefficients; writing the names of the coefficients or variables as a succession of letters (e.g., x and y or a, b, and c) becomes burdensome when N is large. Instead, we adopt the notation that the N variables are denoted $x_1, x_2, \ldots, x_N$ and the coefficients are an indexed array. The system of N equations can be written

$$\begin{array}{ccccccc}
a_{1,1}x_1 & + & a_{1,2}x_2 & + & \cdots & + & a_{1,N}x_N & = & b_1 \\
a_{2,1}x_1 & + & a_{2,2}x_2 & + & \cdots & + & a_{2,N}x_N & = & b_2 \\
\vdots & & \vdots & & \ddots & & \vdots & & \vdots \\
a_{N,1}x_1 & + & a_{N,2}x_2 & + & \cdots & + & a_{N,N}x_N & = & b_N
\end{array} \tag{17.12}$$

This is purely a matter of notation and doesn't impose any relationship among the N^2 coefficients $a_{i,j}$ or the N coefficients b_j. It's convenient to assemble all of the $a_{i,j}$ coefficients together into one matrix, which we call **A**, and all of the b_j coefficients into a column vector, which we call **b**, and the variables x_j into a column vector called **x**. With this notation, Eq. (17.12) can be written concisely in terms of matrix multiplication:

$$\mathbf{A} \cdot \mathbf{x} = \mathbf{b}$$

With the coefficients stored in this matrix format, the MATLAB solution finding program, called `mldivide`, is invoked with just two arguments, **A** and **b**:

```
≫ x = mldivide(A, b);
```

The name `mldivide` seems strange (it stands for "matrix left-divide"), but the operation of computing linear solutions is so fundamental to scientific computing that there is a special syntax for it:

```
≫ x = A \b
```

pronounced "**A** under **b**." The "under" operation is reminiscent of the "b over a" notation b/a. "Under" rather than "over" is used to distinguish the column-vector-divided-by-matrix operation from the usual division operation.[2]

▶

Example: Forces on a Bridge (continued)

We pick up again the truss bridge example introduced in the Chapter 17 introduction. The task is to compute the stresses $S_{i,j}$ in the truss members. Examination of the equations reveals that they are linear in all of the $S_{i,j}$; we can re-write the equations in the introduction in matrix form:

$$\begin{pmatrix} 0 & 0 & \sin\theta_{231} & 0 & \sin\theta_{635} & 0 & 0 \\ 0 & -1 & -\cos\theta_{231} & 0 & \cos\theta_{635} & 1 & 0 \\ -\sin\theta_{213} & 0 & -\sin\theta_{231} & 0 & 0 & 0 & 0 \\ -\cos\theta_{213} & 0 & \cos\theta_{231} & 1 & 0 & 0 & 0 \\ 0 & 0 & 0 & 0 & -\sin\theta_{635} & 0 & -\sin\theta_{453} \\ 0 & 0 & 0 & -1 & -\cos\theta_{635} & 0 & \cos\theta_{453} \\ 0 & 0 & 0 & 0 & 0 & -1 & -\cos\theta_{453} \end{pmatrix} \cdot \begin{pmatrix} S_{1,2} \\ S_{1,3} \\ S_{2,3} \\ S_{2,4} \\ S_{3,4} \\ S_{3,5} \\ S_{4,5} \end{pmatrix} = \begin{pmatrix} Mg \\ 0 \\ 0 \\ 0 \\ 0 \\ 0 \\ 0 \end{pmatrix}$$

(17.13)

To solve these equations, we give specific numerical values to the parameters θ_{ijk}, g, and M. For simplicity, we consider a truss made of equilateral triangles so that $\theta_{ijk} = \pi/3$ for all i, j, and k.

```
≫ theta = pi/3;
≫ M = 2000; % kilos
≫ g = 9.8; % meters/sec^2
```

and then formulate the A matrix and b vector:

```
≫ b = [ M*g; 0; 0; 0; 0; 0; 0]; % a column vector
```

We can make a little shorthand for $\cos\theta$ and $\sin\theta$

```
≫ s = sin(theta); c = cos(theta);
```

---

[2] The "over" operator, actually `mrdivide` (standing for "matrix right-divide"), also works with vectors and matrices, but in a different way than `mldivide`.

and define the matrix

```
>> A = [0,0,s,0,s,0,0;
0,-1,-c,0,c,1,0;
-s,0,-s,0,0,0,0;
-c,0,c,1,0,0,0;
0,0,0,0,-s,0,-s;
0,0,0,-1,-c,0,c;
0,0,0,0,0,-1,-c];
```

The forces are then found using the `mldivide` operator:

```
>> A\b
 ans:  -11316
        5658
       11316
      -11316
       11316
        5658
      -11316
```

The stress in member $(1,2)$ is $-11,316$ with units of $kg \cdot m/s^2$. (We need to keep track of the units ourselves, on the side.) The negative number denotes that the member is in tension. Member $(2,3)$ is in tension (also with magnitude $11,316 \, kg \cdot m/s^2$). Note the symmetry of the stresses, which reflects the symmetry of the truss and the loading.

◀

17.7 A Geometric Interpretation of $A \cdot x = b$

Equation (17.12) can be interpreted in a geometric way that points out some possibilities and difficulties in solving sets of linear equations. Just as the one-dimensional equation $ax = b$ has no solution when $a = 0$, there is an equivalent disqualification based on **A** for solutions to Eq. (17.12). This disqualification algebraically is complicated but has a simple geometrical interpretation in terms of vectors.

Key Term

A *vector* is a point in space. Conventionally (the reasons for this will become clear when we discuss vector addition), a vector is drawn as an arrow. Figure 17.13 shows some vectors in $N = 2$ and $N = 3$ dimensional spaces. The *head of a vector* is the end drawn with the arrow; the *tail of a vector* is the other end.

Key Terms

It's convenient to assign two distinct properties to a vector: a *direction* and a *magnitude*. Multiplying each component of a vector by the same positive scalar factor does not alter the direction. For example, the two vectors $(-1,2)$ and $(-1.5,3)$ point in the same direction; each component of the sec-

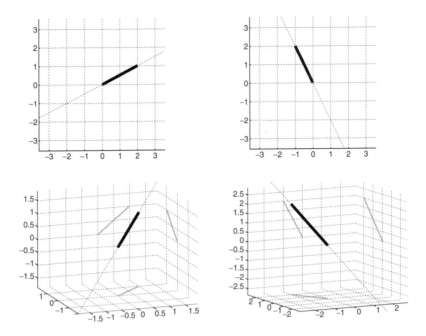

Figure 17.13. Vectors $(2,1), (-1,2), (1,1,1), (-1,2,2)$ For the three-dimensional graphs, the shadow of the vector onto the "walls" of the plotting box is shown.

ond vector is a factor of 1.5 times the corresponding component of the first vector. Multiplication by a negative scalar factor reverses the direction. (See Figure 17.14.)

Key Term   The *dimension* or *cardinality* of a vector is the number of components (this is the value returned by the MATLAB `length` operator). [The magnitude of a vector is, in everyday language, it's length. But this meaning of "length" is quite different from the `length` operator. Computation of the vector magnitude can done with `sum(sqrt(vec.^2))`.]

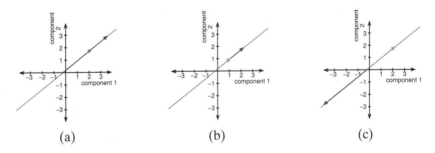

Figure 17.14. Multiplication of a vector by a scalar. The gray vector is multiplied by (a) scalar 2, (b) .5, and (c) -1.5 to produce the black vector.

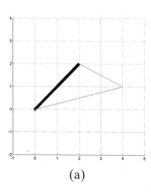
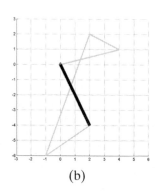
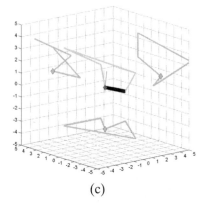

(a)   (b)   (c)

Figure 17.15. Sums of vectors in two and three dimensions. The vectors are shown as light lines, the resultant as a thick, dark line. (a) The two vectors `[4;1]` and `[-2;1]`. (b) The four vectors `[4;1]`, `[-2;1]`, `[-3;-8]`, `[3;2]`. (c) The four vectors `[2;3;1]`, `[-4;0;2]`, `[2;-7;-1]`, `[3;5;-3]`.

Two vectors of the same dimension can be added. When adding two vectors, the components are added one by one. For example, $\begin{pmatrix} 3 \\ 2 \end{pmatrix} + \begin{pmatrix} 1 \\ 7 \end{pmatrix}$ is $\begin{pmatrix} 4 \\ 10 \end{pmatrix}$. Geometrically, adding two vectors is accomplished by picking up the second vector and placing its root at the head of the first vector while maintaining the direction and magnitude. The resultant vector goes from the root of the first vector to the head of the second vector. Adding multiple vectors is similar: assembling a train of vectors root to head. (See Figure 17.15.)

Key Term

Any two vectors have an angle between them. When that angle is 90 degrees, the two vectors are said to be *orthogonal*. Given two vectors A1 and A2, a simple test for orthogonality is to compute sum(A1.*A2). If this sum is zero, the vectors are orthogonal. For example, consider the following several two-dimensional column vectors:

```
≫  A1 = [0;2];
≫  A2 = [1;0];
≫  A3 = [2;1];
≫  A4 = [2;-4];
```

To confirm that A1 is orthogonal to A2, we compute

```
≫  sum(A1.*A2)
   ans: 0
```

Similarly, A3 is orthogonal to A4. But A3 and A2 are not orthogonal since the sum is not zero:

```
≫  sum(A3.*A2)
   ans: 2
```

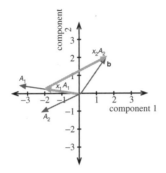

Figure 17.16. The solution to $x_1 A_1 + x_2 A_2 = \mathbf{b}$ is the two scalars, x_1 and x_2, which, when multiplied by the vectors A_1 and A_2, respectively, sum to produce $\mathbf{b}$.

Now we have the necessary tools to draw a geometrical picture of Eq. (17.12).

The coefficients $(b_1, b_2, \cdots, b_N)$ on the right-hand side of Eq. (17.12) constitute an N-dimensional vector which we'll call the *target vector*. Similarly, the left-hand side of Eq. (17.12) can be seen as a sum of scaled vectors.

Key Term

$$
x_1 \begin{pmatrix} a_{1,1} \\ a_{2,1} \\ \vdots \\ a_{N,1} \end{pmatrix} + x_2 \begin{pmatrix} a_{1,2} \\ a_{2,2} \\ \vdots \\ a_{N,2} \end{pmatrix} + \ldots + x_N \begin{pmatrix} a_{1,N} \\ a_{2,N} \\ \vdots \\ a_{N,N} \end{pmatrix} = \begin{pmatrix} b_1 \\ b_2 \\ \vdots \\ b_N \end{pmatrix} \quad (17.14)
$$

Writing Eq. (17.12) this way emphasizes that each of the columns of the matrix $\mathbf{A}$ is a vector; a matrix is a bag of vectors. Since there are N columns in $\mathbf{A}$, there are N vectors. In Eq. (17.12), these N vectors are each scaled by a factor: the first vector by x_1, the second vector by x_2, and so on. By "solving the equation," we mean finding the scaling factors $x_1, x_2, \ldots, x_N$ so that the vector sum on the left-hand side of Eq. (17.14) equals the target vector $\mathbf{b}$ on the right-hand side.

17.8 Least Squares Solutions

The span of a matrix $\mathbf{A}$ is the set of all vectors $\hat{\mathbf{b}}$ that can be constructed by a linear combination of the vectors in $\mathbf{A}$. For instance, if $\mathbf{A}_{\mathrm{one}} = \begin{pmatrix} 3 & 1 \\ 2 & 7 \end{pmatrix}$, then the two vectors in $\mathbf{A}_{\mathrm{one}}$ are $\begin{pmatrix} 3 \\ 2 \end{pmatrix}$ and $\begin{pmatrix} 1 \\ 7 \end{pmatrix}$. Some of the vectors in the span of $\mathbf{A}$ are $\begin{pmatrix} 3 \\ 2 \end{pmatrix}, \begin{pmatrix} 1 \\ 7 \end{pmatrix}, \begin{pmatrix} 4 \\ 9 \end{pmatrix}, \begin{pmatrix} 5 \\ 16 \end{pmatrix}, \begin{pmatrix} 2 \\ -5 \end{pmatrix}, \begin{pmatrix} 2.5 \\ 8 \end{pmatrix}$, and so on. Indeed, any possible two-dimensional vector $\begin{pmatrix} c \\ d \end{pmatrix}$, whatever the values of c and d, is in the span of $\mathbf{A}_{one}$.

Consider now the matrix $\mathbf{A}_{two} = \begin{pmatrix} 1 \\ 2 \end{pmatrix}$. Not every vector $\begin{pmatrix} c \\ d \end{pmatrix}$ falls into the span of $\mathbf{A}_{two}$, only those of the form $\begin{pmatrix} c \\ 2c \end{pmatrix}$. For instance, $\begin{pmatrix} 7 \\ 3 \end{pmatrix}$ is not in the span of $\mathbf{A}_{two}$.

The under operator can generate a solution $\mathbf{x}$ to $\mathbf{A} \cdot \mathbf{x} = \mathbf{b}$. Multiplying $\mathbf{A}$ times $\mathbf{x}$ generates a vector $\hat{\mathbf{b}}$ that is in the span of $\mathbf{A}$:

```
≫  Aone = [3,1;2,7]
   Aone:  3 1
          2 7
≫  Atwo = [1;2]
   Atwo:  1
          2
≫  b = [7;3];
```

Here is the solution to $\mathbf{A}_{one}\mathbf{x} = \mathbf{b}$:

```
≫  xone = Aone \b
   xone:  2.4211
         -0.26316
```

and to $\mathbf{A}_{two}\mathbf{x} = \mathbf{b}$:

```
≫  xtwo = Atwo \b
   xtwo:  1.6
```

For each of these solutions, we can generate the vector $\hat{\mathbf{b}}$ that lies in the span of $\mathbf{A}$:

```
≫  bhatone = Aone*xone % Note: matrix multiplication
   bhatone:  7
             3
≫  bhattwo = Atwo*xtwo % matrix multiplication
   bhattwo:  1.6
             4.8
```

In every case, the vector $\hat{\mathbf{b}}$ satisfies one of two possibilities:

Key Term
1. $\hat{\mathbf{b}}$ is the same as $\mathbf{b}$. In this case, $\mathbf{x}$ is an *exact solution* to $\mathbf{A} \cdot \mathbf{x} = \mathbf{b}$. If there is more than one exact solution possible, the $\mathbf{x}$ returned by the under operator is the one whose magnitude is the smallest of all the exact solutions.

Key Term
2. $\hat{\mathbf{b}}$ is not the same as $\mathbf{b}$. In this case, $\mathbf{x}$ is called a *least squares solution*. This situation arises when $\mathbf{b}$ is not in the span of $\mathbf{A}$; the least squares solution gives a $\hat{\mathbf{b}}$ that is as close as possible to $\mathbf{b}$ while remaining in the span of $\mathbf{A}$.

Least squares solutions are particularly important in fitting linear models.

▶

Example: Linear Fitting of Gasoline Economy

Equation (17.1) gives a model of fuel use. How do we find appropriate values for the parameters?

The gasoline-use data introduced in Table 16.2 give, for each fill-up, the date, the odometer reading, and the number of gallons of gasoline purchased.

```
≫ gasdata = load('gasoline.dat');
```

To put the data in the same form as the model given by Eq. (17.1), we need to extract from the data the variables d, g, c, and h. Variable d is the distance travelled since the last fill-up:

```
≫ d = diff(gasdata(:,2));
```

Variable g is the amount of gasoline that was consumed in traveling that distance, which we take to be the amount purchased at the time of the fill-up:

```
≫ g = gasdata(2:end,3);
```

and for the date t, we take the average of the date of the fill-up and the previous fill-up:

```
≫ t = (gasdata(2:end,1) + gasdata(1:(end-1),1))./2;
```

The quantity t doesn't appear directly in Eq. (17.1), but as the argument to sine and cosine functions. We'll denote these as `ts`, and `tc` respectively:

```
≫ ts = sin(2*pi*t/365.25);
≫ tc = cos(2*pi*t/365.25);
```

This leaves only the variable c: the fraction of driving done in city conditions. No record was made of the driving conditions. This is unfortunate, but also typical of many situations where one is using existing data to address matters that were not considered at the time the data collection was planned. Although we did not measure c, we can try to reconstruct it from the measured data. As it happens, the car involved in the data collection typically is driven every day for short distances in the city and occasionally on the highway for longer distances. When highway driving is being done, the per-day gasoline consumption usually is high. This means that a *proxy variable* for c is the time between fill-ups; when the time between fill-ups is short, a lot of highway driving is being done. We can therefore easily construct a variable c; for example,

Key Term

```
≫ c = min(1,diff(gasdata(:,1)/20));
```

which reflects the assumption that for two fill-ups on the same day or on successive days then almost all of the driving is highway driving, while if there are 20 days or more between fill-ups, then all of the driving is city driving. This is imperfect and depends on the particular way that the car involved is driven, but it is the best we can do with the data available.

Overall, our model is

$$d = (r_{car} + \alpha c + \beta \cos(2\pi t/365.25) + \gamma \sin(2\pi t/365.25)) g \qquad (17.15)$$

Although it's not obvious at first glance, Eq. (17.15) is in the form $\mathbf{b} = \mathbf{A} \cdot \mathbf{x}$. For the sake of staying with the notation $x = A \setminus b$, we make these correspondances explicit:

```
≫  b = d;
≫  A = [g, g.*c, g.*tc, g.*ts]
  ans: 14.2000 0.7100 -9.2864 -10.7425
       15.0000 0.7500 -9.6130 -11.5148
       15.3000 6.8850 -8.7600 -12.5440
       14.6000 13.1400 -5.3799 -13.5727
       15.5000 11.6250 -1.4477 -15.4322
       16.0000 8.0000 1.9391 -15.8821
       and so on for 124 rows altogether
```

The vector $\mathbf{x}$ is comprised of the four parameters r_{car}, α, β, and γ.

Since there are

```
≫  length(b)
  ans: 124
```

elements in b, it's unlikely that the four vectors in $\mathbf{A}$ will add up to give exactly b—the span of $\mathbf{A}$ occupies only a four-dimensional subspace of the 124-dimensional space of data—but we can find the values of the parameters [rcar; alpha; beta; gamma] that give the least squares approximation $\hat{\mathbf{b}}$.

We can now solve equation Eq. (17.15) for the parameter vector [rcar; alpha; beta; gamma]:

```
≫  x = A\b
  x: 17.3463
     0.7601
     -3.7889
     -1.9573
≫  bhat = A*x
  bhat: 303.07
        319.72
        328.37
        310.19
        124 rows altogether
```

The property of the least squares solution is that $\hat{\mathbf{b}}$ is as close as possible to b in the span of $\mathbf{A}$. The difference $\mathbf{b} - \hat{\mathbf{b}}$ is called the *residual*.

Key Term

```
≫  resid = b - bhat
  resid: 6.9316
         14.275
         -47.374
         -84.192
         124 rows altogether
```

The solution x was computed that makes the vector magnitude of the residual as small as possible:

```
≫  sqrt(sum(resid.^2))
  ans: 1307.8
```

Any other proposed solution x would have a larger magnitude. Note that the residual vector and $\hat{\mathbf{b}}$ are orthogonal; the residual gives the direction from $\hat{\mathbf{b}}$ to $\mathbf{b}$:

```
≫  sum(resid.*bhat)
  ans: -7.6882e-009
```

This number isn't exactly zero due to numerical rounding-off.

The under operator makes it technically easy to fit linear models, but it's important to remember that such a fitted model can be only as good as the data used to fit it. Looking at the physical meaning of the solution x reveals a surprise. The order of the values in x corresponds to the order of the columns in A. Since the first column of A is g, the first component of x is r_{car}, the car's ideal highway performance, and is found to be 17.35 miles per gallon. The second component corresponds to the g.*c column of A and therefore is alpha: the city versus highway parameter. According to the result x, city driving increases the performance by 0.76 miles per gallon. That seems very odd, since everyone knows that gasoline economy is worse in city driving than on the highway.

What's wrong here is that our data are not very good. Recall from Chapter 16 that there were outliers in the data from the occasions when a gasoline purchase was not recorded. We identified such outliers based on detailed knowledge of the car, specifically that its range is less than 350 miles, and so any recording with d>350 is suspect. Let's repeat the fit, excluding the outliers:

```
≫  outlier = d > 350;
≫  Anew = A(~outlier,:);
≫  bnew = d(~outlier);
≫  xnew = Anew \bnew
  xnew: 19.5130
        -4.0517
        -2.0163
        -0.9788
```

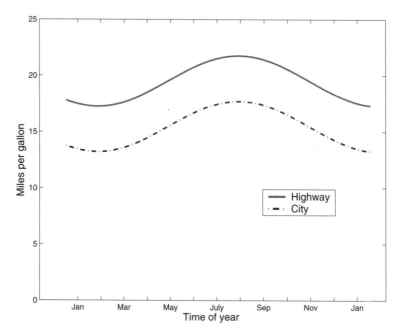

Figure 17.17. Gasoline economy as a function of time of year for pure city and highway driving. This is based on the model of Eq. (17.1) fitted to the data in `gasoline.dat` as described in the text. (Outliers were excluded from the fit.)

The highway performance is now estimated at 19.5 miles per gallon, with a penalty for city driving of 4.1 miles per gallon. The last two parameters describe the seasonal variability in gasoline economy. Figure 17.17 shows highway and city fuel economy, according to the fitted model, as a function of the day of the year.

◀

The linear least-squares technique is extremely powerful; it can be used to solve systems involving many variables, and it requires no initial guess about the values of the parameters. But many models have parameters that appear in a nonlinear way or where it is desired to use a fitting criterion other than least squares. Such nonlinear or non–least-squares fitting problems can be approached by setting up an objective function for optimization by techniques such as the conjugate gradient technique. (A powerful **Key Term** nonlinear technique, for least-squares problems, is the *Levenberg–Marquart method*, which is implemented in many optimization packages.)

Sensitivity of Solutions

The geometric interpretation of solving $\mathbf{A} \cdot \mathbf{x} = \mathbf{b}$ is that we set the value of $\mathbf{x}$ to hit the target $\mathbf{b}$. Recall that the residual is the vector $\mathbf{r} = \mathbf{b} - \hat{\mathbf{b}} = \mathbf{b} - \mathbf{A} \cdot \mathbf{x}$, which describes the minimal displacement between the span of $\mathbf{A}$ and the target vector $\mathbf{b}$. The residual describes how we miss the target. This suggests that the best solution is the one with the smallest residual. But sometimes it is better to miss the target by a larger amount in order to improve other qualities of the solution.

The MATLAB notation $x = A \backslash b$ or equivalently, $x = $ `mldivide(A,b)`, highlights the fact that the solution x is the output of a computation. As described in Chapter 15, any computation has a
Key Term *condition number* that describes how sensitive the output is to changes in the inputs. In the case of $x = A\backslash b$, it's important to remember that either or both A and b may not be very precise. For example, in fitting problems, A and b may be based on measured data, which often contain noise and errors.

Rather than looking at $\mathbf{A} \cdot \mathbf{x} = \mathbf{b}$ in terms of whether an exact solution exists or the size of the residual, it can be useful to consider the effects of the precision of $\mathbf{b}$ and $\mathbf{A}$ and to what extent a solution retains this precision. Suppose the components of $\mathbf{b}$ have a typical size b, and errors or noise in the components of $\mathbf{b}$ are typically of magnitude δb (perhaps as measured
Key Term by the standard deviation). The *relative precision* of $\mathbf{b}$ is $\varepsilon_b = \frac{\delta b}{b}$. A similar relative precision for $\mathbf{A}$ can be defined that we can denote as ε_A. The relative precision of the solution $\mathbf{x}$ depends on ε_b and ε_A according to the bound

$$\varepsilon_x \leq \text{cond}(\mathbf{A})(\varepsilon_A + \varepsilon_b) \tag{17.16}$$

The quantity $\text{cond}(\mathbf{A})$ is called the condition number of the matrix $\mathbf{A}$. The mathematics of the condition number are beyond the scope of this text (see [18]), but the `cond` operator provides a numerical value.

For example, consider the three linear simultaneous equations

$$\begin{array}{rcrcrcl}
3.1x_1 & + & 0.2x_2 & + & 2.9x_3 & = & 4 \\
1.2x_1 & + & 2.5x_2 & - & 0.1x_3 & = & 7 \\
2.2x_1 & + & 3.8x_2 & + & 0.1x_3 & = & 2
\end{array} \tag{17.17}$$

The solution is easily found:

```
≫  A = [3.1 0.2 2.9; 1.2 2.5 -.1; 2.2 3.8 .1];
≫  b = [4;7;2];
≫  x = A\b
  ans:  -77.265
        43.127
        80.999
```

How many significant digits are there in x?

Suppose the coefficients of A are known to two significant digits, and the coefficients of b are known to only one significant digit. This means that the relative precisions are $\varepsilon_A \approx 0.01$ and $\varepsilon_b \approx 0.1$. The condition number of A is

```
≫   cond(A)
    ans: 137.1
```

The errors in A in b might magnified by a factor of almost 100. According to Eq. (17.16), the relative precision of x is $\varepsilon_x \leq 137.1(.01+.1) = 15$. There might not be even a single precise digit in the computed solution! Or, seen another way, we would need to improve our knowledge of the coefficients of A and b to have three significant digits in order for the solution x to have a single significant digit.

The sensitivity of the solution can be seen in another way: modifying the values of A and b randomly to represent the imprecision in the values. This can be accomplished by multiplying each component of A and b by $1 +$ a small random number, where the size of the random number is scaled to ε_A and ε_b. For instance,

```
≫   epsilonA = 0.01;
≫   epsilonb = 0.1;
≫   A.*(1+epsilonA.*randn(size(A)))\(b.*(1+epsilonb.*randn(size(b))))
    ans: -142.27
          78.762
         150.12
```

This simulation can be repeated many times to get an idea about the sensitivity of the solution x to the anticipated imprecision of the values in A and b.

▶

Example: Driving in the City

Let's reconsider the gasoline economy problem. Recall that we defined a proxy variable, c, that stood for the proportion of driving done on the highway. It might be that a better proxy variable would be $c + p_1 c^2 + p_2 c^3$—this would allow a more complicated inference about the relationship between the amount of highway driving and the time between gasoline fill-ups. Finding the best parameters p_1 and p_2 is easy: Just augment the matrix **A** on page 478 with two more columns, c^2 and c^3.

```
≫ A2 = [A, c.^2, c.^3];
≫ A2( outlier,:) \d( outlier,:)
  ans: 21.509
       -12.378
       -1.66
       -0.76701
       4.928
       2.3496
```

The first number is the seasonally averaged miles per gallon for all-highway driving. Our more nuanced approach to the proxy city-driving variable suggests that the all-highway mileage of the car is better. But the consequence of adding in the new variables is a drastic change in the condition number:

```
≫ cond(A)
  ans: 5.3344
≫ cond([A, c.^2, c.^3])
  ans: 579.5
```

Since the relative error of the data in A is about 1%, a condition number this big might produce a result where even the first digit is not reliable.

◄

17.9 Solutions to Systems of Nonlinear Equations

We return to the problem of finding the solution to systems of nonlinear equations. The approach will be to set up a set of indicator functions (one for each nonlinear equation) whose value is zero at a solution. The contours at level zero of each function give the set of points that are solutions to that function. A simultaneous solution is a point that lies at the intersection of the zero contours of all of the functions.

There may be many such solutions, one, or none. For linear functions, where the zero contours are straight lines, it is relatively straightforward to find out which is the case using the techniques in Section 17.6. But this is not true for nonlinear functions where the zero contours can have any shape.

Key Term

Since we have good techniques for solving sets of linear equations, the approach we will take to solving sets of nonlinear equations is *linearization*: We approximate the nonlinear equations by linear equations and solve the approximation. We've already encountered this approach for a single nonlinear function on page 405.

Suppose that there is a solution x_* which we do not know, but that we already have an approximate solution; that is, a point x_0 near to x_*. Although we don't know x_*, we can verify that x_0 is indeed an approximate solution by checking $f_1(x_0) \approx 0$, $f_2(x_0) \approx 0$, $\cdots f_N(x_0) \approx 0$. This situation is illustrated in Figure 17.18 for $N = 2$. The point x_* is at the intersection of the two zero contours; the point x_0 is not.

For each of the N functions $f_i(x)$, we would like to find a linear function whose straight-line zero contour approximates the nonlinear zero contour of $f_i(x)$ near x_*. The intersection of the N-linear zero contours can then be found using the linear equation techniques described in Section 17.6.

In order to use these methods, we need to package the sets of indicator functions whose simultaneous zeros we want to find. There are many ways to do this. For instance, for N functions, we could make an N-element cell array, each element of which contains an `inline` for one function. But note that each of the N functions takes the same input (a vector of N components) and returns a single number. This suggests packaging the N indicator functions as a single computer function that takes an N-element vector as input and returns an N-element vector as output: Each component of the output is the single-number output of the corresponding indicator function.

To illustrate with $N = 2$, consider the pair of nonlinear functions

$$\begin{aligned} f_1(x) &= f_1(x_1, x_2) = x_1^2 + x_2^2 - 4 \\ f_2(x) &= f_2(x_1, x_2) = \tanh(-x_1) + x_2 - 1 \end{aligned} \qquad (17.18)$$

As shown in Figure 17.18, $f_1(x)$ has a zero contour that is a circle of radius 2 centered on the origin, and $f_2(x)$ has a sigmoidal shape. Simultaneous solutions are points where both functions equal zero; that is, where their zero contours intersect.

We can package these two indicator functions into a single computer function:

The two functions in Eq. (17.18)

```
[1]  function res = twoFunExample( x )
[2]  res = zeros(2,1);
[3]  res(1) = x(1).^2 + x(2).^2 - 4;
[4]  res(2) = tanh(-x(1)) + x(2) - 1;
```

Key Term

This takes a vector x as input and returns a vector, giving the values of each of $f_1(x)$ and $f_2(x)$. Such a function, which returns a vector, is called a *vector-valued function*.

```
>>  x0 = [2;3];
>>  twoFunExample(x0)
   ans: 9
        1.036
```

Since the result isn't $\begin{pmatrix} 0 \\ 0 \end{pmatrix}$, the vector $x0$ is not a solution to Eqs. (17.18).

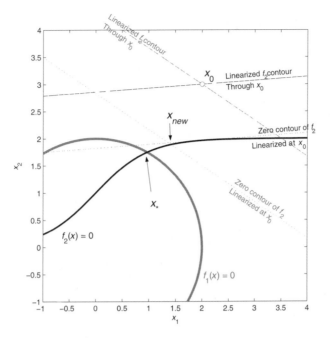

Figure 17.18.   Finding a refined approximation to the system of two nonlinear equations in Eqs. (17.18). The zero contours of $f_1(\mathbf{x})$ and $f_2(\mathbf{x})$ are shown as heavy lines; the simultaneous solution is the intersection of these contours. To find this, we start with a guess $\mathbf{x}_0$, find linear approximations to the zero contours and solve for the intersection of these, giving an improved solution $\mathbf{x}_{new}$.

To improve the guess x0, we can use the techniques for solving linear functions, but we need to approximate the nonlinear functions as linear ones.

Linearization of Functions

For a function $f(\mathbf{x})$, which we write here in the expanded notation $f(x_1, x_2, \cdots, x_N)$, the linear approximation is a function $L(x_1, x_2, \cdots, x_N)$, which is written in terms of parameters $a_1, a_2, \cdots, a_N$ and b:

$$f(\mathbf{x}) \approx L(\mathbf{x}) = a_1 x_1 + a_2 x_2 + \cdots + a_N x_N + b \qquad (17.19)$$

This is always the form of a linear function; the parameters $a_1, \cdots, a_N$ and b are set to specific numerical values to make the linear function approximate $f(\mathbf{x})$ near $\mathbf{x}_0$.

The first thing to notice is that the partial derivatives of $L(\mathbf{x})$ have a particularly simple form:

$$\frac{\partial L}{\partial x_1} = a_1, \ \frac{\partial L}{\partial x_2} = a_2, \ \cdots, \ \frac{\partial L}{\partial x_N} = a_N$$

The definition of the gradient in Eq. (17.3) gives

$$\nabla L(x_1, \cdots, x_N) = (a_1, \cdots, a_N)$$

Unlike a nonlinear function, where the gradient varies with $\mathbf{x}$, the gradient of a linear function is everywhere the same. We can make an exact match between the gradient $\nabla L(\mathbf{x})$ and that of $\nabla f(\mathbf{x}_0)$ simply by setting the coefficients $a_1, \cdots, a_N$ to $\nabla f(\mathbf{x}_0)$—the two functions' gradients will be exactly the same at the point $\mathbf{x}_0$.

The remaining unknown coefficient, b, can be set by matching the value $L(\mathbf{x}_0)$ to that of $f(\mathbf{x}_0)$. From Eq. (17.19) this will be

$$b = f(x_1, \cdots, x_N) - (a_1 x_1 + \cdots + a_N x_N) \tag{17.20}$$

For example, suppose we wish to make a linear approximation to the nonlinear function

$$f(x_1, x_2) = x_1^2 + x_2^2 - 4$$

The first step is to pick specific values $\mathbf{x}_0$ near which we want the approximation to be valid. We select $\mathbf{x} = (2, 3)$; that is, $x_1 = 2$, $x_2 = 3$. To be consistent with the vector notation used earlier, we can define

```
≫  f = inline('x(1).^2 + x(2).^2 - 4');
≫  x0 = [2;3];
```

The gradient of f at x0 can easily be found with the rules of calculus, but we use a numerical approximation just for illustration:

```
≫  a = gradfun(f,x0)
  a:  4
      6
```

This gives the parameters a_1 and a_2 in the linear approximation.

Finding parameter b is a matter of following Eq. (17.20):

```
≫  b = f(x0) - sum(a.*x0)
  b:  -17
```

or, equivalently, but using matrix multiplication notation,

```
≫  b = f(x0) - a'*x0)
  b:  -17
```

Altogether, the linear approximation is

```
>>  flinear = inline('4.*x(1) + 6.*x(2) - 17');
```

This matches both the value of the function f at $\mathbf{x}_0$ and the value of the gradient:

```
>>  f(x0)
   ans:  9
>>  flinear(x0)
   ans:  9
>>  gradfun(f,x0)
   ans:  4
         6
>>  gradfun(flinear(x0))
   ans:  4
         6
```

Key Term

We can repeat this analysis for the other function in the pair, but there is a shortcut. The gradient of a vector-valued function is really a set of gradients, one for each of the values in the vector. Since each gradient is itself an N-element vector, the set of all N gradients can be represented as an $N \times N$ matrix. This matrix is called the *Jacobian matrix*. Computing the Jacobian is a matter of computing the N gradients; the jacobian function is a slight modification of gradfun used to handle vector-valued functions. (In fact, jacobian can be used in place of gradfun.)

For vector-valued functions, the linear approximation can be constructed simultaneously for all of the component functions. The Jacobian matrix greatly simplifies the notation. Using matrix multiplication notation, the linear vector-valued function is

$$\mathbf{L}(\mathbf{x}) = \mathbf{A} \cdot \mathbf{x} + \mathbf{b}$$

where $\mathbf{A}$ is the Jacobian matrix. For example, the vector valued function $f(\mathbf{x})$ in Eq. (17.18), as implemented in twoFunExample, can be linearized around $\mathbf{x}_0$ simply as follows:

```
>>  A = jacobian('twoFunExample', x0)
  A: 4 6
     -0.07 1
>>  b = twoFunExample(x0) - A*x0
  b: -17
     -1.8227
>>  L = inline('[4,6;-.070651 1]*x + [-17; -1.8227]');
```

Admittedly, the preceding line is awkward, since inline doesn't accept variables as parameters; the values need to be typed as characters.

Now the vector-valued function L is a linear match for twoFunExample at x0:

```
≫ L(x0)
  ans: 9
        1.036
≫ twoFunExample(x0)
  ans: 9
        1.036
≫ jacobian(L, x0)
               4      6
  ans:
        -0.070651   1
```

Finding Simultaneous Nonlinear Solutions

Given the ability to find linear approximations to nonlinear functions and the ability to find simultaneous solutions to sets of linear functions, we can construct an algorithm for finding solutions to sets of nonlinear functions. The approach is the same as we saw in Chapter 16: Start with a guess, construct the linear approximation, solve the linear approximation, replace the guess with the linear solution, and iterate. A simple implementation of the methods is given in the programs newtonMethod and secantMethod.

Solving the equations, once they have been properly packaged and once there is an initial guess of the solution, can be attempted using newtonMethod:

```
≫ newtonMethod('twoFunExample', [2,3])
  ans: 0.97033
        1.7488
```

Since there is no possibility of bracketing a solution in two or more variables, there is no guarantee that a solution will be found. Nonetheless, for a sufficiently good initial guess, the iterative methods of a solution are effective.

▶

Example: Global Positioning

As described in Chapter 15, the Global Positioning System (GPS) enables a special radio receiver to compute its position on earth by processing information sent by multiple satellites. Each satellite sends messages giving its position and the time the message is sent; for satellite k, this is the four numbers x_k, y_k, z_k, and t_k, where t_k is the time of transmission.

The receiver records this information along with the time of receipt of the satellite's message. We denote this recorded time of receipt as T_k. Given the time of transmission and the time of receipt, the travel time of the message can be calculated. Since radio messages travel at the speed of light, this travel time can be converted into a distance. Thus, the GPS receiver can calculate the distance d_k to each satellite k and knows the position of that satellite. This provides information about the satellite's location; the receiver is somewhere on a sphere of radius d_k centered at x_k, y_k, z_k. In algebraic terms, the position x, y, z of the receiver is a solution to the equation

$$(x - x_k)^2 + (y - y_k)^2 + (z - z_k)^2 = d_k^2 \qquad (17.21)$$

This is the system of nonlinear equations in three unknowns, x, y, and z; there is one equation for each $k = 1, 2, 3$. Given the data x_k, y_k, z_k and d_k, a simultaneous solution for x, y, z can be computed. Geometrically, this solution is the intersection point of three spheres.

A realistic GPS calculation is more complicated than this and needs to include factors such as atmospheric variation in the speed of radio messages. One complication that we address here stems from the inaccuracy of the receipt time of the message, T_k. The clocks on the satellites are very accurate, but the clock in an inexpensive, hand-held receiver is not. That clock will be off by a certain amount, which we will call b. So the true time of receipt of a message is not T_k, but $T_k - b$, where b is unknown but assumed to be the same for all of the satellite messages used in a given calculation. The equations that need to be solved are

$$(x - x_k)^2 + (y - y_k)^2 + (z - z_k)^2 = c^2((T_k - t_k) - b)^2 \qquad (17.22)$$

There are now four unknowns, x, y, z, and b, so four equations and the information from four satellites are needed to find a solution

The indicator functions are easily constructed by bringing all of the terms in Eq. (17.22) to the right-hand side. The set of four indicator functions can be implemented as a vector-valued function. In the following implementation, `gpsfun`, we have divided the arguments into two groups: The first argument is a vector containing the unknowns (x, y, z, t); the second is a matrix containing the data for all of the satellites:

$$\begin{pmatrix} x_1 & y_1 & z_1 & T_1 - t_1 \\ x_2 & y_2 & z_2 & T_2 - t_2 \\ x_3 & y_3 & z_3 & T_3 - t_3 \\ x_4 & y_4 & z_4 & T_4 - t_4 \end{pmatrix}$$

```
[1] function res = gpsfun(r,s)
[2] % gps(r, s)
[3] % r -- receiver position [x,y,z,b]
[4] % s --- satellite positions in matrix format
[5] % [x1,y1,z1,dt1;
[6] %   x2,y2,z2,dt2;
[7] %   x3,y3,z3,dt3;
[8] %   x4,y4,z4,dt4]
[9] % units: meters and microsecs
[10] res = zeros(4,1);
[11] c2 = (299792458e-6).^2;
[12] for k=1:4
[13]   res(k) = (r(1)-s(k,1)).^2+(r(2)-s(k,2)).^2 ...
[14]   + (r(3)-s(k,3)).^2-c2.*(r(4)-s(k,4)).^2;
[15] end
```

Speed of light (m/μs) is at line [11]. One function for each satellite is at line [12].

To illustrate, here is some realistic information from four satellites with distances given in meters and times in nanoseconds [9].

Satellite	x_k Meters	y_k Meters	z_k Meters	$T_k - t_k$ Nanosecs
1	1.876371950559744e6	-1.064143413406656e7	2.42697646566144e7	0.07234683200e9
2	1.097666464137408e7	-1.308142752230029e7	2.035116937827073e7	0.06730845726e9
3	2.458513954435968e7	-4.335023426659201e6	9.08630032021747e6	0.06738499643e9
4	3.854136195752833e6	7.248575943442946e6	2.526630462778753e7	0.07651971839e9

Putting these data into a matrix named satelliteData, we can solve the GPS equations for the unknown position x, y, z, b of the receiver. But to do so with Newton's method or the method of secants, we need an initial guess as to the solution: Where in the world is the receiver? For these equations, the guess doesn't have to be very good: We put ourselves at the center of the earth (not very realistic but good enough for a start) and assume that the receiver's clock error b is zero.

```
≫  x0 = [0;0;0;0];
≫  x = newtonMethod('gpsfun', x0, satelliteData)
  x: 4.576e+006
     -2.3182e+006
     3.7704e+006
     -2.0005e+006
```

Remembering that the units of x, y, z are in meters, we see that our position x is 4,576.0 km from the center of the earth, y is -2318.2 km, and z is

3770.4 km. The receiver clock is early by by 2 million nanoseconds (this is only 2 thousandths of a second).

The five digits just printed correspond to a precision of 100 meters. More digits are available by printing in a `long` format:

```
≫  format long g
≫  x
  ans:  4576010.71042817
       -2318239.69920796
        3770361.17325821
       -2000493.75732675
```

Taking the printout at face value, it seems that we now have a location that is precise to a small fraction of a micron. But not all of these digits mean something. The input data were specified only to the nearest meter, and the satellite clocks are precise to about one nanosecond [9], so it's natural to assume that the output precision is no better than this. This suggests that the units digit—the digit right before the decimal point—is the last meaningful digit.

But we need to take into account the condition number of the computation. It might be that small changes in the input are amplified (or diminished) in the computation. To explore this, we can add small random perturbations to the input parameters and see how they change the output.

◀

17.10 Exercises

Discussion 17.1:
Generate three independent vectors in a three-dimensional space. Generate many sums by scaling the three vectors by random numbers in the range -5 to 5. Plot out the vectors as points using `plot3`. Rotate the plot to convince yourself that the points fill the three-dimensional space. Now do the same with (a) two vectors, and (b) three vectors, only two of which are linearly independent. Show that in both cases the random points lie on a plane.

Exercise 17.1:
A well-known result from physics is that a rope or chain hung between two points will take on a curve called a catenary. We can simulate this situation on the computer by modeling a chain as a sequence of $N+2$ points, (x_i, y_i) for $i = 0, \ldots, N+1$. We take the first and last points, (x_0, y_0) and (x_{N+1}, y_{N+1}), to be fixed; let's put them at $(0,0)$ and $(0,1)$, respectively. The other points hang below and inbetween the first and last points.

We can compute the shape of the chain by finding the "parameters" $x_1, x_2, \ldots, x_N, y_1, y_2, \ldots, y_N$ that minimize the potential energy of the chain. If we assume that the mass of each point on the chain is m, then the total gravitational potential energy of the chain is

$$E = \sum_{i=1}^{N} mgy_i + \sum_{i=1}^{N+1} k(\sqrt{(x_i - x_{i-1})^2 + (y_i - y_{i-1})^2} - L)^2 \qquad (17.23)$$

where g is the acceleration due to gravity (9.8 m/s^2 at the earth's surface), k is some positive constant, and L is the length of each link in the chain. The constant k specifies how stretchy the chain is. Make k large for a chain that doesn't stretch much.

1. Write a function `drawchain` that takes as an argument the parameter vector $x_1, x_2, \ldots, x_N, y_1, y_2, \ldots, y_N$ and draws a picture of the chain.
2. Write another function `chainenergy` that also takes the parameter vector as an argument and that computes from this the potential energy of the chain.

   Use `fminsearch` or `conjugategradient` to find the minimum energy configuration. You will have to specify an initial condition. Like a real chain, depending on your initial condition, your chain might have kinks in it. These kinks correspond to the local minima of the energy function. (If you have `chainenergy` call `drawchain`, then you will be able to watch the changing shape of the chain as the optimization proceeds. After each time you draw the chain, call the function `drawnow` to force MATLAB to display the image on the screen. Otherwise, only the final image will appear.) (*Hint:* To debug your programs, start with a chain that has only one middle point in it.)
3. Make a few graphs of the shape of a chain for various values of the k-stretchiness parameter.

Exercise 17.2:

The `coffeemodel` program implemented a single exponential model of cooling coffee. A more comprehensive Newton-like model incorporates multiple exponential models. Here is a model for N mechanisms:

$$\theta(t) = \theta_{\text{env}} + \sum_{p=1}^{N} \alpha_p \exp(-k_p t) \qquad (17.24)$$

Assuming that θ_{env} is measured directly, there are $2N$ parameters, $\alpha_1, \cdots, \alpha_N$, and $k_1, \cdots, k_N$.

For $N = 2$, implement the model of Eq. (17.24) and find the four parameters using the coffee-cooling data. Make a graph of the theoretical $\theta(t)$ versus time and compare to the data. As an initial guess for use in the optimization, set k_1 to be roughly what was found for the one-parameter fit, and k_2 to be several times larger, corresponding to the faster cooling from the steaming mechanism. Initially, set all of the α_p to be equal, and keep in mind that the model temperature at $t = 0$ is $\theta_{env} + \sum_{p=1}^{N} \alpha_p$, so you can set α_p to give an initial temperature close to that actually observed.

Repeat this for $N = 3$. In setting an initial guess for k_p, use the values from the $N = 2$ model, adding in a third value that is either larger or smaller.

Exercise 17.3:

A certain flower can take on either an elongated or ovoid form, depending on the temperature. The following table gives some data from an experiment on the probability of the ovoid form as a function of the temperature. At each temperature N flowers were grown.

Temperature	$p_{obs}(t)$	$N_{obs}(t)$
15	0.6	20
12	0.9	10
19	0.4	15
24	0.1	10
28	0.05	20

We wish to fit a model of the following form to these data:

$$p(t) = \frac{1}{2}(1 + \tanh(\alpha(t - \beta)))$$

Note that the function takes on values ranging from 0 to 1; the adjustable parameters for the fit are α and β.

Find the optimal values of α and β to minimize the objective function

$$\sum_{data} \frac{(p - p_{obs}(t))^2}{p(t)(1 - p(t))/N_{obs}(t)}$$

where the sum is over the rows of the table.

Exercise 17.4:

Consider the following data from a physics experiment in which electrons are used to bombard elements of various atomic numbers Z and the energy E of the resulting x-rays are measured.[3]

---

[3] C. W. S. Conover and J. Dudek, *American Journal of Physics*, 64(3):335–338.

Element	Atomic Mass Z	X-ray Energy E (keV)
Mg	12	1.16
Al	13	1.49
Si	14	1.72
S	16	2.28
Ti	22	4.52
V	23	4.96
Cr	24	5.42
Mn	25	5.91
Co	27	6.90
Ni	28	7.49
Cu	29	8.07
Ga	31	9.24
As	33	10.52
Y	39	14.95
Zr	40	15.80
Nb	41	16.60

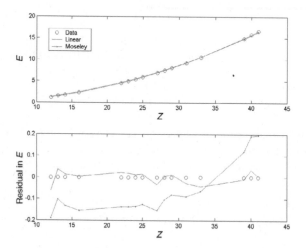

The anticipated relationship between E and Z is called "Moseley's law": $E = B(Z - \sigma)^2$ (a quadratic relationship). Moseley's work in this area helped to develop the concept of atomic mass as distinct from the atomic number and confirmed the shell model of the atom.

One way to look at this problem is as fitting a general quadratic function, $E = \alpha + \beta Z + \gamma Z^2$. This function, which is linear in the parameters, can be fitted by constructing column vectors E and Z containing the data and then assembling the matrix A = [ones(size(Z)), Z, Z.^2] and solving A\E. Once the fit is made, an association can be made: α is σ^2 and B is γ. Unfortunately, the parameter β is unaccounted for in this model, and there is no guarantee that when $Z = \sigma$ the model will give $E = 0$.

Fit the values of the parameters B and σ to the data using a nonlinear least-squares technique and compare the values you get to those found using the linear fit to the general quadratic model.

(The figure shows the two fits compared with the data. On a large scale, both fits seem the same. But the residuals systematically are larger for the nonlinear fit than the linear one. This isn't because the linear fit is better, but because the Moseley model isn't exactly consistent with the data.)

Exercise 17.5:
Write a program to read from the truss-bridge database structure introduced in Chapter 11, and calculate the stress in each member. To do this, you will have to set up a matrix with one line for the horizontal and the vertical forces on each node, except for the nodes that have vertical or horizonal supports. (It is tedious to set up the matrix, but once this is done finding the stresses is a simple matter of invoking mldivide.)

Exercise 17.6:
Linearize the GPS system of equations around the solution to the problem in the text, and find the condition number of the linear system at that point. This can be done using the cond and jacobian programs, evaluating the Jacobian at the solution to the equations.

The condition number of the GPS system becomes worse when the satellites are aligned. To simulate this phenomenon, you can use the randomGPS function, which generates random positions of four satellites and the corresponding message transit times to the point $(0,0,0,0)$. RandomGPS takes as an argument a matrix whose columns are used to generate random positions in three-dimensional space. Depending on the space spanned by these columns, the satellites will be aligned or not. For instance, randomGPS(rand(3,3)) will generate satellite positions randomly in three-dimensional space, while randomGPS(rand(3,2)) will put all of the satellites in random positions on a plane, and randomGPS(rand(3,1)) will put the satellites randomly on a line.

The output of randomGPS is arranged to be a suitable input to gpsfun. For example, here is a set of satellite data:

```
≫    sats = randomGPS(rand(3,3))
   sats:  1.5804 -0.2482 1.3949 7.0798
          0.8671 0.0356 0.4956 3.3335
          -0.8820 -0.3606 -0.8638 4.2901
          -2.4757 -1.5026 -2.5919 12.9642
```

Using Newton's method, we can find the position implied by these data:

```
≫ newtonMethod('gpsfun', [0,0,0,0], sats)
  ans: 1.0e-014 *
        0.1508
       -0.1529
       -0.1593
        0.1548
```

The positions are practically zero, which is the right answer.

Linearize the GPS equations around the solution point when the satellites are randomly placed in three-dimensional space, and find the condition number of the system. Repeat this when the satellites are placed randomly in a plane or along a line.

17.11 Project: Alignment of Images

A tourist stands on a hilltop, surveying the panorama around her. Turning clockwise around, she snaps a series of photos, capturing the horizon in a full circle. She makes sure that the left side of each photo will overlap the right edge of the previous one. Once at home, she will arrange the photos in sequence, using the overlap to align them.

But the developed photos bring a disappointment. No matter how the photographer shifts or rotates the photos, the overlapped areas do not align.

The problem is one of perspective. The adjacent photos were taken with the camera pointing in different directions. Each photo is a projection of a three-dimensional scene onto the two-dimensional firm; in such a projection, lines that are parallel in 3D become, in 2D, convergent to a point on the horizon. The projection distorts shapes and different projections distort shapes differently. Two differently distorted pictures cannot be aligned by shifts of position or rotation.

To accomplish the alignment, we need a more general method. Consider a pair of images. We call one the *target* and the other the *source*. Our goal is to pull pixels from the source and place them in the correct position in the target. Figure 17.19 shows a source and target image plotted with coordinate scales. The source coordinates are labeled (x,y) and the target coordinates (u,v).

We want to extend the target image to the left. For instance, we would like to fill in the pixel at position $(u = -20, v = 400)$. To find a color for this pixel, we use information in the source picture. We need to find where in the (x,y) source coordinates the point $(u = -20, v = 400)$ is located. To do this, we find a pair of functions

$$\begin{aligned} x &= f(u,v) \\ y &= g(u,v) \end{aligned} \qquad (17.25)$$

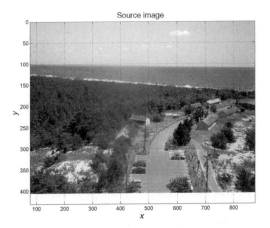
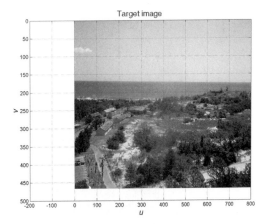

Figure 17.19. Two photos from a panoramic sequence. Some easily identified corresponding pairs of points in the two images have been marked.

These two functions, once we find them, allow us to take any (u,v) coordinate and find the corresponding (x,y) coordinate.

Given the functions in Eq. (17.25), the image-merging algorithm will work like this:

0 Take as arguments a source image s, a target image t, and the coordinate transformations $f(u,v)$ and $g(u,v)$.

1 Pick a set of indices in (u,v) coordinates that we wish to fill in. Call these (u_{new}, v_{new}).

2 Transform (u_{new}, v_{new}) into (x,y) coordinates:

$$
\begin{aligned}
x_{new} &= f(u_{new}, v_{new}) \\
y_{new} &= g(u_{new}, v_{new})
\end{aligned}
$$

3 Read off the pixels $s(x_{new}, y_{new})$ and fill in the target image t with these values. Since in general x_{new} and y_{new} will not be integers, interpolation may be necessary.

The question we face is how to find the functions $f(u,v)$ and $g(u,v)$. One way to do this is by *fitting*. In Figure 17.19, several pairs of corresponding points, one in each image, have been marked. We can use the coordinates of these points as data to fit the transformation functions. By fitting, of course, we mean adjusting parameters in a predefined form of function. What should that form be?

To mimic the manual alignment process, involving rotating and shifting the photo without changing its shape as in Figure 17.20, we can use the functional form

$$
\begin{aligned}
x &= \cos(\theta)u - \sin(\theta)v + c \\
y &= \sin(\theta)u + \cos(\theta)v + d
\end{aligned}
\tag{17.26}
$$

Figure 17.20. The alignment of the panoramic photographs using pure rotation and translation of the left image, as would be done by hand. The rotation is by a small angle in the clockwise direction. In the region of overlap between the two photographs, the target photograph has been drawn to be slightly transparent, so that the discrepancy between the two photographs can be seen. For example, note that the car in the right part of the parking lot is located in different positions in the two photographs.

θ, c, and d are the parameters that are adjusted to accomplish the alignment.

The transformation functions in Eq. (17.26) depend linearly on the variables u and v. This means that a straight line in (u,v) coordinates will remain a straight line after transformation to (x,y) coordinates, exactly what we would expect by a transformation that involves only rotation and shifting. The functions are, however, nonlinear in the parameter θ.

A slightly more general system, allowing for magnifying or reducing the photo before rotating it, is linear both in the variables u, v and in the parameters, which are a, b, c, and d.

$$
\begin{aligned}
x &= au - bv + c \\
y &= bu + av + d
\end{aligned}
\tag{17.27}
$$

Equations (17.27) and (17.26) are "models" of the true transformation that would have mapped the real-world points into the target image. Like almost all models, they are certainly not true; nevertheless they may be useful for accomplishing the task at hand.

Whatever the true transformation is, we know that it can be approximated by a linear form, at least over small distances. The full linear form has six individual parameters a, b, c, d, e, and f.

$$
\begin{aligned}
x &= au + bv + c \\
y &= du + ev + f
\end{aligned}
\tag{17.28}
$$

The full linear transformation, as well as Eqs. (17.27) and (17.26), which are special cases of linear transforms, has the nice property that straight lines stay straight under transformation. Thus, the horizon remains a straight line. However, under these transformations, parallel lines remain parallel, which isn't consistent with the sorts of distortions introduced by perspective. A substantially better approximation—but still an approximation—is given by

$$
\begin{aligned}
x &= a_1u + b_1v + c_1uv + d_1 \\
y &= a_2u + b_2v + c_2uv + d_2
\end{aligned}
\tag{17.29}
$$

To fit this model, take the spts data, which we write as x_i, y_i for $i = 1, \cdots, p$, where there are p data points. Similarly, the tpts data is u_i, v_i for $i = 1, \cdots, p$. Since this is a model that is linear in the parameter, the least-squares fit can be found by solving an equation $\mathbf{A} \cdot \mathbf{x} = \mathbf{b}$. Let the unknown parameter column vector be the eight parameters $\mathbf{x} = (a_1, b_1, c_1, d_1, a_2, b_2, c_2, d_2)$. The column vector $\mathbf{b}$ is $\mathbf{b} = x_1, x_2, \cdots, x_p, y_1, y_2, \cdots, y_p$. The matrix $\mathbf{A}$ is

$$
\mathbf{A} =
\begin{pmatrix}
u_1 & v_1 & u_1v_1 & 1 & 0 & 0 & 0 & 0 \\
u_2 & v_2 & u_2v_2 & 1 & 0 & 0 & 0 & 0 \\
\vdots & \vdots & \vdots & \vdots & \vdots & \vdots & \vdots & \vdots \\
u_p & v_p & u_pv_p & 1 & 0 & 0 & 0 & 0 \\
0 & 0 & 0 & 0 & u_1 & v_1 & u_1v_1 & 1 \\
0 & 0 & 0 & 0 & u_2 & v_2 & u_2v_2 & 1 \\
\vdots & \vdots & \vdots & \vdots & \vdots & \vdots & \vdots & \vdots \\
0 & 0 & 0 & 0 & u_1 & v_p & u_pv_p & 1
\end{pmatrix}
\tag{17.30}
$$

The solution $\mathbf{x}$ can be found in the familiar way. Similar systems can be set up for the other linear models.[4]

Figure 17.21 shows the alignment using the nonlinear transformation of Eq. (17.29) as fitted to the points marked in Figure 17.19. The nonlinear transformation has curved the edges of the source photograph. The agreement between the source and target photographs is very good for areas of the image near the spts and tpts data and much less good far away. In particular, note how the far-left side of the image appears to be distorted compared to the original source image.

All of the models, Eqs. (17.27), (17.28), or (17.29), can be fitted by solving sets of linear equations. To illustrate, we fit Eq. (17.29). The data are the pairs of corresponding points (u_i, v_i) and (x_i, y_i) shown in Figure 17.19.

In this project, you are to build a system for aligning photographs using the four methods described previously. To start, you will need the two photographs to be aligned. (The photographs used in the preceding examples are left.png and right.png.)

---

[4] Since the matrix $\mathbf{A}$ consists of two copies of a 4×4 matrix, it's possible to find a_1, b_1, c_1, d_1 independently of a_2, b_2, c_2, d_2 by solving two 4×4 systems. But this isn't true for the system in Eq. (17.27).

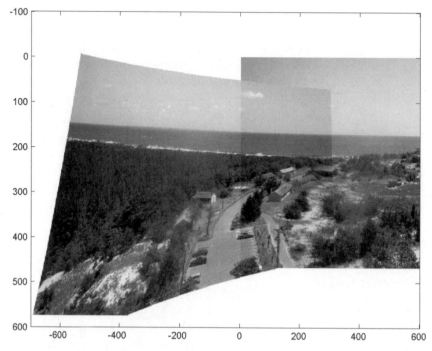

Figure 17.21. Alignment of the two images using the nonlinear model of Eq. (17.29). Compared with the rotation/shift model shown in Figure 17.20, the nonlinear model shows much better alignment within the region of overlap, but there is a distortion at the extreme lower-left region of the image.

The program `identPts` will display the source and target images so that corresponding points can be identified. You can use this program to create two sets of points, one for the target and one for the source image, as indicated in Figure 17.19. The syntax is

```
≫ [spts, tpts] = identPts(source, target);
```

Exercise 17.7:

Write a program `fitLinearModel` to find the parameters of Eq. (17.28) from the data in `spts` and `tpts`.

Exercise 17.8:

Write a program `doLinearInterp` that takes several arguments:

unew and vnew   Lists of new points in the target image coordinate system that you want to find the pixel values of.

s   The source image.

params   The parameters of the transformation.

The program should transform unew and vnew to the (x, y) coordinate system, find the pixel values at those coordinates using interpolation, and return the set unew and vnew along with the pixel values svals.

The interp2 program can perform the interpolation. It returns NaN for when the evaluation point is outside of the range of tabulated points; that is, when (x, y) is outside the range covered by the image.

Note that for color images, the interpolation should be done separately for each of the color planes. (Although the same transformation can be used!)

Exercise 17.9:

The images can be plotted using the image program. For example, given a target image and the transformed image svals (with coordinates unew and vnew), the two images can be overlaid with

```
≫   image(target);
≫   hold on;
≫   image(unew,vnew,svals);
≫   hold off;
```

Using the 'AlphaData' argument to image allows the photographs to be drawn transparently.

Write a program showimages(target, unew, vnew, svals) that produces a nice plot of the aligned images.

Exercise 17.10:

Write a program mergeimages(target, unew, vnew, svals) that works much like showimages but produces an single image with the merged photograph. You will need to figure out how to deal with the fact that the merged photograph will not, in general, be rectangular.

Exercise 17.11:

The transformation of Eq. (17.26) is nonlinear in one of its parameters. Write down a least-squares objective function for fitting the model of Eq. (17.26) with parameters θ, c, and d. Find the optimal parameters using fminsearch or conjugategradient.

There are several ways that this photo-alignment system can be elaborated upon. Some of them are as follows:

1. Modify identpts so that it is possible to zoom in on a little area around a point in order to locate it more precisely.
2. Combine the various transformation functions so that each is used in an area where it gives good results. Suppose that we have $f_1(u, v)$ and $g_1(u, v)$ for the nonlinear transformation of Eq. (17.29) and $f_2(u, v)$ and $g_2(u, v)$ for the rotation model of Eq. (17.27). Then another transformation is given by the functions

$$
\begin{aligned}
f(u, v) &= \alpha f_1(u, v) + (1 - \alpha) f_2(u, v) \\
g(u, v) &= \alpha g_1(u, v) + (1 - \alpha) g_2(u, v)
\end{aligned}
\qquad (17.31)
$$

where α is a number between 0 and 1 that smoothly controls the transition between the two models. Suppose that α is itself a function of u and v: Perhaps α is 1, when u and v are near the data points `tpts`, and 0, when u and v are far away and vary smoothly in between. Such a function might avoid the extrapolation problems of the nonlinear model while allowing nice alignment near the data points.

3. Color alignment. It's possible to find a mapping that transforms RGB colors so that the corresponding pixels in the two images have the same color.

Exercise 17.12:

A problem closely related to alignment is image *warping* or *morphing*, where a transformation is found that maps set of points in one image to the corresponding points in another image. But rather than using these points to fit a transformation, the transformation is interpolated. The `griddata` program, together with the sets of corresponding (u, v) and (x, y) points, can be used to define a transformation from target to source. (For a more sophisticated approach, see [28].)

In order to make a gradual transition from source to target, the $x = f_{\text{interp}}(u, v)$ and $y = g_{\text{interp}}(u, v)$ transformations found by interpolation can be mixed with the identity transformations $x = u$, $y = v$ using a simple linear combination:

$$\begin{aligned}
f(u, v) &= \alpha f_{\text{interp}}(u, v) + (1 - \alpha)u \\
g(u, v) &= \alpha g_{\text{interp}}(u, v) + (1 - \alpha)v
\end{aligned} \tag{17.32}$$

Depending on α, the resulting transformation keeps the original shape of the source or morphs it completely to the target, as shown in Figure 17.22. Write a program `morphimages(source,target,spts,tpts)` that takes a source image, a target image, and the set of corresponding points as arguments and returns the fully morphed image that aligns the corresponding points in the two images.

Write another program `morphgradual(source,target,spts, tpts,n)` that performs the gradual morphing, returning a set of images corresponding to $\alpha = 1/n, 2/n, \cdots, 1$.

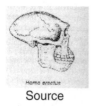 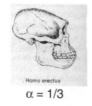  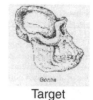

Source $\alpha = 1/3$ $\alpha = 2/3$ $\alpha = 1$ Target

Figure 17.22. Morphing of an image of the skull of *homo erectus* to the skull of a gorilla using the method described in Exercise 17.12. The intermediate images are different sizes of α in Eq. (17.32). (From Fig 12.4, p. 382 from *Evolving: The Theory and Processes of Organic Evolution* by Francisco J. Ayala and James W. Valentine. Copyright © 1979 by The Benjamin/Cummings Publishing Company, Inc. Reprinted by permission of Pearson Education, Inc.).

ASCII **Characters**

These characters should not be used in international interchange without determining that there is agreement between sender and recipient.
— Vint Cerf, Network Working Group, RFC 20, October 16, 1969

The ASCII character set is used by MATLAB to represent character strings. MATLAB stores character strings as sets of 2-byte integers; only the first 127 of these are ASCII values.

Many of the non-alphanumeric characters in the following table refer to specific typographical conventions; for instance, Line Feed (LF), Form Feed (FF), Carriage Return (CR), Horizontal Tabulation (HT), Vertical Tabulation (VT), and Backspace (BS). The Bell (BEL) character reminds us that the teletypewriters in use when ASCII was developed had very limited sound generation capabilities.

Some of the characters are obsolete for general usage and reflect specific protocols for communications among devices: End of Transmission (EOT), Start of Text (STX), End of Text (ETX), Acknowledge (ACK), Negative Acknowledge (NAK), and End of Medium (EM).

The Delete (DEL) character harkens back to the days when perforated paper tape was a common data-storage medium; a byte was a sequence of punch positions in a row across the tape. It's very difficult to fill in punched-out holes in the event of error. So, to correct a byte, all the holes were punched out to indicate that the byte was in error and should be disregarded. Note that DEL is 1111111—all the holes are punched out.

Bit Pattern	ASCII Char.	Num. Val.
0000000	null	0
0000001	soh	1
0000010	stx	2
0000011	etx	3
0000100	eot	4
0000101	enq	5
0000110	ack	6
0000111	bel	7
0001000	bs	8
0001001	ht	9
0001010	lf	10
0001011	vt	11
0001100	ff	12
0001101	cr	13
0001110	so	14
0001111	si	15
0010000	dle	16
0010001	dc1	17
0010010	dc2	18
0010011	dc3	19
0010100	dc4	20
0010101	nak	21
0010110	syn	22
0010111	etb	23
0011000	can	24
0011001	em	25
0011010	sub	26
0011011	esc	27
0011100	fs	28
0011101	gs	29
0011110	rs	30
0011111	us	31
0100000	space	32
0100001	!	33
0100010	"	34
0100011	#	35
0100100	$	36
0100101	%	37
0100110	&	38
0100111	'	39
0101000	(	40
0101001	)	41
0101010	*	42
0101011	+	43

Bit Pattern	ASCII Char.	Num. Val.
0101100	,	44
0101101	-	45
0101110	.	46
0101111	/	47
0110000	0	48
0110001	1	49
0110010	2	50
0110011	3	51
0110100	4	52
0110101	5	53
0110110	6	54
0110111	7	55
0111000	8	56
0111001	9	57
0111010	:	58
0111011	;	59
0111100	<	60
0111101	=	61
0111110	>	62
0111111	?	63
1000000	@	64
1000001	A	65
1000010	B	66
1000011	C	67
1000100	D	68
1000101	E	69
1000110	F	70
1000111	G	71
1001000	H	72
1001001	I	73
1001010	J	74
1001011	K	75
1001100	L	76
1001101	M	77
1001110	N	78
1001111	O	79
1010000	P	80
1010001	Q	81
1010010	R	82
1010011	S	83
1010100	T	84
1010101	U	85
1010110	V	86

Bit Pattern	ASCII Char.	Num. Val.
1010111	W	87
1011000	X	88
1011001	Y	89
1011010	Z	90
1011011	[	91
1011100	\	92
1011101	]	93
1011110	^	94
1011111	_	95
1100000	`	96
1100001	a	97
1100010	b	98
1100011	c	99
1100100	d	100
1100101	e	101
1100110	f	102
1100111	g	103
1101000	h	104
1101001	i	105
1101010	j	106
1101011	k	107
1101100	l	108
1101101	m	109
1101110	n	110
1101111	o	111
1110000	p	112
1110001	q	113
1110010	r	114
1110011	s	115
1110100	t	116
1110101	u	117
1110110	v	118
1110111	w	119
1111000	x	120
1111001	y	121
1111010	z	122
1111011	{	123
1111100			124
1111101	}	125
1111110	~	126
1111111	del	127

Matrix
Operations

Key Term

The basic operator for matrices and vectors is *matrix multiplication*. This is the *
operator, as opposed to the . * operator for scalar multiplication. MATLAB
enforces the distinction between row and column vectors based on their shape.

```
≫  m = [1 2; 4 3]
  m: 1 2
     4 3
≫  x = [1 2
  x: 1 2
≫  y = [30; 40]
  y: 30
     40
≫  x*y
  ans: 110
≫  x*m
  ans: 9 8
≫  m*y
  ans: 110
       240
≫  m*m
  ans: 9 8
       16 17
```

MATLAB enforces the requirement that the number of columns in the left argument of * equals the number of rows in the right element. For example, m*x makes no sense:

```
≫  m*x
  ???: Error using ==> *
       Inner matrix dimensions must agree.
```

The transpose operator can be used to swap rows for columns.

```
≫  transpose(x)
  ans: 1
       2
```

There is a shorthand special syntax for transposing: x'. This works for matrices as well; for example,

```
≫  m'
  ans: 1 4
       2 3
```

The standard matrix operators are readily available:

Matrix inverse: The matrix inverse operator is inv:

```
    ≫  minv = inv(m)
      minv:  -0.6000 0.4000
              0.8000 -0.2000
```

As expected, a matrix times its inverse gives the identity matrix:

```
    ≫  minv*m
      ans: 1 0
           0 1
```

Solving $Ax = b$: Many people believe that the way to solve $Ax = b$ for x is to compute $A^{-1} \cdot b$. This is inefficient, may be inaccurate, and won't give an answer if A doesn't have an inverse even though there still might be some x that satisfies $Ax = b$. A better method of solution uses the "under" operator \. To illustrate,

```
     ≫  A = [1 2 3; 1 2 1]
```

The inverse of A doesn't exist, but there are some values of **b** for which there is a solution; for example,

```
     ≫  b = [3;2]
```

The solution is

```
≫  x = A \b
  x:  0
     .7500
     .5000
```

as can be confirmed by multiplication:

```
≫  A*x
  ans:  3
        2
```

Eigenvalues and **Eigenvectors:**

```
≫  eig(m)
  ans:  -1
         5
```

When two return values are provided, the eigenvectors are the columns of the first one and the eigenvalues (as a diagonal matrix) as the second:

```
≫  [vecs,vals] = eig(m);
≫  vals
  ans: -0.7071 -0.4472
        0.7071 -0.8944
≫  vecs
  ans: -1 0
        0 5
```

Singular-value decomposition:

```
≫  [u,s,v] = svd(m)
≫  u
  ans: 0.3827 0.9239
       0.9239 -0.3827
≫  s
  ans: 5.3983 0
       0 0.9262
≫  v
  ans: 0.7555 -0.6552
       0.6552 0.7555
```

APPENDIX C

Binary
Flat Files

When writing pure numerical data to a file, it is possible to avoid including formatting information in a file. Since the binary numerical types are of fixed length (for instance, a single-precision floating point number is 32 bits), there is no need to mark the boundaries between entries. Avoiding such formatting data can save space. It can also make the files harder to read, since the file itself doesn't document its own layout.

Reading and writing files using the binary file operators involves considerably more choices than reading and writing spreadsheet files: the number of elements to read, the type of elements, whether conversion between different machine formats is to be performed, and how to format the output. Binary files created on one type of computer can be hard to read on another type of computer, since the arrangements of bits in a number may be different.

As a simple illustration, we use the file counting.bin, which contains the ten numbers 1 to 10 stored as signed, 2-byte integers. This binary format is often used by laboratory equipment for storing signals. To read the file, we must first "open" the file:

```
>> fid = fopen('counting.bin', 'r');
```

The first argument is the name of the file; as always, if the complete pathname isn't given, the current working directory is used as the path stem. The second argument identifies the mode in which we want to use the file: for reading 'r', or for writing 'w'. There are several other modes for appending, mixed reading and writing, and so on. An additional argument can be given to identify conversions to perform between machine formats on different types of computers. Fopen returns a file identifier that is needed for subsequent operations on the file.

To read the contents of the file, the `fread` operator is used.

```
≫ fread(fid, Inf, 'int16')
   ans: 1
        2
        3
        4
        5
        6
        7
        8
        9
        10
```

The second argument tells how many items to read from the file; `Inf` signifies to read the entire file. The last argument describes the binary format to use in interpreting the bits in the file. The string `'int16'` stands for a 16-bit signed integer. There are many other possibilities given in the documentation for `fread`.

To finish the process, the file is closed.

```
≫ fclose(fid)
```

There are many possible variations on this. For instance, it's sometimes useful to read only part of the file, or to arrange the output into a different shape. Here are the commands for reading the first six elements of the file, arranged as a 2×3 matrix. Since the file has been closed, it must be opened again.

```
≫ fid = fopen('counting.bin', 'r');
```

Then reading can commence:

```
≫ fread(fid, [3,2], 'int16')
   ans: 1 4
        2 5
        3 6
```

Not all of the elements have been read from the file, and since it has not yet been closed, we can read in the next element

```
≫ fread(fid,1,'int16')
   ans: 7
```

and continue. We set the size argument to `Inf` to signal that the remainder of the file should be read:

```
≫ fread(fid,Inf,'int16')
   ans: 8
        9
        10
```

Further attempts to read from the file will return an empty matrix:

```
>>    fread(fid,Inf,'int16')
  ans:  Empty matrix:  0-by-1
```

So it's time to close the file:

```
>>    fclose(fid)
```

Writing to binary files is done with the `fwrite` operator.

Exercise C.1:

Use the binary flat file operator `fwrite` to store as signed 16-bit integers the sequence `-3:3` in a file named `negAndPos.bin`. (*Hint:* You will have to use `fopen` and `fclose` as well.) Produce an m-file for reading the file three times, using first signed 16-bit integers, then unsigned 16-bit integers, and finally single bits.

APPENDIX D

Program Index

Programs marked with a * are m-files available at www.macalester.edu/~kaplan.

abs, 384
all, 57, 58
alpha, 452
alphabetcompare*, 272, 273
and, 57
angle, 384
any, 57, 58
asciicompare*, 272, 273
axes, 227, 236

cat, 367, 369
ceil, 130, 166
char, 386
class, 365
coffeemodel*, 463
collectLeaves*, 313
colon, 51
colorbar, 452
colormap, 186
combinations*, 304
compsort*, 283
cond, 495

conjugategradient*, 462, 465, 492, 501
contour, 451
countchars*, 169
countCharsInFiles*, 169, 315
csvread, 98, 463
cumprod, 48
cumsum, 48

datestr, 237, 239
dbclear, 198
dbcont, 198
dbdown, 198
dbquit, 198
dbstack, 198
dbstatus, 198
dbstep, 198
dbstop, 198
dbup, 198
derivfun*, 411
diff, 222, 239
dlmread, 99
drawbridge*, 290, 291

APPENDIX E

Data Files

These files are available at www.macalester.edu/~kaplan.

Solutions to Selected Exercises

Chapter 1

Exercise 1.4

If x is already between 1 and 100, look up the output on the graph. Otherwise, we exploit the fact that $\sqrt{10^{2n}x} = 10^n \sqrt{(x)}$. We need to find the n such that $1 \le 10^{2n} \le x$ so that we can use the graph. To do this, set up a counter n initialized to zero. There are two possibilities:

1. If x is less than 1, replace x by $100 \times x$ and subtract one from n. Keep doing this until x reaches the range 1 to 100.
2. Otherwise: if the original x was greater than 100, replace x by $x/100$ and add one to n. Keep doing this until x reaches the range 1 to 100.

Then look up the square root of x on the graph and multiply this by 10^n.

Exercise 1.6

Assume we already have an algorithm for finding the smallest number in a list.

1. Find the smallest number in the list.
2. If n is 1, this is the output of the computation.
3. Otherwise: remove the smallest number from the list, reduce n by one, and continue with step (1).

Chapter 2

Discussion 2.4

Scanning the characters in `1.3e(1+2)` from left to right, the tokenizer tries to build up a number. In MATLAB, the character following the e must be either a digit or a + or - sign. Since the (character is not admissible here, the tokenizer gives up and indicates that it was not able to construct a complete numerical token. The obvious thing that a human would do [evaluate `(1+2)` to produce 3 and inserting this in place of `(1+2)`] doesn't respect the strict sequence followed by the interpreter: first tokenize the expression, then evaluate it.

Discussion 2.6

We use a notation where innermost boxes are done first:

1. $\boxed{\boxed{\boxed{1-2}-3}-4}$ all operations are of equal precedence, so the order is left to right.

2. $\boxed{1-\boxed{2\ ^{\wedge}3}-4}$ the exponentiation has higher precedence than the addition and subtraction.

Exercise 2.2

```
rdivide( minus(3,plus(5,times(2,8),4)
```

Exercise 2.4

```
(a+b)+(c-d)
(a+sqrt(a))./((a.^b)+b)
(a./(b./c)).^(d - (f.^g))
```

Chapter 3

Discussion 3.1

```
≫  '3+2' + 7
   ans: 58 50 57
```

Since `'3+2'` is a character string with three elements, the ASCII values are 51, 43, and 50.

Discussion 3.4

Lower-case letters have an ASCII value that is 32 greater than upper-case letters. 32 corresponds to the 6th bit. When the bit is on the letter is lower case, `toupper` subtracts 32 from any ASCII character in the range `'a'` to `'z'`, leaving all other letters alone.

Discussion 3.6

An individual might write 10 characters per second. Doing this every second of a 100-year life gives `100*365.25*24*3600*10` or 3×10^{10} characters per person. With 10^{11} people, we have 3×10^{21} characters. The number of bits needed to give a unique identifier to every character is `log2(3e21)`, just over 71 bits.

Discussion 3.7

Sorts in decending order.

Exercise 3.1

```
sum(2.^(0:4))
```

Exercise 3.4

Once you have defined a vector v, the statement `sum(v)./length(v)` will compute the mean. To compute the standard deviation:

```
sqrt(sum((v-sum(v)./length(v)).^ 2)./...
     (length(v)-1))
```

Exercise 3.6

```
>> xl=2; xr=4; yb=1; yt=3;
```

We need to trace the *x*- and *y*-coordinates of the path around the rectangle:

```
plot([xl, xr, xr, xl, xl],[yb, yb, yt, yt, yb])
xlim([0,5]); ylim([0,5])
```

Exercise 3.8

"Unless" can be translated as "and not."

```
>> ((a+b)>N) & ~(b<0)
```

Exercise 3.10

```
monthsWithR=logical([1,1,1,1,0,0,0,0,1,1,1,1])
monthsWithR(m)
```

Exercise 3.12

There are 256 leap years from 1780 to 2832. Years 2004, 2008, ... are leap years, but 1900 and 2100 are not since they are evenly divided by 100. 2000 is a leap year since, although divisible by 100, it is also divisible by 400.

```
years = 1780:2832;
leap = (rem(years,4)==0 ...
~rem(years,100)==0) ...
  | rem(years,400)==0
sum(leap)
```

Chapter 4

Discussion 4.2

With the 4-column matrix `elmstreet` defined:

1. Number of residents: 22. `sum(elmstreet(:,1))`
2. Number of children: 9. `sum(elmstreet(:,3))`
3. Average number of children per household: 1.28. `mean(elmstreet(:,3))`
4. Average number of children per household in households with children. `mean(elmstreet( elmstreet(:,3)>0 ,3))`

Exercise 4.2

1. Odd indexed elements of x: `x(1:2:end)`
2. First half of x including middle element: `x(1:ceil(end/2))`
3. Last half: `x((1+ceil(end/2)):end)`

Exercise 4.6

1. Odd indices followed by even indices: `x( [1:2:end, 2:2:end])`
2. Perfect shuffle. This is not so easy. Pull out the first and last halves as separate arrays:

   ```
   ≫   Fh = x(1:ceil(end/2))
   ≫   Lh = x((1+ceil(end/2)):end)
   ```

   Then assign these values into a placeholder array

   ```
   ≫   nx = x;
   ≫   nx(1:2:end)  = fh;
   ≫   nx(2:2:end)  = Lh;
   ```

   Where nx contains the desired result.

Chapter 5

Exercise 5.3

Create file `swap.m` with this content:

```
tmp = a;
a = b;
b = tmp;
```

Exercise 5.4

The data file `counting.bin` is in the data directory (see Appendix E).

```
fid = fopen('counting.bin', 'r');
one   = fread(fid, 16, 'ubit1');
two   = fread(fid, 16, 'ubit1');
three = fread(fid, 16, 'ubit1');
% and so on
```

Chapter 6

Discussion 6.2

The error message stems from the fact that the variable a defined on the command line is hidden when the in-line expression a.*x is evaluated.

Exercise 6.4

```
function res = numToBoolean(x)
% numToBoolean(x) ---
% converts numerical x to boolean
% zero -> false, nonzero -> true
res = x ~= 0;
```

Exercise 6.6

```
function res = BooleanToTF(bool)
% BooleanToTF(bool)
% Convert boolean to TF character array
res = char('F'+zeros(size(bool)));
res(bool) = 'T';
```

```
function res = TFToBoolean(TFstring)
% TFToBoolean(TFstring) ---
% converts TF string to 0/1 boolean
% anything that's not T or t will become 0
res = logical(zeros(size(TFstring)));
res(upper(TFstring)=='T') = logical(1);
```

Exercise 6.8

```
function [s,d] = sumdiff(a,b);
s = a+b;
d = a-b;
```

Exercise 6.10

```
function [h,m,s] = daysToHMS(days)
% daysToHMS(days)-converts days
%          to hours, mins, secs.
tmp = days.*24; % in hours;
h = floor(tmp);
tmp = (tmp - h).*60; % remainder in minutes
m = floor(tmp);
s = (tmp - m).*60; % remainder in seconds;
```

Exercise 6.12

In the statement

```
>> [a,a] = max([5,4,8,2,7])
```

two successive output assignments are made. The first output will be the argument's maximum, 8. The second assignment is the index of this maximum value, namely, 3. Since the same variable a is used for both output assignments, a will end up with the value of the last assignment. This is not specified as part of the language and may vary from computer to computer.

Exercise 6.14

```
f = inline('strrep(x,''if'',''when'')')
```

Chapter 7

Exercise 7.2

There are many ways to write max4. Here is one with seemingly complex logic:

```
function res = max4(a,b,c,d)
if a >= b
    if a >= c & a >= d
        res = a;
    elseif c >= d
        res = c;
    else
        res = d;
    end
else
    if b >= c & b >= d
        res = b;
    elseif c >= d
        res = c;
    else
        res = d;
    end
end
```

A much simpler approach successively develops the answer:

```
function res = max4(a,b,c,d)
if a >= b
    res = a; % best so far
end
if c > res % compare to best so far
```

```
        res = c; % new best
    end
    if d > res % compare to best so far
        res = d; % new best
    end
```

The tools of iteration, in Chapter 8, make it easier to write this style of program.

Exercise 7.4

```
function res = max5(a,b,c,d,e)
if a >= b
    res = max4(a,c,d,e); % b is not it
else
    res = max4(b,c,d,e); % a is not it
end
```

Exercise 7.6

```
function res = absolutevalue(x)
if imag(x)==0 % no imaginary part
    if x < 0 % just the ordinary absolute value
        res = -x;
    else
        res = x;
    end
else
    res = sqrt(real(x).^2 + imag(x).^2);
end
```

The mathematically inclined might prefer to use complex conjugates

```
function res = absolutevalue(x)
res = sqrt(x.*conj(x));
```

The built-in function `abs` is more economical than either version of `absolutevalue`.

Exercise 7.8

This can be written in a `switch-case` style, which is readable but verbose:

```
function res = protons(atom)
switch(atom)
    case 'H'
        res = 1;
    case 'He'
        res = 2;
```

```
        case 'Li'
            res = 3;
        % remainder go here
        otherwise
            error('Not an atomic symbol');
    end
```

An alternative uses arrays:

```
function res = protons(atom)
symbols = {'H','He','Li','Be',...
           'B','C','N','O','F','Ne'};
nprotons = [1,2,3,4,5,6,7,8];
ind = strcmpi(atom, symbols);
if ~any(ind)
    error([atom, ' not an atomic symbol']);
else
    res = nprotons(ind);
end
```

Exercise 7.10

```
function res = roottype(a,b,c)
discrim = b.*b - 4.*a.*c;
if discrim == 0
    res = 'degenerate';
elseif discrim < 0
    res = 'complex';
else
    res = 'real';
end
```

Exercise 7.12

Although it's tempting to write a simple function like

```
function res = YMDtoDecimalYear(y,m,d)
res = y + m./12 + d./365;
```

this is not right, since not all months are $\frac{1}{12}$ of the year and there is no accounting for leap years.

A complete consideration takes into account the length of each month and the leap year:

```
function res = YMDtoDecimalYear(y,m,d)
% days to the beginning of each month
daysToMonth = cumsum([0,31,28,31,30,...
        31,30,31,31,31,31,30]);
```

```
if rem(y,4)==0 & ...
        ~(rem(y,100)==0 & ~(rem(y,400)==0))
    res = y + (daysToMonth(m)+(m>2)+d)./366;
else
    res = y + (daysToMonth(m)+d)./365;
end
```

The calendar functions built into MATLAB use serial days instead of serial years.

Chapter 8

Discussion 8.2
The 16-digit integer 1111111111111111 is represented exactly in computer arithmetic. But the 17-digit integer 11111111111111111 is represented only approximately with 11111111111111110, which is even.

Exercise 8.2

```
function res = squarerootV3(x)
if x == 0
    res = 0;
elseif x == 1
    res = 1;
elseif x < 0
    res = 1i.*squarerootV3(-x);
else
    res = squarerootV2(x);
end
```

Exercise 8.4
This function can be written in a looping/accumulator style:

```
function res = myfind(bool)
res = [];
for k=1:length(bool)
    if bool(k)
        res = [res, k];
    end
end
```

or in an indexing style:

```
function res = myfind(bool)
res = 1:length(bool);
res = res(bool);
```

Exercise 8.6

```
function res = alphabeticallyBefore(frst,scnd)
first = lower(frst); % convert to lower case
second = lower(scnd);
for k=1:(min(length(first), length(second)))
    if first(k) ~= second(k)
        if first(k) < second(k)
            res = 1;
        else
            res = 0;
        end
        return; % we're finished
    end
    % if we got here, they must be identical
    % up to the length of the shorter one
    if length(first) < length(second)
        res = 1; % first is the shorter
    else
        res = 0; % same or second is shorter
    end
end
```

Exercise 8.8

Here is a function that works for integer $n \geq 0$ with the least significant bit in the rightmost position. Note that we haven't checked that this condition holds (or even that n is an integer).

```
function res = intToBits(n)
res = [];
while n>0
    lastBit = rem(n,2);
    % Least Significant Bit on right
    res = [lastBit, res];
    n = (n-lastBit)./2;
end
```

The `bitsToInt` program takes a bit vector as an argument with the least significant bit on the right.

```
function res = bitsToInt(bits)
res = 0;
for k=1:length(bits)
    res = 2.*res + bits(k);
end
```

A more complete function would verify that the contents of `bits` are only zeros and ones.

Exercise 8.10

```
function [biggest, nextBiggest] = biggestTwo(v)
if length(v) < 1
    biggest = [];
    nextBiggest = [];
elseif length(v) < 2
    biggest = v(1);
    nextBiggest = [];
else
    biggest = max(v(1:2));
    nextBiggest = min(v(1:2));
    for k=3:length(v)
        if v(k) >= biggest
            nextBiggest = biggest;
            biggest = v(k);
        elseif v(k) >= nextBiggest
            nextBiggest = v(k);
        end
    end
end
```

Chapter 9

Exercise 9.2

```
function res = asktime
persistent lasttime;
persistent count;
if isempty(count) % initialize persistents
    count = 0;
    lasttime = now; % get time (in serial days)
end
answers = {'Didn''t you just ask?', ...
  'Same time it was last time you asked.', ...
  'Go buy a watch!'};
thistime = now; % get the actual time
fivemins = 1./(24.*12); % in units of days
% decide whether to give an useful answer
if count < 2 | ((thistime-lasttime)>fivemins)
   res=datestr(thistime);
else
   res=answers{ceil(length(answers).*rand(1))};
end
count = count+1;
lasttime = thistime;
```

Exercise 9.4

Since variables defined in one environment are out of scope in the function's invocation environment, the values of a and b cannot be changed through a function invocation. Using a reference, however, would give the function access to the location of the data stored in a and b, and thus would allow the value to be changed even though the environment in which a and b were defined is out of scope.

Exercise 9.6

```
function [scores, values] = keepbest(one,two)
persistent v; % values
persistent s; % scores
switch one
    case 'setup'
        v = cell(1,two);
        s = zeros(1,two) - Inf;
    case 'report'
        % don't need to do anything
    otherwise
        % see whether the new one is better than
        % any of the old ones
        L = min(find(one > s));
        % if so, insert the new one in the list
        if ~isempty(L) & L <= length(s)
            v((L+1):end) = v(L:(end-1));
            v{L} = two;
            s((L+1):end) = s(L:(end-1));
            s(L) = one;
        end
end
values = v(s ~= -Inf);
scores = s(s ~= -Inf);
```

Chapter 11

Discussion 11.2

All but the last item in the list are in correct alphabetical order in either the ASCII ordering or the conventional alphabetical ordering. Since 'George' is in the first half of the list, the search works regardless of which of the two comparison functions are used.

Exercise 11.2

```
function res = streetPrinceton(item)
res = ~isempty(findstr(item{4},'Princeton'));
```

```
function res = exchange231(item)
exchange = item{3}(1:3); % first 3 chars
res = strcmp(exchange, '231');
```

Exercise 11.4

```
function res = lengthcompare(one,two)
% Compares two strings alphabetically,
% returning 1 for a match
%             2 if <one> is before <two>
%             3 if <one> is after <two>
a = length(one);
b = length(two);
if a==b
    res = 1;
elseif a < b
    res = 2;
else
    res = 3;
end
```

Exercise 11.6

Here are four criterion functions that apply to this problem. Note that all of them state a positive condition; for instance, the exchange being 642 or the first name being Alice.

```
one = inline('strcmp(x{3}(1:3), ''642'')');
two = inline( ...
    '~isempty(strfind(x{4},''Fairview''))');
three = inline('strcmp(x{1},''Cummings'')');
four = inline('strcmp(x{2}(1:5),''Alice'')');
```

Putting the telephone data in cell array data:
Live on Fairview with last name Cummings:

```
inds = intersect(selectFromTable(data,one),...
        selectFromTable(data,two))
data(inds,:)
```

Live on Fairview or have last name Cummings:

```
inds = union(selectFromTable(data,one),...
        selectFromTable(data,two))
data(inds,:)
```

Live on Fairview or have last name Cummings but not both:

```
inds = setxor(selectFromTable(data,one),...
       selectFromTable(data,two))
data(inds,:)
```

642 exchange but not on Fairview:

```
inds = setdiff(selectFromTable(data,one),...
       selectFromTable(data,two))
data(inds,:)
```

Last name Cummings but first name not Alice:

```
inds = setdiff(selectFromTable(data,three),...
       selectFromTable(data,four))
data(inds,:)
```

Chapter 12

Exercise 12.2
Both fib and factorialrecursive will fail if their arguments are negative or are not integers because the base cases will never be reached.

Exercise 12.4

```
function res = fibmemo(n)
persistent fibs; % storage for past results
if isempty(fibs) %! Initialize storage
    fibs = zeros(1000,1);
end

if n==1 %! Base case
    res = 0;
elseif n == 2 %! Base case
    res = 1;
elseif length(fibs) >= n &  fibs(n) > 0
    % We already know the answer
    res = fibs(n);
else % We don't know, so figure it out
    res = fibmemo(n-1) + fibmemo(n-2);
    % and store it for the future
    fibs(n) = res;
end
```

Chapter 13

Exercise 13.2
(*Caution:* If you are wearing earphones, make sure the volume on your computer is adjusted properly to avoid dangerously loud sounds.) Many people can hear sounds down to about 50 Hz, but computer systems don't always faithfully produce such low-frequency sounds. To produce a sound at 100 Hz, lasting one second, do this:

```
[t,fs] = puretone(100,1);
sound(t,fs)
```

If you don't hear anything and aren't sure if the problem is in your ear or in the computer, try a 1000 Hz sound. Anyone with normal hearing should hear that.

Exercise 13.4
The notes in a scale correspond to $n = 0, 2, 4, 5, 7, 9, 11, 12$.

```
function [res,sf] = playscale(basefreq,dur)
if nargin < 2
    dur = 0.5;
end
res = [];
n = [0,2,4,5,7,9,11,12];
for k=1:length(n)
    [note,sf] = puretone( ...
            basefreq.*(2.^(n(k)/12)), dur);
    res = [res,note];
end
sound(res,sf);
```

Try `playscale(440)`. The sound has clicks that result from the abrupt ending of each tone made by `puretone`. A more natural note has a smoother ending.

Chapter 14

Discussion 14.2
In a low resolution image of the sort frequently found as thumbnail pictures on the Internet, the distance between samples is often similar to the size of stripes in clothing. This leads to aliasing and beating. Depending on the orientation of the stripes, the aliasing might be horizonal, vertical, or some combination of these. The result is that uniformly striped clothing can appear different, depending on which way the clothing is curving around the body.

Exercise 14.1
The `imagequant` function is very simple for images scaled between 0 and 1:

```
function res = imagequant(im,n)
res = round(im.*n)./n;
```

Chapter 15

Discussion 15.2

All comparisons involving NaN return false. This is appropriate behavior considering that NaN can result from a computation where the result is indeterminate; two indeterminate numbers should not be regarded as equal. Consider the mathematical statement $\frac{x}{y} = \frac{x^2}{y}$ which is true only when $x = 1$. If NaN == NaN, then the statement x./y == (x.^2)./y would be true whenever y==0.

Exercise 15.3

A base-10 number like 0.3333 can be written as an integer mantissa, 3333×10^M. In this case, $M = -4$. The size of M sets the number of digits in the integer mantissa.

The same principles apply in base 2. There is an integer mantissa N and an integer exponent M, organized as a base-2 scientific notation: $N \times 2^M$. If we set M to be -3, then 2^M is $\frac{1}{8}$. So, we look for an integer N such that $\frac{N}{8}$ is close to 0.3. There is no answer that works exactly, but $N = 2$ or $N = 3$ are the closest.

If we set M to be -4, then 2^M is $\frac{1}{16}$, and we get an good approximation to 0.3 with $M = 5$. Similarly, if M is set to be -5, then 2^M is $\frac{1}{32}$, and the best approximation is $N = 10$. For any choice of M, you can find the best corresponding integer N and write it as a binary number.

Exercise 15.4

The cause of the problem is the precision lost when adding or subtracting two numbers of different sizes.

When x is very small, then 1 and x^2 have very different sizes. Since the mantissa of a floating point number has a precision of only about 16 digits, when x^2 has a size smaller than 10^{-16}, none of the few digits of x^2 can contribute to the sum $1 - x^2$ and this sum is effectively 1. Thus, the computed function becomes $0/x^2$ or 0. When x^2 is a bit bigger than 10^{-16}, only one or two digits of x^2 appear in the sum: The numerical quantity $1 - \cos x$ therefore grows in a staircase fashion as the digits change, rather than in the smooth manner expected mathematically. In contrast, the denominator x^2 can be computed quite precisely, since it does not involve adding numbers. This gives the smooth, curved shape to each of the segments in the staircase.

Bibliography

[1] M. M. Waldrop, *The Dream Machine: JCR Licklider and the Revolution that Made Computing Personal*, Viking, 2001

[2] H. H. Goldstine, *A History of Numerical Analysis from the 16th Century through the 19th Century*, Springer-Verlag, 1977

[3] K. Devlin, *The Maths Gene*, Widenfeld and Nicolson, London, 2000, p. 39

[4] R. Kaplan, *The Nothing that Is: A Natural History of Zero*, Oxford University Press, 1999

[5] From the US Bureau of the Census "World POPClock Projection," `http://www.census.gov/cgi-bin/ipc/popclockw`, on April 3, 2001

[6] J. Langhorne and W. Langhorne (trans.), *Plutarch's Lives*, translated from the original Greek, William & Joseph Neal, Baltimore, Maryland, 1834

[7] The *New York Times*, as printed in the St. Paul *Pioneer Press*, June 7, 2000, p. E1

[8] The Unicode Consortium, *The Unicode Standard*, Addison-Wesley Developers Press, 2000

[9] G. Nord, D. Jabon, and J. Nord, "The Global Positioning System and the Implicit Function Theorem," *SIAM Review*, 1998, 40(3):692–696

[10] Garmin Corporation, *GPS Guide for Beginners*, `www.garmin.com`

[11] S. Mori, C. Y. Suen, and K. Yamamoto, "Historical Review of OCR Research and Development," *Proc. of the IEEE*, 1992, 80(7):1029–1058

[12] C. E. Shannon, "The Mathematical Theory of Communication," *Bell System Technical Journal*, July and October 1948. Reprinted in C. E. Shannon and W. Weaver, *The Mathematical Theory of Communication*, University of Chicago Press, 1963

[13] F. Pascal, *Understanding Relational Databases with Examples in SQL-92*, Wiley, 1993

[14] L. Atkinson, *Core MySQL*, Prentice Hall, 2002

[15] S. Singh, *The Code Book*, Anchor Books, 1999

[16] P. J. Pratt and J. J. Adamski, *Database Systems Management and Design*, 3rd ed. Boyd and Fraser, 1994

[17] W. H. Press, S. A. Teukolsky, W. T. Vettingling, and B. P. Flannery, *Numerical Recipes in C: The Art of Scientific Computing*, 2nd ed., Cambridge University Press, 1992

[18] M. T. Heath, *Scientific Computing: An Introductory Survey*, 2nd ed., McGraw-Hill, 2002

[19] S. C. Chapra and R. P. Canale, *Numerical Methods for Engineers*, 4th ed. McGraw-Hill, 2002

[20] D. F. Andrews and A. M. Herzberg, *Data: A Collection of Problems from Many Fields for the Student and Research Worker*, Springer-Verlag, 1985

[21] F. J. Ayala and J. W. Valentino, *Evolving: The Theory and Processes of Organic Evolution*, Benjamin Cummings, 1979, p. 382

[22] P. Marchand, *Graphics and GUIs with MATLAB*, CRC Press, 1996; S. J. Chapman, *MATLAB Programming for Engineers*, Brooks/Cole, 2000

[23] J. R. Jensen, *Introductory Digital Image Processing*, 2nd ed., Prentice Hall, 1995, p. 185

[24] S. Wolfram, *A New Kind of Science*, Wolfram Media, Inc., 2002

[25] P. Prusinkiewicz and A. Lindenmayer, *The Algorithmic Beauty of Plants (The Virtual Laboratory)*, Springer-Verlag, 1990

[26] Physionet, `physionet.org`, file n50 at 336 to 342 seconds

[27] R. Feynman, *QED*, Princeton University Press, 1988

[28] T. Beier and S. Neely, "Feature-Based Image Metamorphosis" *Computer Graphics,* 1992, 265(2):35–42

Index